The ROV Manual: A User Guide for Observation-Class Remotely Operated Vehicles

The ROV Manual: A User Guide for Observation-Class Remotely Operated Vehicles

Robert D. Christ and Robert L. Wernli Sr

ELSEVIER
AMSTERDAM • BOSTON • HEIDELBERG • LONDON • NEW YORK • OXFORD
PARIS • SAN DIEGO • SAN FRANCISCO • SINGAPORE • SYDNEY • TOKYO
Butterworth-Heinemann is an imprint of Elsevier
BH

Butterworth-Heinemann is an imprint of Elsevier
Linacre House, Jordan Hill, Oxford OX2 8DP
30 Corporate Drive, Suite 400, Burlington, MA 01803

First edition 2007

British Library Cataloguing in Publication Data
A catalogue record for this book is available from the British Library

Library of Congress Cataloging-in-Publication Data
A catalog record for this book is available from the Library of Congress

ISBN: 978-0-7506-8148-3

For information on all Butterworth-Heinemann publications visit our web site at http://books.elsevier.com

Typeset by Charon Tec Ltd (A Macmillan Company), Chennai, India
www.charontec.com

Printed and bound in the UK

07 08 09 10 10 9 8 7 6 5 4 3 2 1

Contents

Foreword

When I returned to college to become a mechanical engineer in 1970, I had never heard the term ROV. The only remotely operated vehicles I knew of were essentially satellites that orbited the Earth. With the collapse of the space industry in the early 1970s, my attention was drawn toward the next frontier – the ocean. I was recruited by the Naval Undersea Center in 1973, primarily due to the allure of those interesting underwater robots being developed there like the *CURV III* (Cable-Controlled Underwater Recovery Vehicle). I was hooked!

My interest grew, the technology advanced and several ROV companies were born. This booming industry, especially around my home in San Diego, led to the creation of the first remotely operated vehicle conference, ROV '83, which I chaired. The theme of 'A Technology Whose Time Has Come' was timely and helped launch a conference series that continues today under the title of Underwater Intervention.

One of the first products of that conference series was the *Operational Guidelines for ROVs*, a ground-breaking publication by the ROV Committee of the Marine Technology Society (MTS). I had the pleasure of producing the book along with Frank Busby and a committee of experts who also helped launch the ROV conference series.

There were some other specialized publications that were produced to address ROV use and maintenance, but none that had the scope of the earlier Guidelines. The MTS ROV Committee asked me to update the Guidelines, and I, along with Jack Jaeger, who handled the production, took on the task. The result – *Operational Effectiveness of Unmanned Underwater Systems* – covered the entire scope of ROVs, from the history, through design and operation, and ending with a look into the future. The 700-plus page book, published as a CD-ROM in 1999, set another milestone in undersea vehicle documentation. My thanks to all those who contributed material for the book.

In 2005, I retired from my job at the Navy laboratory in San Diego and decided to begin my next career as a consultant on underwater systems and author of not only technical publications, but also fiction. My first novel – *Second Sunrise* – was an award winner and the sequel will be out in 2007. Hopefully, this will be the beginning of a series of undersea techno-thrillers that will not only entertain, but also educate my readers.

Now, to this publication – *The ROV Manual*. Once again, I was contacted to review and provide a critique of an original manuscript written by Robert Christ. I know the amount of effort that goes into preparing such a book and provided my thoughts on what looked like an excellent publication that addressed the specialized observation class of ROVs. To make a long story short, we agreed to work together on the manual and drive it to completion, which made our publisher more than happy.

For my part, I've added a few chapters and helped Bob edit and complete the manuscript. Bob has added his own real-life experiences throughout the book, which provide excellent anecdotes for the novice and expert alike. To the readers of this book, I hope it fulfills your needs and sparks a desire for an exciting career in underwater robotics.

Robert L. Wernli Sr

Preface

Principles of operating ROV equipment are similar throughout the size ranges of submersible systems. Although this manual covers a wide range of underwater technologies, the vehicle classification and size range covered within this manual is the free-swimming, tethered, surface-powered, observation-class ROV system with submersible sizes from the smallest of micro-ROV systems up to a submersible weight of 200 pounds (91 kilograms).

The purpose of this manual is to put forth a basic 'How To' for usage of observation-class ROV systems (along with some theory) in a variety of underwater tasks. With the addition of underwater duties to today's law enforcement profession's mission requirements, the requirement to view underwater items of interest is apparent. The use of ROVs and similar technologies provides the public safety, military, maritime security, archeological, and commercial diver with an array of options that offer reductions in risk (as opposed to using divers), potential time savings, and reduced cost.

This manual is a manufacturer non-specific document for ROV deployment that also contains standard operating procedures (SOPs), training materials, and qualification standards for qualifying personnel to operate ROVs. The material will augment manufacturer's equipment-specific instructions for use. All ROV systems share a similar set of operating parameters, environment, and methods of use. This manual also seeks to clarify the general techniques for optimizing ROV deployment to achieve operational requirements. Although there are many contributors to this manual, by industry experts as well as product manufacturers, this manual in no way endorses any specific product or manufacturer.

As with any new class of vehicle, the deployment of ROV systems involves specific tactics, techniques, and procedures to fully utilize the equipment's potential and provide satisfactory results for the customer. Although an ROV is a powerful tool, it is still a machine subject to operator experience, equipment capabilities, and environmental factors that affect its utility. ROVs are a compromise of size and power balanced against operational requirements. A large ROV system may accomplish more open-water tasks than a smaller system due to its ability to muscle to a location (through currents, distance offset, and around obstructions) with increased payload and additional sensors. But a larger vehicle with its support system may not fit aboard a vessel of opportunity (e.g. a 25-foot Rigid Hull Inflatable Boat or other small deployment platform), thus increasing the amount/type of resources needed to support the use of a larger ROV. A micro-ROV system may not possess adequate thrust to pull its own tether to the inspection site and is more likely to be affected by environmental conditions. However, these smaller systems are more mobile and can be operated in confined spaces with fewer resources. Regardless of the ROV size, visibility will vastly affect the amount of time and level of difficulty of the inspection task.

This manual focuses on the two lowest common denominators: (1) The technology and (2) The application. There is often misconception as to the utility and application of this technology. The goal of this manual is to introduce the basic technologies required, how they relate to specific requirements, and to help identify the equipment necessary for a cost-effective and successful operation.

Acknowledgments

Special thanks to the following people for their invaluable assistance in producing this volume:

Andrew Bazeley, President, Tecnadyne Advanced Product Development
Andy Goldstein, Vice President, Desert Star Systems LLC
Arnt Olsen, Sales Manager, Kongsberg Maritime Inc.
Bill Kirkwood, Associate Director of Engineering, Monterey Bay Aquarium Research Institute
Bill Nagy, Anteon Corporation
Blades Robinson, Director, International Association of Dive Rescue Specialist
Boris Rozman, Director, Indel Ltd.
Brett T. Seymour, Photographer, NPS Submerged Resources Center
Buddy Mayfield, President, Outland Technology, Inc.
Charles L. Strickland, President, Photosea Systems Inc.
Chris Gibson, Marketing Director, VideoRay LLC
Chuck Daussin, Vice President, Outland Technology, Inc.
Colin Dobell, President, Inuktun Services, Ltd.
Dan Fjellroth, Editor, ROV.NET
Darren Moss, Technical Support (Robotics Products), Benthos, Inc.
David Brown, Marketing Communications Manager, Sonardyne International
Don Rodocker, President, SeaBotix, Inc.
Doug Wilson, Ph.D., Imagenex Technology Corporation
Dr Douglas E. Humphreys, President, Vehicle Control Technologies, Inc.
Dr Robert H. Rines, President, Academy of Applied Science
Drew Michel, ROV Committee Chairman, Marine Technology Society
Edward O. Belcher, President, Sound Metrics Corp.
Jack Fishers, President, JW Fishers Manufacturing
Jacob Smith, Electronics Technician, US Coast Guard MSST San Pedro
Jeff Patterson, Senior Engineer, Imagenex Technology Corporation
Jesse Rodocker, Vice President, SeaBotix, Inc.
Joel Russell, Editor, Best Publishing Company
John T. Johnson, Vice President, Oceaneering International
Julee Gooding, Owner, Maximum Presence
Kenneth McDaniel, Anteon Corporation
Kyrill V. Korolenko, Chief Scientist, Naval Undersea Warfare Center – Newport
Lane Stephens, Owner, Entirely by Design
Larry E. Murphy, Chief, NPS Submerged Resources Center
Lev Utchakov [deceased], Director, Indel Ltd.
Marco Flagg, President, Desert Star Systems LLC
Martin Fairall, Electronics Technician, US Coast Guard MSST Seattle
Maurice Fraser, Sales Engineer, Tritech International

Mitch Henselwood, Sales Manager, Imagenex Technology Corporation
Nick Smedley, Sales Manager, Sonardyne International
Richard Faulk, President, Above and Below the H2O
Richard Marsh, President, Tritech International
Ronan Gray, Oceanographic Engineer, DeepSea Power and Light
Russell Landry, President, Capital Consultants, Inc.
Scot Trip, Senior Analyst, US Coast Guard R&D Center
Scott Bentley, President, VideoRay LLC
Steve Fondriest, President, Fondriest Environmental, Inc.
Steve Van Meter, Hazardous Robotics Specialist, NASA Kennedy Space Center
Terry Knight, Co-Founder, Inuktun Services, Ltd.
Willie Wilhemson, President, Imagenex Technology Corporation

Illustration credits

US Navy	Figures 1.3, 1.4, 1.5, 9.1
Hydro Products	Figure 1.6
JAMSTEC	Figure 1.7
Schilling Robotics	Figure 1.8
Steve Van Meter	Figure 1.9
VideoRay LLC	Figures 3.8, 3.10, 3.11, 3.16, 3.17, 3.18, 3.19, 3.20, 3.21, 3.22, 3.23, 3.29
Tecnadyne	Figure 3.9
Inuktun Services Ltd.	Figure 3.13, 11.6(d)
Outland Technology	Figure 3.14
Sonardyne	Figures 4.21, 4.22(c), 4.23(c), 4.26, 8.3
Desert Star LLC	Figures 4.25, 4.27, 4.28, 4.32, 4.35, 4.36, 4.37, 4.38, 8.4
Imagenex Technology Corp.	Figures 5.2, 5.15, 5.16, 5.17, 5.29, 5.30, 5.32, 5.34(a), 5.35(a)
Deep Sea Power & Light	Figures 7.6, 7.7
Academy of Applied Science	Figure 5.21(a)
Sound Metric Corp.	Figures 5.25, 5.26, 5.27, 5.58
Charles Strickland (Photosea Systems)	Figure 7.2
ECA SA	Figures 9.2, 9.7, 9.8, 9.9
Hydroid	Figure 9.3
BAE Systems	Figures 9.4, 9.5, 9.6
US National Park Services	Figure 11.5, Cover Photo
SeaBotix, Inc.	Figure 11.6(a through c), 11.7(d through f)
JW Fishers Manufacturing	Figure 11.7(a through c)

History and dedication

My father, C.J. Christ, learned in the late 1960s that a German U-boat (U-166) was sunk on 1, August 1942 by a US Coast Guard aircraft just south of our hometown of Houma, Louisiana. Thus began a 35+ year quest to find the wreck. Through that odyssey, a host of people cycled through our lives with similar interests in this project that was fueled by my father's infectious enthusiasm. Oceanographic pioneer Demitri Rebikoff stayed at our home for a month while staging for tests of his Rebikoff *Remora* early diver propulsion vehicle. The brilliant Dr Harold Edgerton from MIT brought down one of his new inventions by the name of 'side-looking sonar' (later termed 'side-scan sonar') that we spent interminable hours dragging behind small boats in the Ship Shoal area south of Last Island, Louisiana.

Instead of playing baseball each weekend, my Dad brought me offshore while he made dive after dive on the old WW2 wreck sites. Later, I began to make these dives myself to these rusted hunks of metal lying inert in the mud on the Outer Continental Shelf of the Mississippi river estuary.

As I grew older, those rusting hulks turned into living beings as my father had me accompany him on interview after interview with the survivors of those sinkings. This fueled an intense interest in history, taking me further into the technical aspects of underwater work. In the late 1990s, one of Dad's strongest long-term supporters of the U-166 project, Johnny Johnson of Oceaneering International, sent me offshore to work on large ROV systems in support of the Oil and Gas industry to 'learn how the real work is done'.

While performing a pipeline survey for British Petroleum during 2001 in 5000 feet of seawater 45 miles south of the Mississippi river delta (over 140 miles east of the recorded sinking location of the U-166), C&C Technologies from Lafayette, Louisiana found a blip on a sonar readout that resembled a German Type IXC U-Boat. The 60-year-old enigma of the U-166 had been solved – with my Dad known as the 'Grandfather of the U-166 Project'.

I dedicate this book to its inspiration, C.J. Christ – Pilot, Diver, Historian, Father, and Friend.

Robert D. Christ

C.J. Christ, Adm. Karl Donitz, and Capt. Peter 'Ali' Kramer in Germany, c. 1979 (from the C.J. Christ Archives).

Introduction

While working in the offshore Oil and Gas Industry in the 1990s for Oceaneering International, I noticed that the majority of the work being performed by large work-class ROV (WCROV) systems was simply taking pictures of work sites. The majority of human activity in the world's waters happens in the area known as the 'littoral' waters, generally described as from the surface to a depth of 600 feet. To deploy a WCROV system to the work site, heavy industrial machinery is required in order to simply launch the vehicle into the water. To support the crane and equipment needed for the submersible, a large 'platform of opportunity' is needed, such as a large supply or salvage boat. To staff this vessel, engineers, deckhands, vessel operators, galley hands, etc. are needed in addition to the ROV crew. All of this just to deploy a camera to a place where a meaningful picture can be taken. It seemed to be a waste of resources, in many of the operational situations, to use a multimillion dollar WCROV when a smaller system could have performed the same operation much more cost-effectively.

While on an unrelated trip to Russia in 1999, I was introduced to a Russian company tied to the Shirshov Institute, which was producing a micro-ROV system designed for structural penetration of wrecks and other light-duty tasks. Thus began my odyssey of observation-class ROV usage in underwater tasks.

ROV systems will not completely replace divers in the near future due to the lack of sensory feedback needed to do intricate tasks. But the ROV, in many cases, can replace putting a human in harm's way. This allows a safer and more cost-effective means of performing the mundane tasks of searching and monitoring. It takes less time, less effort, less risk, and (as a result) less money to drop a self-propelled camera into the water, go to the work site to look around and perform a task. By using an ROV system, the diver can be moved to a remote location that minimizes the risk to personnel (i.e. remote from the hazards of temperature, hyperbarics, moving machinery, and other underwater hazards).

The idea for this manual came about while I was trying to learn how to operate a Remotely Operated Vehicle (ROV). The larger operators of ROV equipment had operations manuals with standard operating procedures and standards, but the information on how to apply this knowledge was disparate and in many different locations. Most of the learning was provided as on-the-job training programs. Later, as I moved into smaller ROV systems, the problem became even more acute since there was no chain of command present with the necessary depth of knowledge of these skills (as with a larger ROV crew running working-class systems). In short, I had several failed projects due to my inexperience and lack of understanding of the operation and capability of ROV systems.

There is a body of knowledge with tricks and techniques that exist for operating ROV systems. Unfortunately, this knowledge set resides with the commercial ROV operators who do not normally share these tidbits of information, since their competitive advantage relies upon their knowledge and experience.

There are some excellent books written by an English gentleman by the name of Chris Bell published through an oil and gas publishing company. The focus of those manuals is on the basic science of ROV operations, since most of the entry-level folks purchasing that type of manual are destined for employment on large ROV systems. Entry-level oilfield ROV technicians learn all of the components of the equipment before operating the vehicle in the field. With the observation-class ROV system, however, there is usually no experienced supervisor or other trained crew members upon which to rely. Thus, the technician has to figure out the entire process on the spot. In this text, we focus on the 'How To' aspect of ROV operations, since we assume the operator, as the lead technician, is using a fully functional ROV system in a field location. Accordingly, a general knowledge of electricity/electronics, a basic understanding of physics, and a familiarity with the operation of general machinery is necessary.

At the end of my tenure in 2003 with the micro-ROV manufacturer VideoRay (a company which I co-founded in 1999), I made a series of recommendations to a large customer of mine (the United States Coast Guard) for the implementation of the ROV program into their fleet. Among the recommendations was a strong suggestion that they write a 'How To' manual including recommended standard operating procedures as well as a basic manufacturer non-specific training manual for ROVs. The recommendation turned into a contract – thus the genesis of this manual.

In 1979, the NOAA Office of Ocean Engineering produced a survey of ROVs (US Government Printing Office Stock #003-017-00465-1) to include the then-current technology. While reading through that manual, I noticed practically all of the same issues still exist today regarding the application of this type of technology. My operational experience with a range of systems (sizes, classifications, and applications) has shown that there is a commonality with the vehicles and environments. All vehicles share the same operating principles. Conveying them in one text is the focus of this manual.

For efficient use of this manual, read through the sections relevant to all systems then branch into more specific applications. However, read through all sections as time permits, since they contain information that could help solve operational problems in the field.

ROVs are a very useful tool to conduct a wide range of underwater tasks. Some of the projects encountered will be fascinating and satisfying. Others will be frustrating and embarrassing. In 2002, I spent an entire day attempting to recover an ROV system stuck in a doorway deep within the wreck of the *USS Arizona* in Pearl Harbor. The National Park Service project supervisor (Larry Murphy) kept threatening to use bolt cutters on the tether. Several high-profile officials were watching. That was frustrating! In 2003, I spent an incredible eight hours doing an internal wreck penetration/survey of a Swedish C-47 aircraft (shot down by a Russian MiG in 1952) at 425 feet of sea water. That was exhilarating!

Little of the Earth's oceans have been explored for the simple reason that the environment is so inhospitable to us humans. ROV technology is now maturing to the point where this exploration may now be conducted while we remain topside in a much safer and more comfortable environment. A whole new chapter in the exploration of the seas is unfolding each and every day with the advent of underwater robotic technology. The choice of delving into this technology places us on the wave of explorers 'going where no one has gone before'. We live in exciting times indeed.

Enjoy your work!

Robert D. Christ

Chapter 1

A Bit of History

1.1 INTRODUCTION

The strange thing about history is that it never ends. In the case of Remotely Operated Vehicles (ROVs), that history is a short one, but very important nonetheless, especially for the observation-class ROVs.

Two critical groups of people have driven ROV history: (1) Dedicated visionaries and (2) exploiters of technology. Those who drove the development of ROVs had a problem to solve and a vision, and they did not give up the quest until success was achieved. There were observation-class vehicles early in this history, but they were far from efficient. In time, however, the technology caught up with the smaller vehicles, and those who waited to exploit this technology have led the pack in fielding smaller, state-of-the-art ROVs.

This section will discuss what an ROV is, address some of the key events in the development of ROV technology, and address the breakthroughs that brought observation-class ROVs to maturity.

1.2 WHAT IS AN ROV?

Currently, underwater vehicles fall into two basic categories (Figure 1.1): Manned Underwater Vehicles and Unmanned Underwater Vehicles (UUVs). The US Navy often uses the definition of UUV as synonymous with Autonomous Underwater Vehicles (AUVs), although that definition is not a standard across industry.

According to the US Navy's UUV Master Plan (2004 edition, section 1.3), an 'unmanned undersea vehicle' is defined as a:

> *Self-propelled submersible whose operation is either fully autonomous (pre-programmed or real-time adaptive mission control) or under minimal supervisory control and is untethered except, possibly, for data links such as a fiber-optic cable.*

The civilian moniker for an untethered underwater vehicle is the AUV, which is free from a tether and can run either a pre-programmed or logic-driven course. The difference between the AUV and the ROV is the presence (or absence) of direct hard-wire communication between the vehicle and the surface. However, AUVs can also be linked to the surface for direct communication through an acoustic modem, or (while on the surface) via an RF (radio frequency) and/or optical link. But in this book, we are concerned with the ROV.

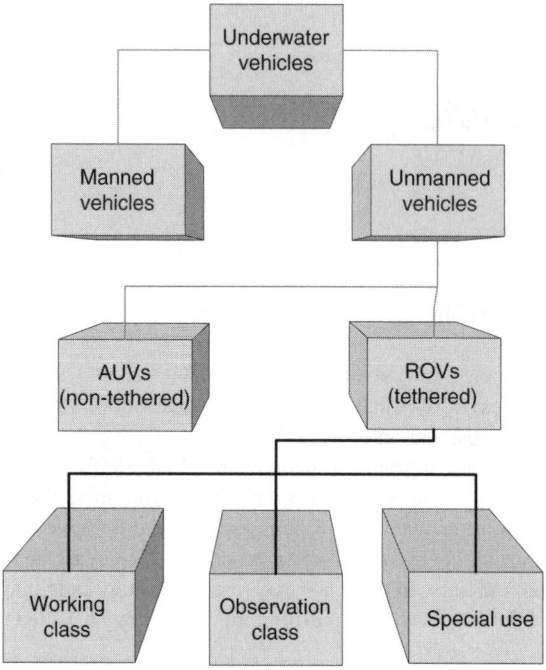

Figure 1.1 *Underwater vehicles to ROVs.*

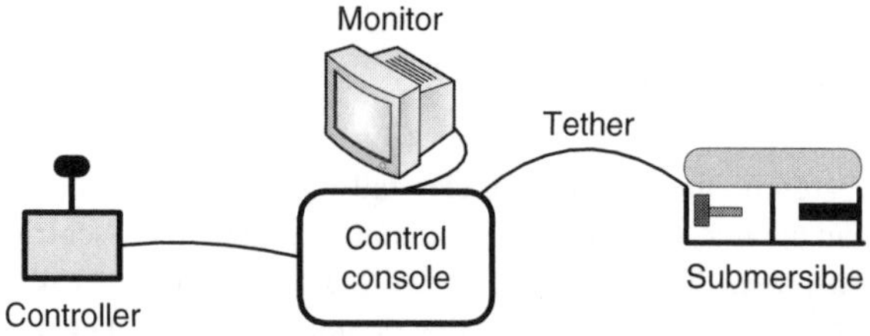

Figure 1.2 *Basic ROV system components.*

Simplistically, an ROV is a camera mounted in a waterproof enclosure, with thrusters for maneuvering, attached to a cable to the surface over which a video signal is transmitted (Figure 1.2). Practically all of today's vehicles use common consumer industry standards for commercial off-the-shelf (COTS) components. The following section will provide a better understanding of the scope of this definition.

1.3 IN THE BEGINNING

One way to discuss the historical development of ROVs is to consider them in terms of the cycle of life – from infancy to maturity. Anyone who has raised a child will quickly

Figure 1.3 *The Navy's* CURV II *vehicle.*

understand such a categorization. In the beginning the ROV child was 'nothing but a problem: Their bottles leaked, their hydraulics failed, sunlight damaged them, they were too noisy and unreliable, were hard to control and needed constant maintenance. Beginning to sound familiar?' (Wernli, 1998).

Some have credited Dimitri Rebikoff with developing the first ROV – the *POODLE* – in 1953. However, the vehicle was used primarily for archeological research and its impact on ROV history was minimal – but it was a start.

Although entrepreneurs like Rebikoff were making technology breakthroughs, it took the US Navy to take the first real step to an operational system. The Navy's problem was the recovery of torpedoes that were lost on the seafloor. Replacing a system that essentially grappled for the torpedo, the Navy (under a contract awarded to VARE Industries, Roselle, New Jersey) developed a maneuverable underwater camera system – a Mobile Underwater Vehicle System. The original VARE vehicle, the *XN-3*, was delivered to the Naval Ordnance Test Station (NOTS) in Pasadena, California, in 1961. This design eventually became the Cable-Controlled Underwater Research Vehicle (*CURV*).

The Navy's *CURV* (and its successor – *CURV III*) made national headlines twice:

- The *CURV* retrieved a lost atomic bomb off the coast of Palomares, Spain in 1966, from 2850 feet (869 meters) of water, even though working beyond its maximum depth. The *CURV*'s sister vehicle, *CURV II*, is shown in Figure 1.3.
- *CURV III*, which had become a 'flyaway' system, was sent on an emergency recovery mission from San Diego to a point offshore, near Cork, Ireland in 1973. With little air left for the two pilots of the *PISCES III* manned submersible, which was trapped on the bottom in 1575 feet (480 meters) of water, the *CURV III* attached a recovery line that successfully pulled the doomed crew to safety.

Figure 1.4 *Co-author R. Wernli (right) directs the launch of the US Navy's WSP/PIV.*

With such successes under its belt, the Navy expanded into more complex vehicles, such as the massive Pontoon Implacement Vehicle (PIV), which was developed to aid in the recovery of sunken submarines, shown with the integrated Work Systems Package (WSP) (Figure 1.4).

At the other end of the scale, the US Navy developed one of the very first small-size observation ROVs. The *SNOOPY* vehicle, which was hydraulically operated from the surface, was one of the first portable vehicles (Figure 1.5).

This version was followed by the *Electric SNOOPY*, which extended the vehicle's reach by going with a fully electric vehicle. Eventually sonars and other sensors were added and the childhood of the small vehicles had begun.

Navy-funded programs helped Hydro Products (San Diego, CA) get a jump on the ROV field through the development of the *TORTUGA*, a system dedicated to investigating the utility of a submarine-deployed ROV. These developments led to Hydro Products' RCV line of 'flying eyeball' vehicles (Figure 1.6).

These new intruders, albeit successful in their design goals, still could not shake that lock on the market by the manned submersibles and saturation divers. In 1974, only 20 vehicles had been constructed, with 17 of those funded by various governments. Some of those included:

- France – *ERIC* and *Telenaute* and ECA with their *PAP* mine countermeasure vehicles.
- Finland – *PHOCAS* and Norway – the *SNURRE*
- UK – British Aircraft Corporation (*BAC-1*) soon to be the *CONSUB 01*; *SUB-2*, *CUTLET*
- Heriot-Watt University, Edinburgh – *ANGUS* (001, 002, and 003)
- Soviet Union – *CRAB-4000* and *MANTA* vehicles.

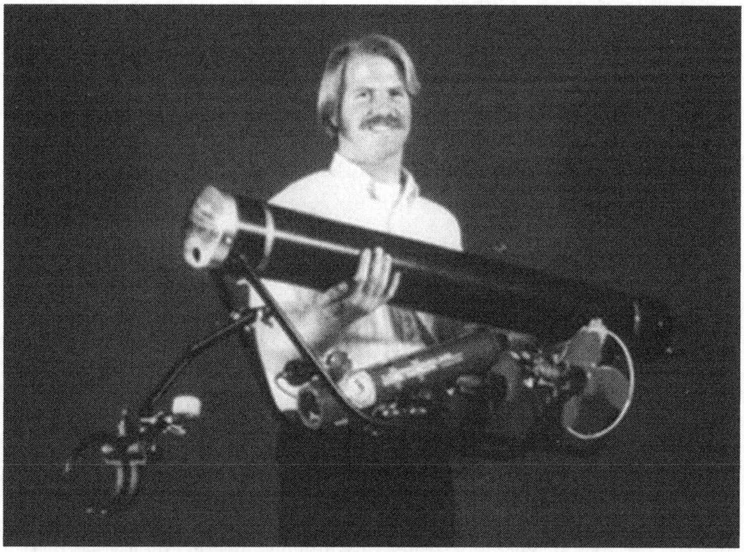

Figure 1.5 *US Navy's hydraulic* SNOOPY.

Figure 1.6 *Hydro Products' RCV 225 and RCV 150 vehicles.*

It could be said that ROVs reached adolescence, which is generally tied to a growth spurt, accented by bouts of unexplained or irrational behavior, around 1975. With an exponential upturn, the number of vehicles grew to 500 by the end of 1982. And the funding line also changed during this period. From 1953 to 1974, 85 percent of the vehicles built were government funded. From 1974 to 1982, 96 percent of the 350 vehicles produced were funded, constructed, and/or bought by private industry.

The technological advancements necessary to take ROVs from adolescence to maturity had begun. This was especially true in the electronics industry, with the miniaturization of the onboard systems and their increased reliability. With the ROV beginning to be accepted by the offshore industry, other developers and vehicles began to emerge:

- USA – Hydro Products – the *RCV 125*, *TORTUGA*, *ANTHRO*, and *AMUVS* were soon followed by the *RCV 225*, and eventually the *RCV 150*; AMETEK, Straza Division, San Diego – turned their Navy funded *Deep Drone* into their *SCORPIO* line; Perry Offshore, Florida – started their RECON line of vehicles based on the US Navy's *NAVFAC SNOOPY* design.
- Canada – International Submarine Engineering (ISE) started in Canada (*DART*, *TREC*, and *TROV*).
- France – Comex Industries added the *TOM-300*, C.G. Doris produced the *OBSERVER* and *DL-1*.
- Italy – Gay Underwater Instruments unveiled their spherical *FILIPPO*.
- The Netherlands – Skadoc Submersible Systems' *SMIT SUB* and *SOP*.
- Norway – Myers Verksted's *SPIDER*.
- Sweden added SUTEC's *SEA OWL* and Saab-Scandia's *SAAB-SUB*.
- UK – Design Diving Systems' *SEA-VEYOR*, Sub Sea Offshore's *MMIM*, Underwater Maintenance Co.'s *SCAN*, Underwater and Marine Equipment Ltd.'s *SEA SPY*, *AMPHORA*, and *SEA PUP*, Sub Sea Surveys Ltd.'s *IZE*, and Winn Technology Ltd.'s *UFO-300*, *BOCTOPUS*, *SMARTIE*, and *CETUS*.
- Japan – Mitsui Ocean Development and Engineering Co., Ltd. had the *MURS-100*, *MURS-300* and *ROV*.
- Germany – Preussag Meerestechnik's *FUGE*, and VFW-Fokker GmbH's *PINGUIN B3* and *B6*.
- Other US – Kraft Tank Co. (*EV-1*), Rebikoff Underwater Products (*SEA INSPECTOR*), Remote Ocean Systems (*TELESUB-1000*), Exxon Production Research Co. (*TMV*), and Harbor Branch Foundation (*CORD*).

From 1982 to 1989 the ROV industry grew rapidly. The first ROV conference, ROV '83, was held with the theme 'A Technology Whose Time Has Come!' Things had moved rapidly from 1970, when there was only one commercial ROV manufacturer. By 1984 there were 27. North American firms (Hydro Products, AMETEK, and Perry Offshore) accounted for 229 of the 340 industrial vehicles produced since 1975.

Not to be outdone, Canadian entrepreneur Jim McFarlane bought into the business with a series of low-cost vehicles – *DART*, *TREC*, and *TROV* – developed by International Submarine Engineering (ISE) in Vancouver, British Columbia. But the market was cut-throat; the dollar to pound exchange rate caused the ROV technology base to transfer to the UK in support of the Oil and Gas Operations in the North Sea. Once the dollar/pound exchange rate reached parity, it was cheaper to manufacture vehicles in the UK. Slingsby Engineering, Sub Sea Offshore, and the OSEL Group cornered the North Sea market and the once dominant North American ROV industry was soon decimated. The only North American survivors were ISE (due to their diverse line of systems and the can-do attitude of their owner) and Perry, which wisely teamed with their European competitors to get a foothold in the North Sea.

However, as the oil patch companies were fighting for their share of the market, a few companies took the advancements in technology and used them to shrink the ROV to a new class of small, reliable, observation-class vehicles. These vehicles, which were easily portable when compared to their larger offshore ancestors, were produced at a cost that civil organizations and academic institutions could afford.

The *MiniRover*, developed by Chris Nicholson, was the first real low-cost, observation-type ROV. This was soon followed by Deep Ocean Engineering's (DOE's) *Phantom* vehicles. Benthos (now Teledyne Benthos) eventually picked up the *MiniRover* line and, along with DOE, cornered the market in areas that included civil engineering, dam and tunnel inspection, police and security operations, fisheries, oceanography, nuclear plant inspection, and many others.

The 1990s saw the ROV industry reach maturity. Testosterone-filled ROVs worked the world's oceans; No job was too hard or too deep to be completed. The US Navy, now able to buy vehicles off the shelf as needed, turned its eyes toward the next milestone – reaching the 20 000-foot (6279 meters) barrier. This was accomplished in 1990, not once, but twice:

- *CURV III*, operated by Eastport International for the US Navy's Supervisor of Salvage, reached a depth of 20 105 feet (6128 meters).
- The Advanced Tethered Vehicle (*ATV*), developed by the Space and Naval Warfare Systems Center, San Diego broke the record less than a week later with a record dive to 20 600 feet (6279 meters).

It didn't take long and this record was not only beaten by Japan, but obliterated. Using JAMSTEC's *Kaiko* ROV (Figure 1.7), Japan reached the deepest point in the Mariana Trench – 35 791 feet (10 909 meters) – a record that can be tied, but never exceeded.

The upturn in the offshore oil industry is increasing the requirement for advanced undersea vehicles. Underwater drilling and subsea complexes are now well beyond diver depth, some exceeding 3000 meters deep. Due to necessity, the offshore industry has teamed with the ROV developers to ensure integrated systems are being designed that can be installed, operated, and maintained through the use of remotely operated vehicles. ROVs, such as Schilling Robotics UHD ROV (Figure 1.8), are taking underwater intervention to a higher technological level.

In the late 1990s (Wernli, 1998), it was estimated that there were over 100 vehicle manufacturers, and over 100 operators using approximately 3000 vehicles of various sizes and capabilities. According to the 2006 edition of *Remotely Operated Vehicles of the World*, there are over 450 builders and developers of ROVs (including AUVs), and over 175 operators. As far as the number of actual vehicles in the field … well, tracking that number will be left to the statisticians. However, the number of small, observation-class vehicles being used today probably comes close to the all-inclusive 3000 vehicles worldwide in 1998, which provides a perfect segue into the next section.

Figure 1.7 Kaiko – *the world's deepest diving ROV.*

1.4 TODAY'S OBSERVATION-CLASS VEHICLES

One can spend a lifetime running numbers and statistics; However, it can be easily said that the technology is here to reach into the shallower depths in a cost-effective manner and complete a series of critical missions. Whether performing dam inspections, body recoveries, fish assessment or treasure hunting, technology has allowed the development of the advanced systems necessary to complete the job – and stay dry at the same time.

Technology has moved from vacuum tubes, gear trains and copper/steel cables to microprocessors, magnetic drives and fiber-optic/Kevlar cables. That droopy-drawered infant has now graduated from college and can work reliably without constant maintenance. As computers have moved from trunk-sized 'portable' systems to those that fit in a pants pocket, observation-class ROVs have moved from

Figure 1.8 *Schilling Robotics UHD.*

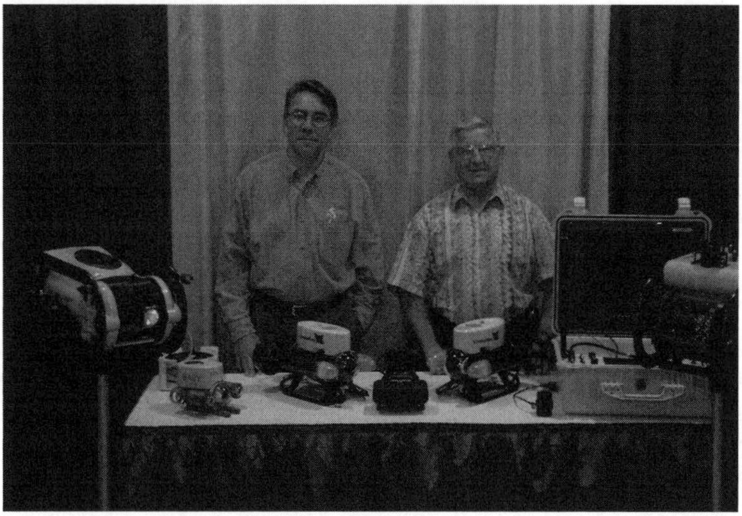

Figure 1.9 *Authors Bob Christ (L) and Robert Wernli (R) with examples of today's observation class vehicles.*

portable, i.e. a team of divers can carry one, to handheld vehicles that can complete the same task. Examples of observation class vehicles in use today are shown in Figure 1.9. Table 1.1 provides a summary of those vehicles that weigh in less than 70 kg (150 lb) and have 25 or more in the field.

Table 1.1 *Observation-class vehicles – small vehicles that weigh less than approx. 70 kg (150 lb) with over 25 sold.*

Name	*Company*	*Wt. (kg) in air*	*Depth (m)*	*Built*
AC-ROV	AC-CESS CO, UK	3	75	75
Firefly	Deep Ocean Engineering, USA	5.4	46	30
H300	ECA Hytec, France	65	300	24
Hyball	SMD Hydrovision Ltd., UK	41	300	185
Little Benthic Vehicle (LBV)	SeaBotix, Inc., USA	10–15	150–1500	300
Navaho	Sub-Atlantic (SSA alliance), UK	42	300	35
Offshore Hyball	SMD Hydrovision Ltd., UK	60	300	50
Outland 1000	Outland Technology Inc., USA	17.7	152	39
Phantom 150	Deep Ocean Engineering, USA	14	46	27
Phantom XTL	Deep Ocean Engineering, USA	50	150	81
Prometeo	Elettronica Enne, Italy	48–55	–	36
RTVD-100MKIIEX	Mitsui, Japan	42	150	310*
Seaeye 600 DT	Seaeye Marine Ltd., UK	65	300	63
Seaeye Falcon	Seaeye Marine Ltd., UK	50	300–1000	72
Stealth	Shark Marine Technologies Inc., Canada	40	300	50+
VideoRay†	VideoRay LLC, US	4–4.85	0–305	550+

* Includes RTV-100.

† Includes Deep Blue, Explorer, Pro 3XE, and Pro 3XE GTO.

Information from: *Remotely Operated Vehicles of the World*, 7th Edition, Clarkson Research Services Ltd., 2006/2007, ISBN 1-902157-75-3.

The following chapters will provide an overview of observation-class ROVs, related technologies, the environment they will work in, and words to the wise about how to – and how not to – use them to complete an underwater task. How well this is done will show up in the History chapter of future publications.

Chapter 2
ROV Design

This chapter describes the different types of underwater systems, the basic theory behind vehicle design/communication/propulsion/integration, and explains the means by which a typical ROV gets everyday underwater tasks performed.

2.1 UNDERWATER VEHICLES TO ROVs

The previous chapter provided an introduction to what an ROV is and is not, along with a brief history of how these underwater robots arrived at their present level of worldwide usage. Since these vehicles are an extension of the operator's senses, the communication with them is probably the most critical aspect of vehicle design.

The communication and control of underwater vehicles is a complex issue and sometimes occludes the lines between ROV and AUV. Before addressing the focus of this text, the issues involved will be investigated further.

The basic issues involved with underwater vehicle power and control can be divided into the following categories:

- Power source for the vehicle
- Degree of autonomy (operator controlled or program controlled)
- Communications linkage to the vehicle.

2.1.1 Power Source for the Vehicle

Vehicles can be powered in any of the following three categories: Surface-powered, vehicle-powered, or a hybrid system.

- *Surface-powered* vehicles must, by practicality, be tethered, since the power source is from the surface to the vehicle. The actual power protocol is discussed more fully later in this text, but no vehicle-based power storage is defined within this power category.
- *Vehicle-powered* vehicles store all of their power-producing capacity on the vehicle in the form of a battery, fuel cell, or some other means of power storage needed for vehicle propulsion and operation.
- *Hybrid system* involves a mixture of surface and submersible supplied power. Examples of the hybrid system include the battery-powered submersible with a surface-supplied charger (through a tether) for recharging during times of less-than-maximum power draw; a surface-powered vehicle with an on-board power source for a transition from ROV to AUV (some advanced capability torpedo

designs allow for swim-out under ship's power to transition to vehicle power after clearing the area) and other variations to this mix.

2.1.2 Degree of Autonomy

According to the National Institute of Standards and Technology (Huang, 2004), unmanned vehicles may be operated under several modes of operation, including fully autonomous, semi-autonomous, tele-operation, and remote control.

- *Fully autonomous* – A mode of operation of an unmanned system (UMS) wherein the UMS is expected to accomplish its mission, within a defined scope, without human intervention. Note that a team of UMSs may be fully autonomous while the individual team members may not be, due to the need to coordinate during the execution of team missions.
- *Semi-autonomous* – A mode of operation of a UMS wherein the human operator and/or the UMS plan(s) and conduct(s) a mission and requires various levels of human–robot interaction (HRI).
- *Tele-operation* – A mode of operation of a UMS wherein the human operator, using video feedback and/or other sensory feedback, either directly controls the motors/actuators or assigns incremental goals, waypoints in mobility situations, on a continuous basis, from off the vehicle and via a tethered or radio/acoustic/optic/other linked control device. In this mode, the UMS may take limited initiative in reaching the assigned incremental goals.
- *Remote control* – A mode of operation of a UMS wherein the human operator, without benefit of video or other sensory feedback, directly controls the actuators of the UMS on a continuous basis, from off the vehicle and via a tethered or radio-linked control device using visual line-of-sight cues. In this mode, the UMS takes no initiative and relies on continuous or nearly continuous input from the user.

2.1.3 Communications Linkage to the Vehicle

The linkage to the vehicle can come in several forms or methods depending upon the distance and medium through which the communication must take place. Such linkages include:

- Hard-wire communication (either electrical or fiber optic)
- Acoustic communication (via underwater analog or digital modem)
- Optical communication (while on the surface)
- Radio frequency (RF) communication (while on or near the surface).

What is communicated between the vehicle and the operator can be any of the following:

- *Telemetry* – The measurement and transmission of data or video through the vehicle via tether, RF, optical, acoustic, or other means

- *Tele-presence* – The capability of an unmanned system to provide the human operator with some amount of sensory feedback similar to that which the operator would receive if inside the vehicle
- *Control* – The upload/download of operational instructions (for autonomous operations) or full tele-operation.
- *Records* – The upload/download of mission records and files.

ROVs receive their power, their data transmission or their control (or all three) directly from the surface through direct hard-wire communication (i.e. the tether). In short, the difference between an ROV and an AUV is the tether (although some would argue that the divide is not that simple).

2.1.4 Special-Use ROVs

Some of the special-use remotely operated vehicles come in even more discriminating packages:

- *Rail cameras* – Work on the drilling string of oil and gas platforms/drilling rigs (a pan and tilt camera moving up and down a leg of the platform to observe operations at the drill head with or without intervention tooling)
- *Bottom crawlers* – Lay pipe as well as communications cables and such while heavily weighted in the water column and being either towed or on tracks for locomotion
- *Towed cameras* – Can have movable fins that allow 'sailing' up or down (or side to side) in the water column behind the towing vehicle
- *Swim-out ROVs* – Smaller free-swimming systems that launch from larger ROV systems.

Although this text covers many of the technologies associated with all underwater vehicles, the subject matter will focus on the free-swimming, surface-powered, teleoperated (or semi-autonomous) observation-class remotely operated vehicle with submersible weights of less than 200 pounds (91 kg).

2.2 AUTONOMY PLUS: 'WHY THE TETHER?'

In order to illustrate where ROVs fit into the world of technology, an aircraft analogy will be discussed first, then the vehicle in its water environment.

Autonomy with regard to aerial vehicles runs the full gamut from man occupying the vehicle while operating it (e.g. a pilot sitting in the aircraft manipulating the controls for positive navigation) to artificial intelligence on an unmanned aerial vehicle making unsupervised decisions on navigation and operation from start to finish (Figure 2.1). However, where the human sits (in the vehicle or on a separate platform) is irrelevant to the autonomy discussion, since it does not effect how the artificial brain (i.e. the controller) thinks and controls.

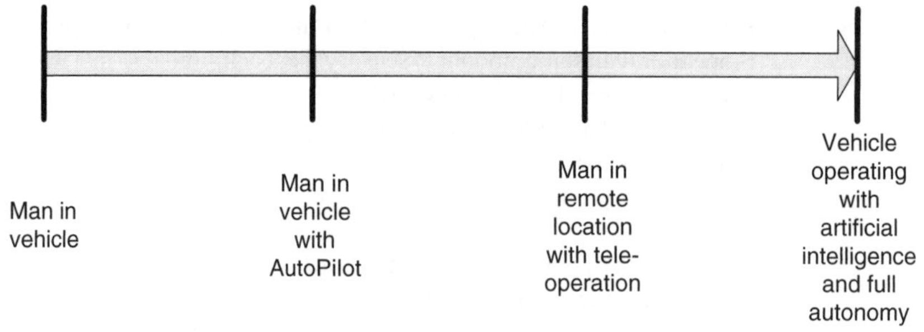

Figure 2.1 *Degrees of autonomy.*

2.2.1 An Aircraft Analogy

To set the stage with an area most are familiar with, the control variations of an aircraft will be defined as follows:

- *Man in vehicle* – Pilot sitting aboard the aircraft in seat manning controls.
- *Man in vehicle with AutoPilot* – Pilot sitting aboard aircraft in seat with AutoPilot controlling the aircraft's navigation (pilot supervising the systems).
- *Man in remote location with tele-operation* – Technician sitting in front of control console on the ground (or another aerial platform) with radio frequency link to the unmanned aircraft while the technician is manipulating the controls remotely.
- *Vehicle operating with artificial intelligence and full autonomy* – No human supervisor directly controlling the vehicle. The vehicle controls are pre-programmed with the vehicle making objective decisions as to the conduct of that flight from inception to termination.

Predator UAVs have recently been retrofitted with weapons, producing the new designation UCAV (unmanned combat aerial vehicle). The most efficient technology would allow that UCAV (without human intervention) to find, detect, classify, and deliver a lethal weapon upon the target, thus eliminating the threat. And here is the crux – is a responsible commander in the field comfortable enough with the technology to allow a machine the decision of life and death? This may be an extreme example, but (for now) a human must remain in the decision loop. To continue this example, would any passenger (as a passenger) fly in a commercial airliner without a pilot physically present in the cockpit? This may not happen soon, but one can safely predict that unpiloted airliners are in our future. Unattended trolleys are currently used in many airports worldwide.

2.2.2 Underwater Vehicle Variations

Now, the aircraft analogy will be reconsidered with underwater vehicle control in mind.

- *Man in vehicle* – Manned submersible pilot sitting aboard the vehicle underwater in the pilot's seat manning the controls.

- *Man in vehicle with AutoPilot* – Same situation with AutoPilot controlling the submersible's navigation (pilot supervising the systems).
- *Man in remote location with tele-operation* – Technician sitting in front of control console on the surface (or other submerged platform) with tether or other data link to the submersible while the technician is manipulating the controls remotely.
- *Vehicle operating with artificial intelligence and full autonomy* – No human supervisor directly controlling the vehicle. The vehicle controls of the Autonomous Underwater Vehicle (AUV) are pre-programmed, with the vehicle making objective decisions as to the conduct of that dive from inception to termination.

During operation Iraqi Freedom, Mine Countermeasure (MCM) AUVs were used for mine clearance operations. The AUV swam a pre-programmed course over a designated area to search and detect mine-like objects on the bottom. Other vehicles (or marine mammals or divers) were then sent to these locations to classify and (if necessary) neutralize the targets.

The new small UUVs are going through a two-stage process where they Search (or survey), Classify, and Map. The Explosive Ordinance Disposal (EOD) personnel then return with another vehicle (or marine mammal or human divers) to Reacquire, Identify and Neutralize the target. Essentially, the process is to locate mine-like targets, classify them as mines if applicable, then neutralize them. What if the whole process can be done with one autonomous vehicle? And again the crux – is the field commander comfortable enough with the vehicle's programming to allow it to distinguish between a Russian KMD-1000 Bottom-type influence mine and a manned undersea laboratory before destroying the target? For the near term, man will remain in the decision loop for the important operational decisions. But again, one can safely predict that full autonomy is in our future.

2.2.3 Why the Tether?

Radio frequency (RF) waves penetrate only a few wavelengths into water due to water's high attenuation of its energy. If the RF is of a low frequency, the waves will penetrate further into water due to longer wavelengths. But with decreasing RF frequencies, data transmission rates suffer. In order to perform remote inspection tasks, live video is needed at the surface so that decisions by humans can be made on navigating the vehicle and inspecting the target. Full tele-operation (under current technology) is possible only through a high-bandwidth data link.

With the UAV example above, full tele-operation was available via the RF link (through air) between the vehicle and the remote operator. In water, this full telemetry is not possible (with current technologies) through an RF link. Acoustic in-water data transmission (as of 2006) is limited to less than 100 kilobytes per second (insufficient for high-resolution video images). A hard-wire link to the operating platform is needed to have a full tele-operational in-water link to the vehicle. Thus, the need exists for a hard-wire link of some type, in the foreseeable future, for real-time underwater inspection tasks.

2.2.4 Tele-operation Versus Remote Control

An ROV pilot will often operate a vehicle remotely with his/her eyes directly viewing the vehicle while guiding the vehicle on the surface to the inspection target. This navigation of the vehicle through line of sight (as with the remote control airplane) is termed Remote Control (RC) mode. Once the inspection target is observed through the vehicle's camera or sensors, the transition is made from RC operational mode to tele-operation mode. This transition is important because it changes navigation and operation of the vehicle from the operator's point of view to the vehicle's point of view. Successful management of the transition between these modes of operation during field tasks will certainly assist in obtaining a positive mission completion.

Going back to the UAV analogy, many kids have built and used RC model aircraft. The difference between an RC aircraft and a UAV is the ability to navigate solely by use of onboard sensors. A UAV can certainly be operated in an RC mode while the vehicle is within line-of-sight of the operator's platform, but once line-of-sight is lost, navigation and control are only available through tele-operation or pre-programming.

The following is an example of this transition while performing a typical observation-class ROV inspection of a ship's hull: The operator swims the vehicle on the surface (Figure 2.2) via RC to the hull of the vessel until the inspection starting point is gained with the vehicle's camera, then transitions to navigation via the vehicle's camera.

2.2.5 Degrees of Autonomy

An open-loop control system is simply a condition on a functioning machine whereby the system has two basic states: 'On' or 'Off'. The machine will stay On/Off for as long as the operator leaves it in that mode. The term 'open-loop' (or essentially 'no loop') refers to the lack of sensor feedback to control the operation of the machine. An example of an open-loop feedback would be a simple light switch that, upon activation, remains in the 'On' or 'Off' condition until manually changed.

Beginning with pure tele-operation (which is no autonomy), the first step toward full autonomy is the point at which the vehicle begins navigation autonomously within given parameters. This is navigation through 'closed-loop feedback'.

Closed-loop feedback is simply control of an operation through sensor feedback to the controller. A simple example of a closed-loop feedback system is the home air-conditioning thermostat. At a given temperature, the air-conditioner turns on, thus lowering the temperature of the air surrounding the thermostat (if the air-conditioning is ducted into that room). Once the air temperature reaches a certain pre-set value, the thermostat sends a signal to the air-conditioner (closing the control signal and response loop) to 'turn off', completing this simple closed-loop feedback system.

The most common first step along this line for the ROV system is the auto heading and auto depth functions. Any closed-loop feedback control system can operate on an ROV system, manipulating control functions based upon sensor output. Operation of the vertical thruster as a function of constant depth (as measured by the variable water pressure transducer) is easily accomplished in software to provide auto depth capability. For example, consider an auto depth activation system on an ROV at 100 feet (30 meters) of seawater. The approximate (gauge) pressure is 3 atmospheres or

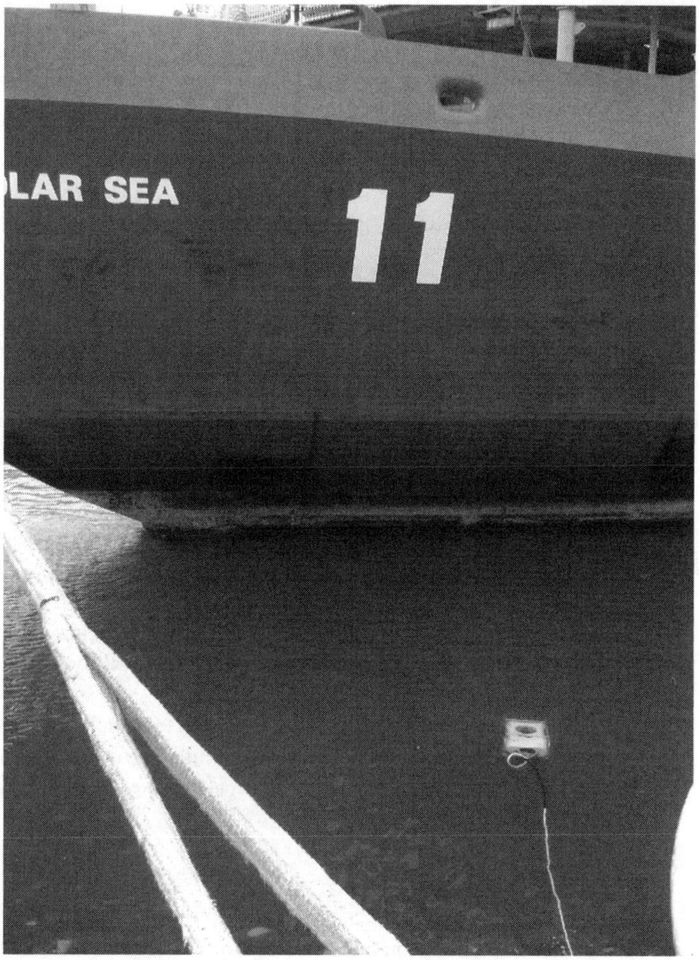

Figure 2.2 *Surface swim in RC mode.*

45 psig (3 bar). As the submersible sinks below that pressure (as read by the pressure transducer on the submersible), the controller switches on the vertical thruster to propel the vehicle back toward the surface until the 45 psig (3 bar) reading is reacquired (the reverse is also applicable).

The same applies to auto altitude, where variation of the vertical thruster maintains a constant height off the bottom based upon echo soundings from the vehicle's altimeter. Similarly, auto standoff from the side of a ship for hull inspections can be based upon a side-looking acoustic sensor, where variation of the sounder timing can be used to vary the function of the lateral thruster.

Any number of closed-loop variables can be programmed. The submersible can then be given a set of operating instructions based upon a matrix of 'if/then' commands to accomplish a given mission. The autonomy function is a separate issue from communications. A tethered ROV can be operated in full autonomy mode just as an untethered AUV may be operated in full autonomy mode. The only difference between a fully autonomous ROV and a fully autonomous AUV is generally considered to be the presence/absence of a hard-wire communications link, i.e. a tethered AUV is actually an ROV.

2.3 THE ROV

The following sections will address the most critical areas of ROV design and operational considerations.

2.3.1 What is the Perfect ROV?

This section, and those to follow, will investigate the premise that vehicle geometry does not affect the motive performance of an ROV (over any appreciable tether length) nearly as much as the dimensions of the tether.

Accordingly, the perfect ROV would have the following characteristics:

- Minimal tether diameter (for instance, a single strand of unshielded optical fiber)
- Powered from the surface having unlimited endurance (as opposed to battery operated with limited power available)
- Very small in size (to work around and within structures)
- Have an extremely high data pipeline for sensor throughput.

ROV systems are a trade-off of a number of factors, including cost, size, deployment resources/platform, and operational requirement. But the bottom line, which will become obvious in the following sections, is that the tether design can help create, or destroy, the perfect ROV.

2.3.2 ROV Classifications

ROV systems come in three basic categories:

- *Observation class* – Observation-class ROVs are normally a 'flying eye' designed specifically for lighter usage with propulsion systems to deliver a camera and sensor package to a place where it can provide a meaningful picture or gather data. The newer observation-class ROVs enable these systems to do more than just see. With its tooling package and many accessories, the observation-class ROV is able to deliver payload packages of instrumentation, intervention equipment, and underwater navigational aids, enabling them to perform as a full-function underwater vehicle.
- *Work class* – Work-class systems generally have large frames (measured in multiple yards/meters) with multi-function manipulators, hydraulic propulsion/actuation, and heavy tooling meant for larger underwater construction projects where heavy equipment underwater needs movement.
- *Special use* – Special-use ROV systems describe tethered underwater vehicles designed for specific purposes. An example of a special-use vehicle is a cable burial ROV system designed to plow the sea floor to bury telecommunications cables.

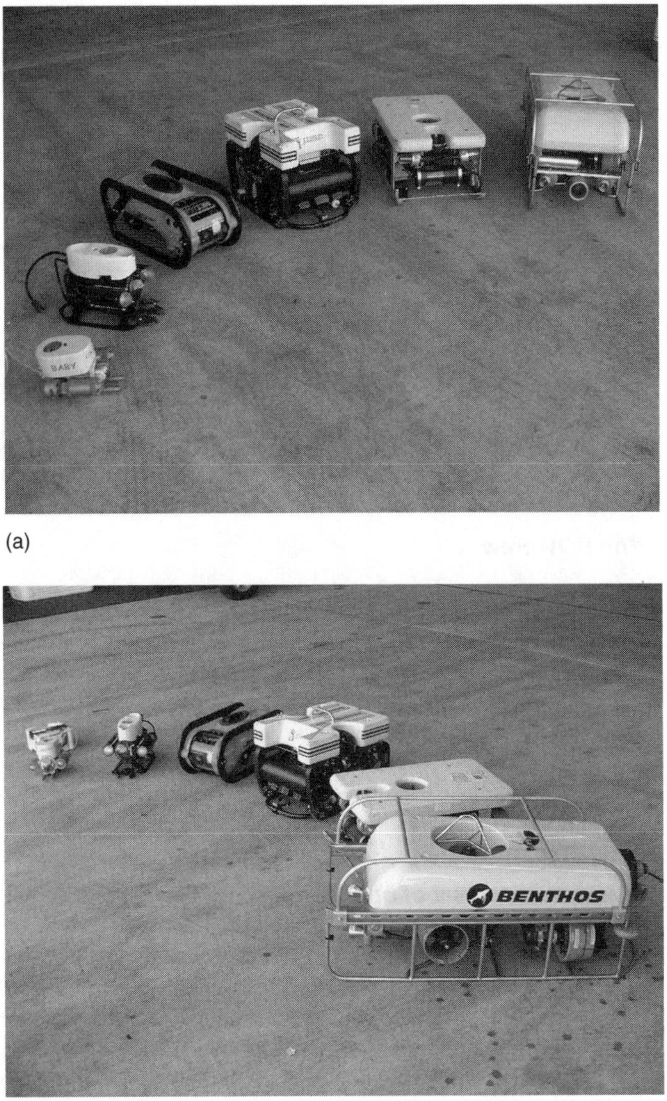

(a)

(b)

Figure 2.3 *Various system sizes of observation-class ROV systems.*

2.3.3 Size Considerations

2.3.3.1 *Vehicle size versus task suitability*

As discussed in the previous section, ROVs range from the small observation-class to large, complex work-class vehicles. The focus of this book is on the smaller observation-class systems. The various sizes of observation-class ROV systems have within them certain inherent performance capabilities (Figure 2.3). The larger systems

Table 2.1 *Task versus ROV system size.*

Tasking	*Best size*
External pipeline inspection	Large
External hull inspection	Medium/large
Internal wreck survey	Small
Open-water scientific transect	Medium/large
Calm-water operations	All sizes

have a higher payload and thruster capability, allowing better open-water operations. The smaller systems are much more agile in getting into tight places in and around underwater structures, making them more suitable for enclosed structure penetrations. The challenge for the ROV technician is to find the right system for the job.

Table 2.1 provides examples of tasks along with the 'best' system size selection (size definitions are discussed in more detail in Chapters 8 and 12).

2.3.3.2 *The ROV crew*

For a larger work-class system, the ROV crew consists of a supervisor as well as two or more team members possessing specialized knowledge in areas such as mechanical or electrical systems. On the smaller systems, the supervisor does not normally have the range of personnel due to costing issues. For deployment of an observation-class ROV system, the ROV crew should comprise, at the very least, an operator as well as a tether tender. For performing an inspection task, it is quite helpful to also have a third person to take prolific notes as well as a second set of eyes to view the job for content and completeness.

2.3.3.3 *Platform or vessel of opportunity*

The operations platform for larger ROV systems could range from a drilling rig deck to the moon pool of a specialized, dynamically positioned Diver Support Vessel (DSV) outfitted specifically for ROV operations. With observation-class systems, any number of work platforms may be used depending upon the work environment and the equipment being deployed. As a minimum, the work platform will require the following characteristics:

1. A water ingress point footprint large enough to deploy the submersible safely.
2. A comfortable platform from which to operate the video and electronics console.
3. A direct line of communication from the ROV pilot to the operator of the mobile deployment platform.
4. A sufficient power supply to run all equipment for the duration of the operation.

2.3.4 Buoyancy and Stability

As discussed further in Chapter 3, any vehicle has movement about six degrees of freedom (Figure 2.4): Three translations (surge, heave, and sway along the longitudinal,

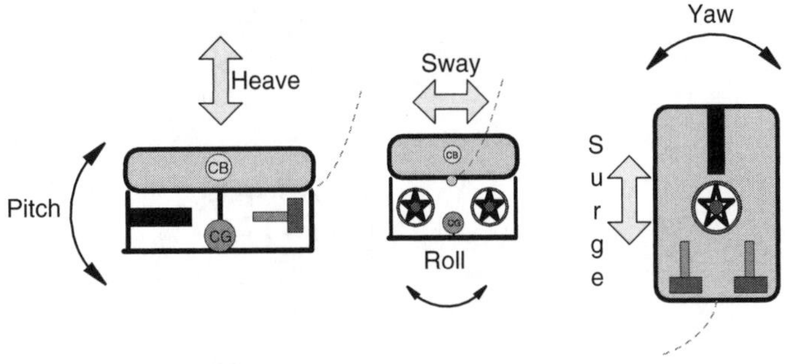

Figure 2.4 *Vehicle degrees of freedom.*

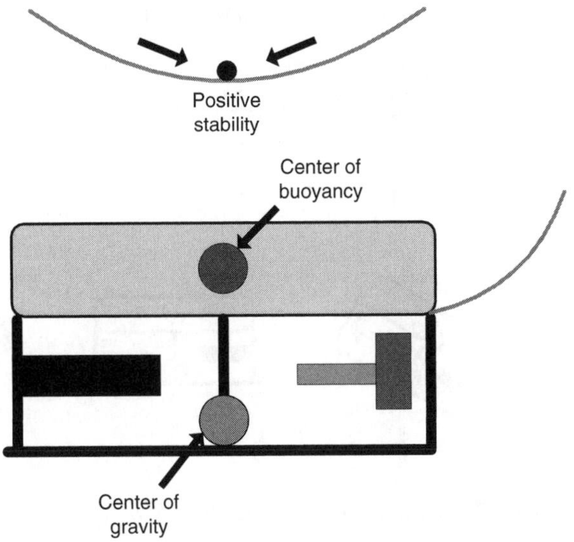

Figure 2.5 *Positive ROV stability.*

vertical, and transverse (lateral) axes respectively) and three rotations (roll, yaw, and pitch about these same respective axes). This section will address the interaction between vehicle static and dynamic stability and these degrees of freedom.

ROVs are not normally equipped to pitch and roll. The system is constructed with a high center of buoyancy and a low center of gravity to give the camera platform maximum stability about the longitudinal and lateral axes (Figure 2.5). Most small ROV systems have fixed ballast with variable positioning to allow trimming of the system nose-up/nose-down or for roll adjustment/trim. In the observation class, the lead (or heavy metal) ballast is normally located on tracks attached to the bottom frame to allow movement of ballast along the vehicle to achieve the desired trim.

2.3.4.1 *Hydrostatic equilibrium*

According to Archimedes' principle, any body partially or totally immersed in a fluid is buoyed up by a force equal to the weight of the displaced fluid. If somehow one

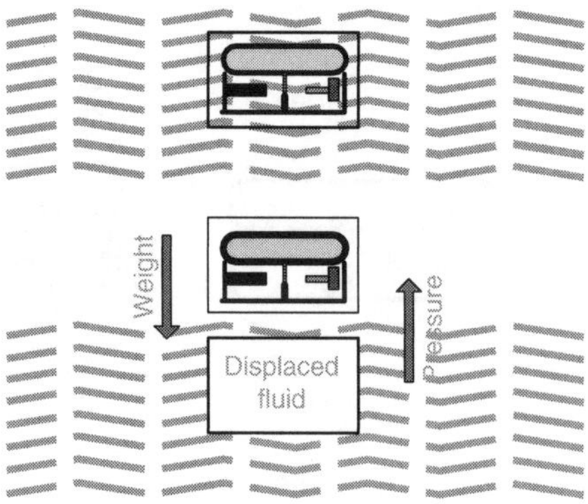

Figure 2.6 *Hydrostatic equilibrium of ROV.*

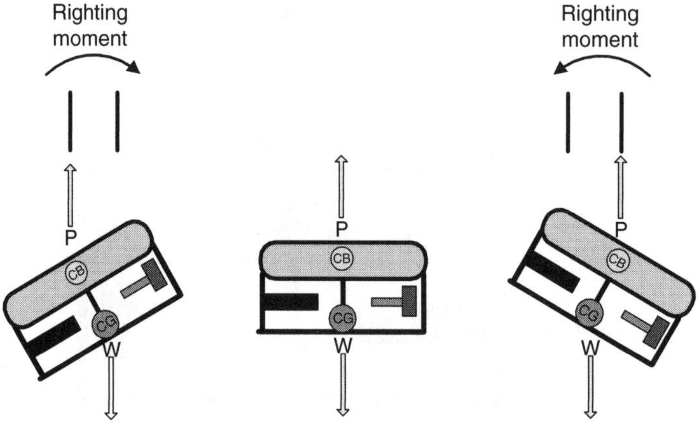

Figure 2.7 *Righting forces acting on an ROV.*

could remove the body and instantly fill the resulting cavity with fluid identical to that surrounding it, no motion would take place: The body weight would exactly equal that of the displaced fluid.

The resultant of all of the weight forces on this displaced fluid (Figure 2.6) is centered at a point within the body termed the 'center of gravity' (CG) (see Figures 2.5 and 2.7). This is the sum of all the gravitational forces acting upon the body by gravity. The resultant of the buoyant forces countering the gravitational pull acting upward through the CG of the displaced fluid is termed the 'center of buoyancy' (CB). There is one variable in the stability equation that is valid for surface vessels with non-wetted area that is not considered for submerged vehicles. The point where the CB intersects the hull centerline is termed the metacenter and its distance from the CG is termed the metacentric height (usually written as GM). For ROV considerations, all operations are with the vehicle submerged and ballasted very close to neutral buoyancy, making only the separation of the CB and the CG the applicable reference metric for horizontal stability.

Per Van Dorn (1993):

The equilibrium attitude of the buoyant body floating in calm water is determined solely by interaction between the weight of the body, acting downward through

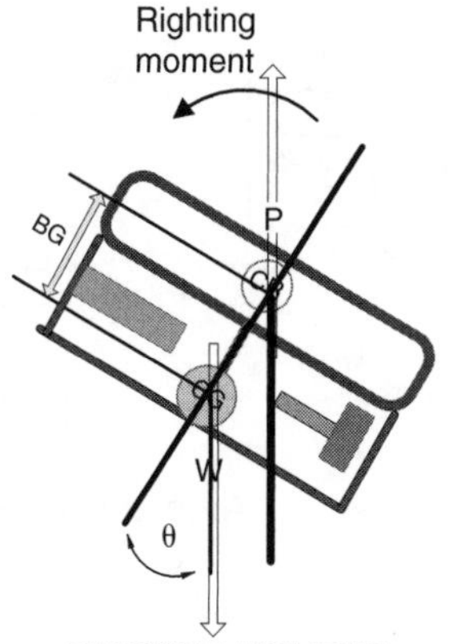

Figure 2.8 *ROV righting moment.*

its CG, and the resultant of the buoyant forces, which is equal in magnitude to the weight of the body and acts upward through the CB of the displaced water. If these two forces do not pass through the same vertical axis, the body is not in equilibrium, and will rotate so as to bring them into vertical alignment. The body is then said to be in static equilibrium.

2.3.4.2 *Transverse stability*

Paraphrasing Van Dorn (1993) to account for ROVs, having located the positions of the CG and the upright CB of the vehicle, one can now investigate the transverse (lateral) stability. This is done without regard to external forces, merely by considering the hull of the vehicle to be inclined through several angles and calculating the respective moments exerted by the vertically opposing forces of gravity and buoyancy. These moments are generated by horizontal displacements (in the vehicle's reference frame) of the CB relative to the CG, as the vehicle inclines, such that these forces are no longer collinear, but are separate by same distance d, which is a function of the angle of inclination, Θ. The magnitude of both forces remains always the same, and equal to the vehicle's weight W, but their moment ($W \times d$) is similarly a function of Θ. If the moment of the buoyancy (or any other) force acts to rotate the vehicle about its CG opposite to the direction of inclination, it is called a righting moment; If in the same direction, it is called a heeling moment.

Referring to Figure 2.8, as the BG becomes smaller, the righting moment decreases in a logarithmic fashion until static stability is lost.

2.3.4.3 *Water density and buoyancy*

It is conventional operating procedure to have vehicles positively buoyant when operating to ensure they will return to the surface if a power failure occurs. This positive buoyancy would be in the range of 1 lb (450 grams) for small vehicles and 11–15 lb

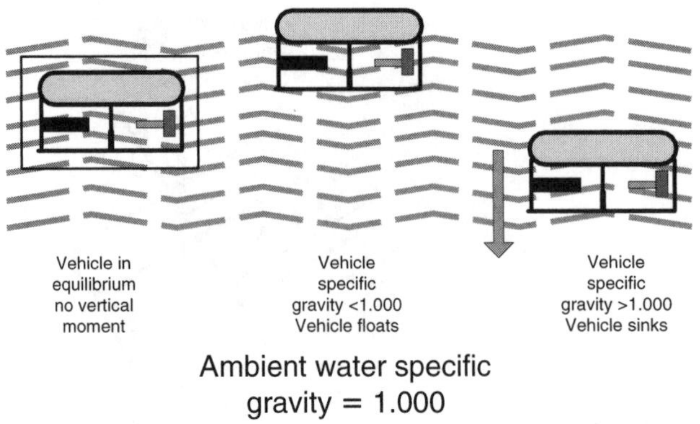

Figure 2.9 *Effect of water specific gravity on ROV buoyancy.*

(5–7 kg) for larger vehicles, and in some cases, work-class vehicles will be as much as 50 lb (23 kg) positive. Another reason for this is to allow near-bottom maneuvering without thrusting up, forcing water down, thus stirring up sediment. It also obviates the need for continual thrust reversal. Very large vehicles with variable ballast systems that allow for subsurface buoyancy adjustments are an exception. The vehicles operated by most observation-class operators will predominantly have fixed ballast.

As more fully described in Chapter 6, the makeup of the water in which the submersible operates will determine the level of ballasting needed to properly operate the vehicle. The three major water variables affecting this are temperature, salinity, and pressure.

More than 97 percent of the world's water is located in the oceans. Many of the properties of water are modified by the presence of dissolved salts. The level of dissolved salts in sea water is normally expressed in grams of dissolved salts per kilogram of water (historically expressed in imperial units as parts per thousand or 'PPT' with the newer accepted unit as the practical salinity unit or 'PSU'). Open ocean seawater contains about 35 PSU of dissolved salt. In fact, 99 percent of all ocean water has salinity of between 33 and 37 PSU.

Pure water has a specific gravity of 1.00 at maximum density temperature of about 4°C (approximately 39°F). Above 4°C, water density decreases due to molecular agitation. Below 4°C, ice crystals begin to form in the lattice structure, thereby decreasing density until the freezing point. It is well known that ice floats, demonstrating the fact that its density is lower than water.

At a salt content of 24.7 PSU, the freezing point and the maximum density temperature of sea water coincide at −1.332°C. In other words, with salt content above 24.7 PSU, there is no maximum density of sea water above the freezing point.

Most ROVs have a fixed volume. When transferring a submersible from a freshwater environment (where the system was neutrally ballasted) to a higher density salt-water environment, the ROV pilot will notice that the system demonstrates a more positive buoyancy, much like an ice cube placed into a glass of water. In order to neutralize the buoyancy of the system, ballast weights will need to be added to the submersible until neutral buoyancy is re-achieved. The converse is also true by going from salt water into fresh water or between differing temperature/salinity combinations with the submersible (Figure 2.9).

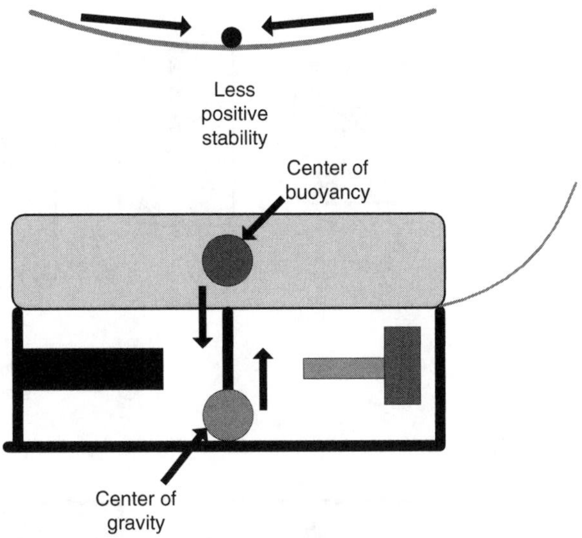

Figure 2.10 *ROV with ballast moved up.*

Water is effectively (for our purposes) incompressible. At deeper depths, water will be at a higher density, slightly affecting the buoyancy of the submersible. The water density buoyancy shift at the deeper operating depths is partially offset due to the compression of the air-filled spaces of the submersible. This balance is more or less dependent upon the system design and the amount of air-filled space within the submersible.

2.3.5 Dynamic Stability

As with a child's seesaw, the further a weight is placed from the fulcrum point, the higher the mechanical force, or moment, needed to 'upset' that weight (the term 'moment' is computed by the product of the weight times the arm or distance from the fulcrum). It is called 'positive stability' when an upset object inherently rights itself to a steady state. When adapting this to a submersible, positive longitudinal and lateral stability can be readily achieved by having weight low and buoyancy high on the vehicle. This technique produces an intrinsically stable vehicle on the pitch and roll axis. In most observation-class ROV systems, the higher the stability the easier it is to control the vehicle. With lower static stability, expect control problems (Figure 2.10).

External forces, however, do act upon the vehicle when it is in the water, which can produce apparent reductions in stability. For example, the force of the vertical thruster when thrusting down appears to the vehicle as an added weight high on the vehicle and, in turn, makes the center of gravity appear to rise, which destabilizes the vehicle in pitch and roll. The center of buoyancy and center of gravity can be calculated by taking moments about some arbitrarily selected point.

Other design characteristics also affect the stability of the vehicle along the varying axis. The so-called 'aspect ratio' (total mean length of the vehicle versus total mean width of the vehicle) will determine the vehicle's hull stability (Figure 2.11), as will thruster placement (Figure 2.12).

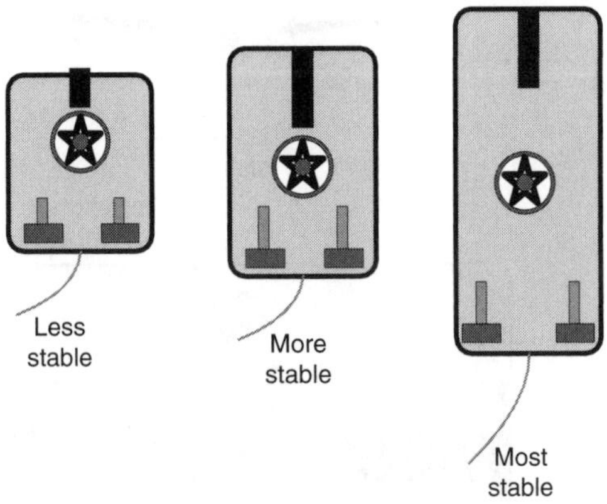

Figure 2.11 *Vehicle geometry and stability.*

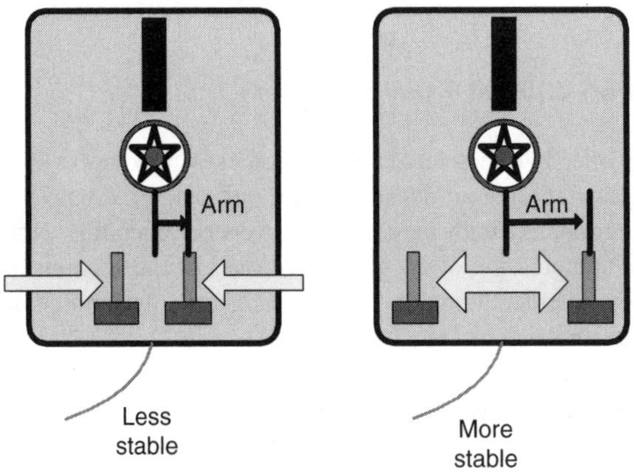

Figure 2.12 *Thruster placement and stability.*

Most attack submarine designers specify a 7:1 aspect ratio as the optimum for the maneuvering-to-stability ratio (see Moore and Compton-Hall, 1987). For ROVs, the optimal aspect ratio and thruster placement will be dependent upon the anticipated top speed of the vehicle, along with the need to maneuver in confined spaces.

2.3.5.1 *Mission-related vehicle trim*

Two examples of operational situations where observation-class ROV trim could be adjusted to assist in the completion of the mission follow:

1. If an ROV pilot requires the vertical viewing of a standpipe with a camera tilt that will not rotate through 90°, the vehicle may be trimmed to counter the lack in camera mobility (Figure 2.13).
2. If the vehicle is trimmed in a bow-low condition while performing a transect or a pipeline survey (Figure 2.14), when the thrusters are operated the vehicle will

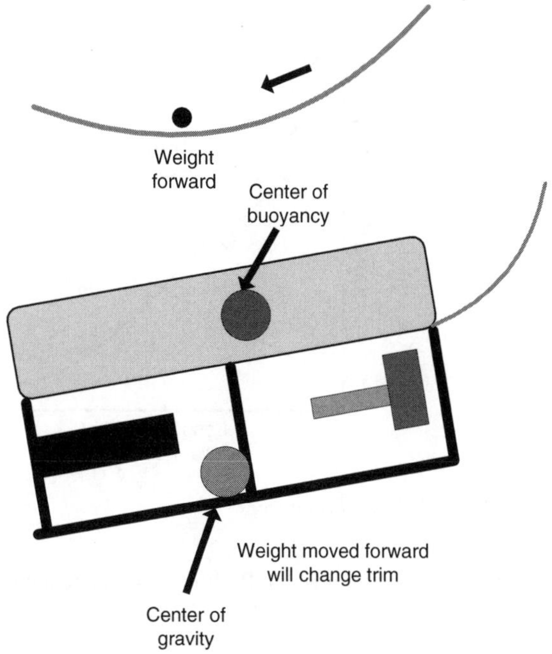

Figure 2.13 *Vehicle trim with weight forward.*

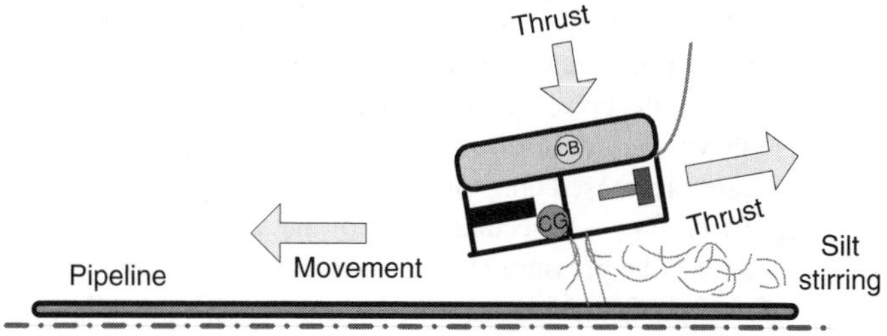

Figure 2.14 *Movement down a pipeline with vehicle out of trim.*

tend to drive into the bottom, requiring vertical thrust (and stirring up silt in the process). The vehicle ballast could be moved aft to counter this condition.

2.3.5.2 *Point of thrust/drag*

Another critical variable in the vehicle control equation is the joint effect of both the point of net thrust (about the various axis) and the point of effective total drag.

The drag perspective will be considered first. One can start with the perfect drag for a hydrodynamic body (like an attack submarine) then work toward some practical issue of manufacturing a small ROV.

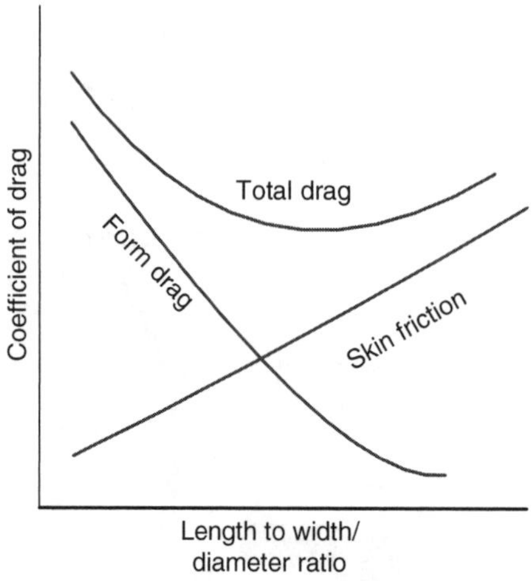

Figure 2.15 *Vehicle drag curves.*

As stated best in Burcher and Rydill (1994), there are two basic types of drag with regard to all bodies:

1. *Skin friction drag* – Friction drag is created by the frictional forces acting between the skin and the water. The viscous shear drag of water flowing tangentially over the surface of the skin contributes to the resistance of the vehicle. Essentially this is related to the exposed surface area and the velocities over the skin. Hence, for a given volume of vehicle hull, it is desirable to reduce the surface area as much as possible. However, it is also important to retain a smooth surface, to avoid roughness and sharp discontinuities, and to have a slowly varying form so that no adverse pressure gradients are built up, which cause increased drag through separation of the flow from the vehicle's hull.
2. *Form drag* – A second effect of the viscous action of the vehicle's hull is to reduce the pressure recovery associated with non-viscous flow over a body in motion. Form drag is created as the water is moved outward to make room for the body and is a function of cross-sectional area and shape. In an ideal non-viscous flow there is no resistance since, although there are pressure differences between the bow and stern of the vehicle, the net result is a zero force in the direction of motion. Due to the action of viscosity there is reduction in the momentum of the flow and, whilst there is a pressure build up over the bow of the submersible, the corresponding pressure recovery at the stern is reduced, resulting in a net resistance in the direction of motion. This form drag can be minimized by slowly varying the sections over a long body, i.e. tending toward a needle-shaped body even though it would have a high surface-to-volume ratio.

As shown in Figure 2.15, there is an optimum aspect ratio whereby the total drag formed from both form drag and skin friction is minimized. Assuming a smoothly shaped contour forming a cylindrical hull, that aspect is somewhere in the range of a 6:1 aspect ratio (length-to-diameter ratio). The practicalities of building a cost-effective underwater vehicle (including the engineering headaches of procuring and

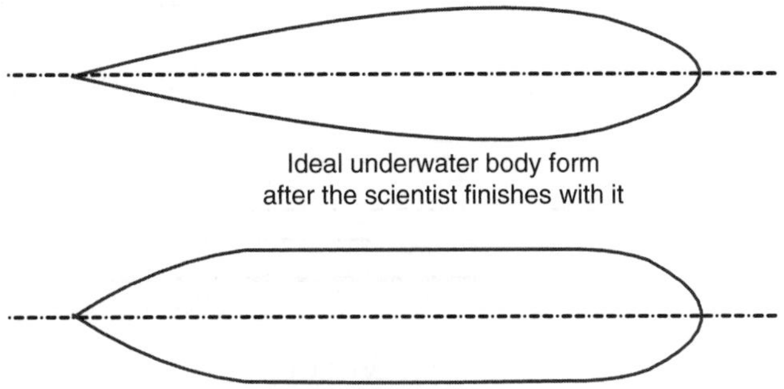
Ideal underwater body form
after the scientist finishes with it
Parallel mid-section after
the procurement folks butcher it

Figure 2.16 *Underwater vehicle body forms.*

forming constantly changing form factors) always get in the way of obtaining the perfect underwater design. Figure 2.16 shows the ideal submarine form, then a slightly modified form factor popular in the defense industry. From this perfect form, the various aspects of the drag computation can be analyzed.

Skin friction drag

The drag dynamics of submerged vehicles were worked out during manned submersible research done in the 1970s by the office of the Oceanographer of the Navy. According to Busby (1976):

> *Skin friction is a function of the viscosity of the water. Its effects are exhibited in the adjacent, thin layers of fluid in contact with the vehicle's surface, i.e. the boundary layer (Figure 2.17). The boundary layer begins at the surface of the submersible where the water is at zero velocity relative to the surface. The outer edge of the boundary layer is at water stream velocity. Consequently, within this layer is the velocity gradient and shearing stresses produced between the thin layers adjacent to each other. The skin friction drag is the result of stresses produced within the boundary layer. Initial flow within the boundary layer is laminar (regular, continuous movement of individual water particles in a specific direction) and then abruptly terminates into a transition region where the flow is turbulent and the layer increases in thickness. To obtain high vehicle speed, the design must be toward retaining laminar flow as long as possible, for the drag in the laminar layer is much less than that within the turbulent layer.*
>
> *An important factor determining the condition of flow about a body and the relative effect of fluid viscosity is the 'Reynolds number'. This number was evolved from the work of Englishman Osborne Reynolds in the 1880s. Reynolds observed laminar flow become abruptly turbulent when a particular value of the product of the distance along a tube and the velocity, divided by the viscosity, was reached. The Reynolds number expresses in non-dimensional form a ratio between inertia forces and viscous forces on a particle, and the transition from the laminar to the turbulent area occurs at a certain critical Reynolds number value. This critical Reynolds number value is lowered by the effects of surface imperfections and*

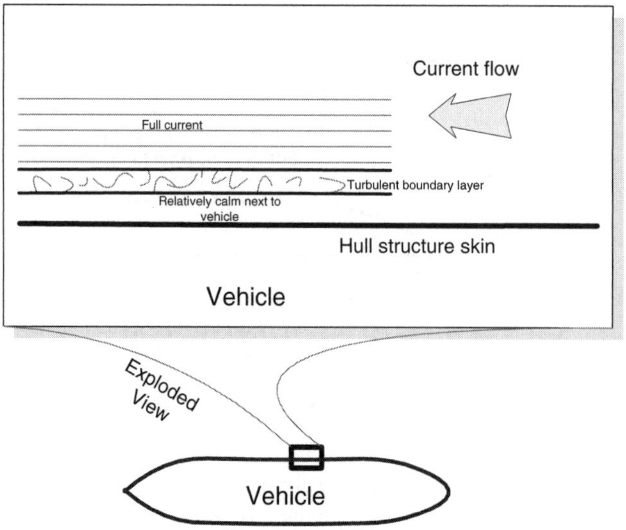

Figure 2.17 *Ideal form with skin surface detail.*

regions of increasing pressure. In some circumstances, sufficient kinetic energy of the flow may be lost from the boundary layer such that the flow separates from the body and produces large pressure or form drag.

The Reynolds number can be calculated by the following formula:

$$Re = pVl/m = Vl/v,$$

where:
p = density of fluid (slugs/ft^3)
V = velocity of flow (ft/s)
m = coefficient of viscosity (lb-s/ft^2)
$v = m/p$ = kinematic viscosity (ft^2/s)
l = a characteristic length of the body (ft).

An additional factor is roughness of the body surface, which will increase frictional drag. Naval architects generally add a roughness-drag coefficient to the friction-drag coefficient value for average conditions.

Form drag

A variation on a standard dynamics equation can be used for ROV drag curve simulation. With an ROV, the two components causing typical drag to counter the vehicle's thruster output are the tether drag and the vehicle drag (Figure 2.18). The function of an ROV submersible is to push its hull and pull its tether to the work site in order to deliver whatever payload may be required at the work site. The only significant metric that matters in the motive performance of an ROV is the net thrust to net drag ratio. If that ratio is positive (i.e. net thrust exceeds net drag), the vehicle will make headway to the work site. If that ratio is negative, the vehicle becomes a very high-tech and very expensive boat anchor.

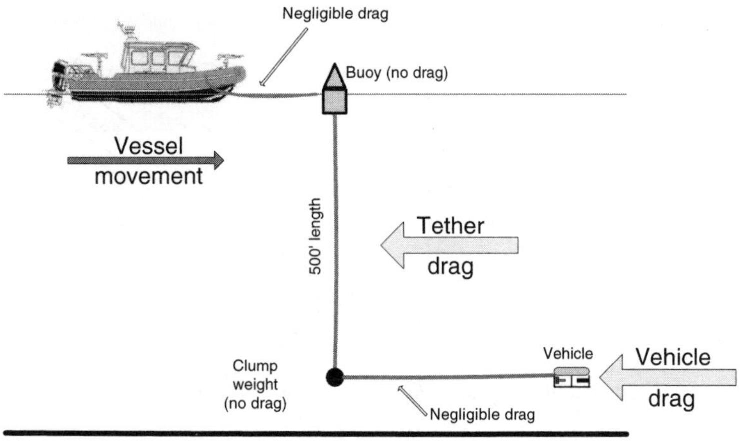

Figure 2.18 *System drag components.*

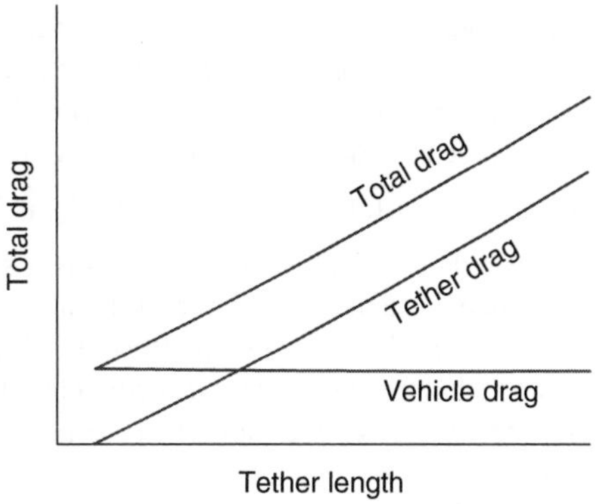

Figure 2.19 *Component drag at constant speed.*

ROV thrusters must produce enough thrust to overcome the drag produced by the tether and the vehicle. The drag on the ROV system is a measurable quantity derived by hydrodynamic factors that include both vehicle and tether drag. The drag produced by the ROV is based upon the following formula:

$$\text{Vehicle drag} = 1/2 \times \sigma A V^2 \, C_\text{d},$$

where

σ = (density of sea water)/(gravitational acceleration), where density of sea water = 64 lb/ft^3 (1025 kg/m^3) and gravitational acceleration = 32.2 ft/s^2 (9.8 m/s^2).

A = Characteristic area on which C_d (the drag coefficient) is non-dimensionalized. For an ROV, A is defined as the cross-sectional area of the front or the vehicle. In some cases, the ROV volume raised to the 2/3 power is used.

V = Velocity in feet per second – (1 knot) = 1.689 feet/second = 0.51 meters/second.

C_d = Non-dimensional drag coefficient. This ranges from 0.8 to 1 when based on the cross-sectional area of the vehicle.

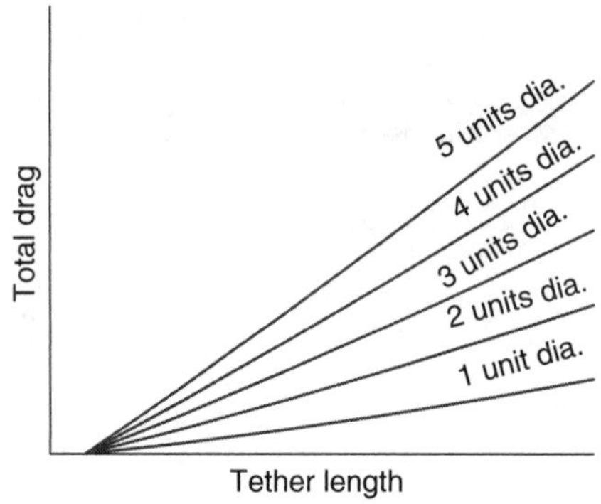

Figure 2.20 *Linear tether drag at constant speed with varying diameter.*

Total drag of the system is equal to the vehicle drag plus the tether drag (Figure 2.19).

In the case of cables, the characteristic area, A, is the cable diameter in inches divided by 12, times the length perpendicular to the flow.

The C_d for cables ranges from 1.2 for unfaired cables; 0.5–0.6 for hair-faired cable; and 0.1–0.2 for faired cables. Since the cylindrical form has the highest coefficient of drag, the use of cable fairings to aid in drag reduction can have a significant impact.

Accordingly, the total drag of the system is defined as:

$$\text{Total drag} = 1/2\,\sigma A_v V^2 C_{dv} + 1/2\sigma A_u V_u^2 C_{du} \text{ (where v = vehicle; u = umbilical).}$$

A simple calculation can be performed if it is assumed that the umbilical cable is hanging straight down and that the tether from the end of the umbilical (via a clump or TMS) to the vehicle is horizontal with little drag (Figure 2.18). For this calculation, it will be assumed that the ship is station keeping in a 1-knot current (1.9 km/h) and the vehicle is working at a depth of 500 feet (152 m). The following system parameters will be used:

Unfaired umbilical diameter = 0.75 inch (1.9 cm)
A, the characteristic area of the vehicle = 10 square feet (0.93 square meters)

Based on the above, the following is obtained:

Vehicle drag = $1/2 \times 64/32.2 \times 10 \times (1.689)^2 \times 0.9 = 25.5$ pounds (11.6 kg)
Umbilical drag = $1/2 \times 64/32.2 \times (0.75/12 \times 500) \times (1.689)^2 \times 1.2$
= 106.3 pounds (48.2 kg)

Note: Computations will be the same in both imperial and metric if the units are kept consistent. This simple example shows why improvements in vehicle geometry do not make significant changes to system performance. The highest factor affecting ROV performance is tether drag.

The following discussion will consider the drag of individual components.

Drag computations for the vehicle assume a perfectly closed frame box. Drag computations for the tether are in the range of a cross-section of ROV systems sampled during recent field trials of small observation-class systems.

By varying the tether diameter, the relationships in Figure 2.20 can be developed.

Figure 2.21 shows that by varying the speed with a constant length of tether, the vehicle will display a similar curve, producing a drag curve that is proportional to velocity squared.

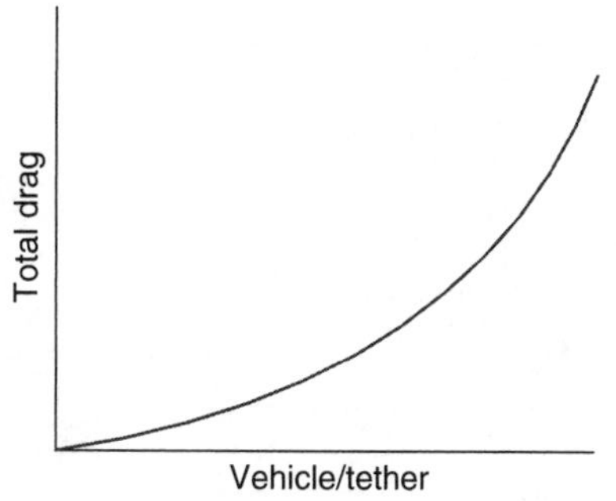

Figure 2.21 *Linear tether drag at varying speed with constant diameter.*

Table 2.2 *Specifications of ROVs evaluated.*

System and parameter	*Large ROV A*	*Small ROV A*	*Small ROV B*	*Large ROV B*	*Small ROV C*	*Medium ROV A*
Depth rating (ft)	500	330	500	1150	500	1000
Length (in)	24	10	14	39	21	18.6
Width (in)	15	7	9	18	9.65	14
Height (in)	10	6	8	18	10	14
Weight in air (lb)	39	4	8	70	24	40
Number of thrusters	4	3	3	4	4	4
Lateral thruster	Yes	No	No	Yes	Yes	No*
Approx. thrust (lb)	25	2	5	23	9	12
Tether diameter (in)	0.52	0.12	0.44	0.65	0.30	0.35
Rear camera	No	No	Yes	No	No	No
Side camera	No	No	No	Yes	No	No
Generator req. (kW)	3	1	1	3	1	3

* Medium ROV A possesses lateral thrusting capabilities due to offset of vertical thrusters.

The power required to propel an ROV is calculated by multiplying the drag and the velocity as follows:

$$\text{Power} = \text{Drag} \times V/550.$$

The constant 550 is a conversion factor that changes feet/pounds/seconds to horsepower. As discussed previously, the drag of a vehicle is proportional to the velocity of the vehicle squared. Accordingly, the propulsion power used is proportional to the velocity cubed. To increase the forward velocity by 50 percent, for example from 2 knots to 3 knots, the power increases by $(3/2)^3$, or $(1.5)^3$, which is 3.4 times more power. To double the speed, the power increases by $(2)^3$, or eight times. Increased speed requirements have a severe impact on vehicle design.

Table 2.2 lists some observation-class systems tested during United States Coast Guard (USCG) procedures trials (without specific names and using figures within each vehicle manufacturer's sales literature) with their accompanying dimensions.

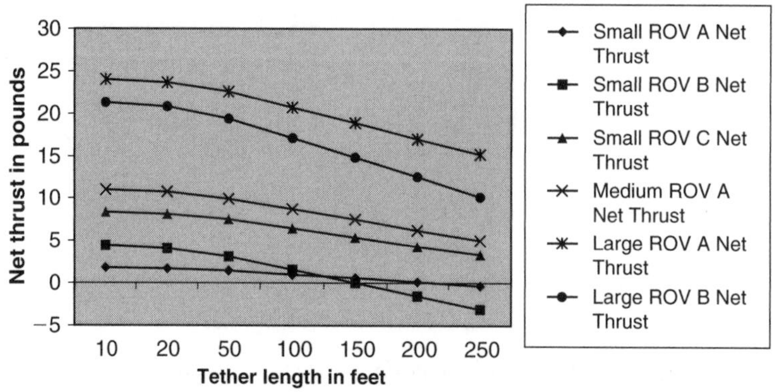

Figure 2.22 *Drag curves of systems tested at 0.5 knot.*

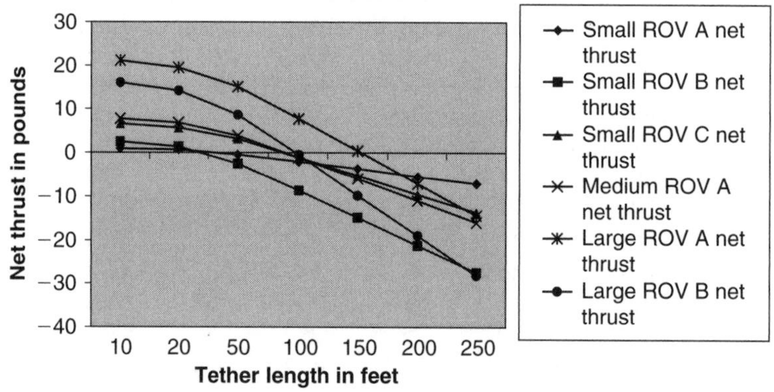

Figure 2.23 *Drag curves of systems tested at 1.0 knot.*

At a given current velocity (i.e. 1 knot), the drag can be varied (by increasing the tether length) until the maximum thrust is equal to the total system drag. That point is the maximum tether length for that speed that the vehicle will remain on station in the current. Any more tether in the water (i.e. more drag) will result in the vehicle losing way against the current. Eventually (when the end of the tether is reached) the form drag will turn the vehicle around, causing the submersible to become the high-tech equivalent of a sea anchor.

The charts in Figures 2.22–2.25 show the approximate net thrust (positive forward thrust versus total system drag) curves at 0.5, 1, 1.5, and 2.0 knots for the ROVs described in Table 2.2.

The net thrust is shown with the horizontal line representing zero net thrust. All points below the zero thrust line are negative net thrust, causing the vehicle to lose headway against an oncoming current. Note: These tether lengths represent theoretical cross-section drag for a length of tether perfectly perpendicular to the oncoming water (see the example, Figure 2.18). Vehicle drag assumes a perfectly closed box frame with the dimensions from Table 2.2 for the respective system.

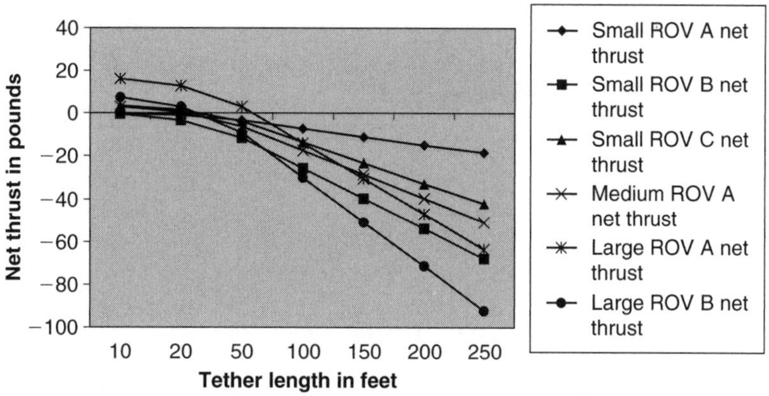

Figure 2.24 *Drag curves of systems tested at 1.5 knots.*

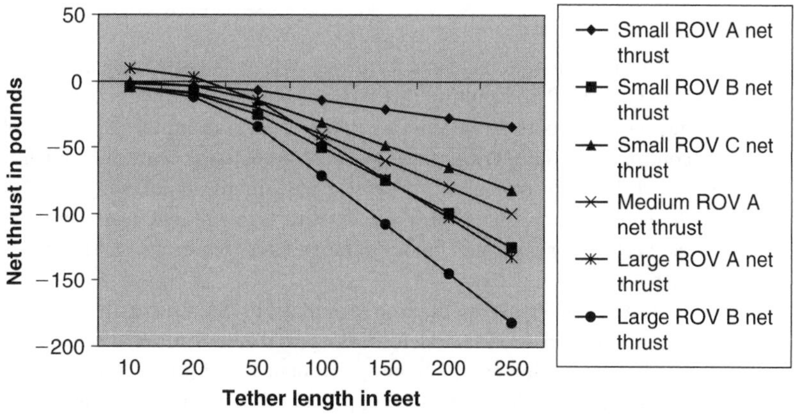

Figure 2.25 *Drag curves of systems tested at 2.0 knots.*

With a tether in a perfectly streamlined configuration (i.e. tether following directly behind the vehicle) the tether drag profile changes and is significantly reduced (Figure 2.26).

The obvious message from this data is that the tether drag on the vehicle is the largest factor in ROV deployment and usage. The higher the thrust-to-drag ratio and power available, the better the submersible pulls its tether to the work site.

2.3.5.3 *Tether effects*

Tether pull point

Stability testing was performed on a small ROV system at Penn State University's Advanced Research Lab in their water tunnel. The water flow was slowly brought up while observing the vehicle's handling characteristics as well as its computed, versus actual, zero net thrust point.

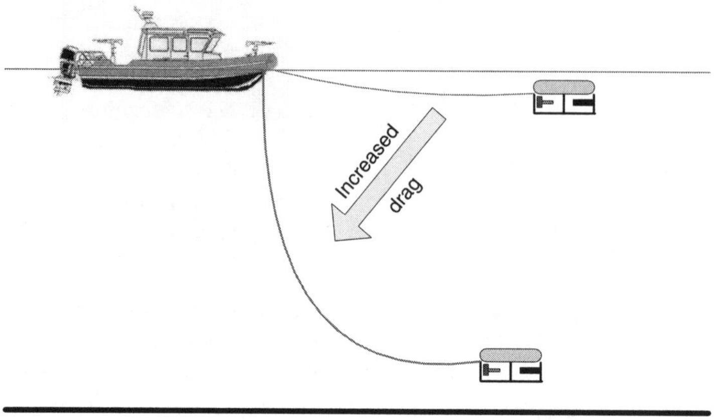

Figure 2.26 *Tether profile drag increases to the perpendicular point.*

This particular vehicle had a tether pull point significantly above the line of thrust (see Figure 2.27), resulting in a 'bow up' turning moment. As the speed ramped up during the tests, with little tether in the water, the vehicle was still able to maintain control about the vertical plane by counteracting the 'bow up' tendency with vertical thrust-down. However, at a constant speed with the tether being lengthened, the tether drag produced an increasingly higher tether turning moment, eventually overpowering the vertical thruster and shooting the submersible to the surface in an uncontrolled fashion.

If the tether is placed in close proximity to the thruster, parasitic drag will occur due to the skin friction and form drag from the thruster discharge flow across the tether. When selecting the tether placement, it is best to design the tether pull point (Figure 2.28) as close to the center point of thrust to balance any turning moment due to the tether pull point.

Tether pull/lay

Some vehicle manufacturers place the tether pull point atop the vehicle. The benefit to this placement is the tether will not lay as easily in the debris located on the bottom, allowing a cleaner tether channel from the vehicle to the surface. This is beneficial if the vehicle is operated in minimal currents with little or no horizontal offset. If either a horizontal offset or a current (or both) is encountered, the vehicle may experience difficulty through partial (if not total) loss of longitudinal and/or lateral stability.

Hydrodynamics of vehicle and tether

The most typical arrangement for an observation-class system involves a clean tether (i.e. without clump weight) following the vehicle to the work site. The tether naturally settles behind the vehicle and slopes in the current as it feeds toward the surface. As the vehicle speed ramps up, the flow drag on the tether correspondingly melds the

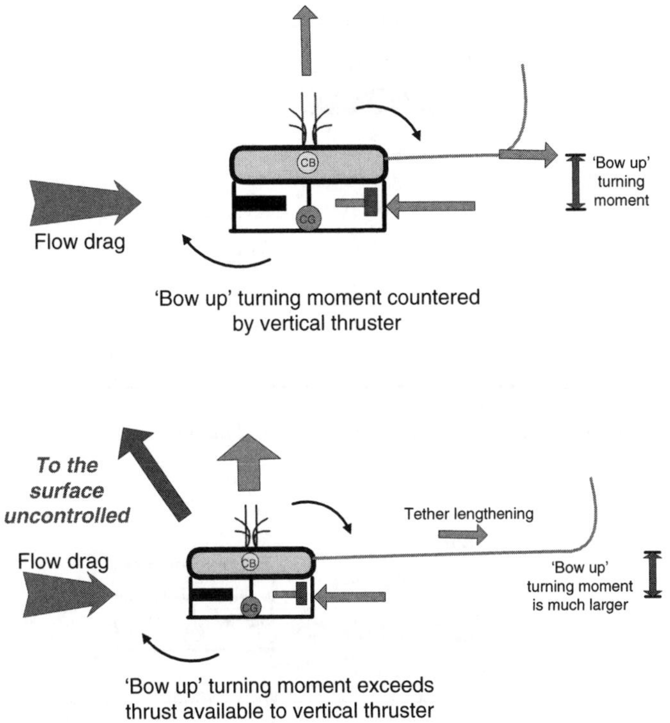

Figure 2.27 *Vehicle stability considerations.*

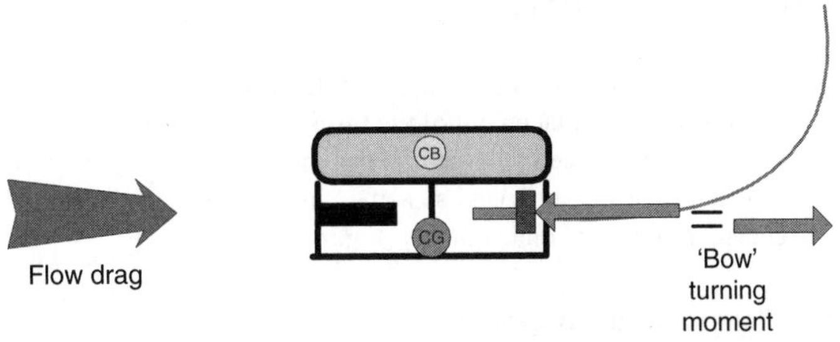

Figure 2.28 *Minimal bow turning moment with tether on thrust line.*

tether into its wake, forming a 'sail' of sorts behind the vehicle (Figure 2.29). A small reduction in the drag due to reduced angle of incidence to the oncoming water flow is more than offset with the additional form and flow drag of the excessive tether in the water. There is an old technique used by surface-supplied commercial divers to counter the excessive umbilical in the water – grab hold of something on the bottom while the tender takes up the slack. The same technique can be applied to ROVs by placing the vehicle on a stationary item on the bottom then having the tender pull the excessive tether back on deck.

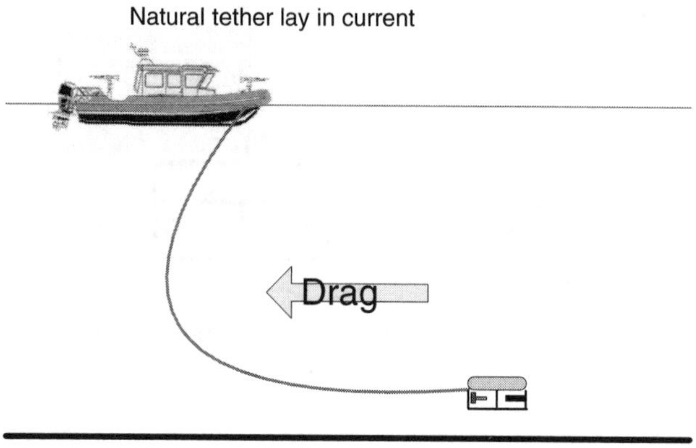

Figure 2.29 *Natural tether lay behind the vehicle as the speed ramps up.*

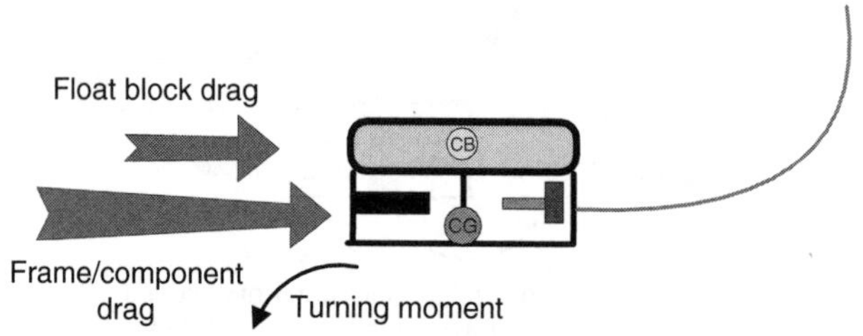

Figure 2.30 *Bow turning moment due to asymmetrical drag as speed ramps up.*

2.3.5.4 *Thrusters and speed*

As the speed of the vehicle ramps up, the low speed stability due to static stability is overtaken by the placement of the thrusters with regard to the center of total drag. In short, for slow-speed vehicles the designer can get away with improper placement of thrusters. For higher speed systems, thruster placement becomes a more important consideration in vehicle control (Figure 2.30).

Propeller efficiency and placement

Propellers come in all shapes and sizes based upon the load and usage. Again using the aircraft correlation, on fixed-pitch aircraft propellers for small aircraft there are 'climb props' and 'cruise props'. The climb prop is optimized for slower speeds, allowing better climb performance while sacrificing on cruise speeds. Conversely, a cruise prop has better range and speed during cruise, but climb performance suffers. Likewise, a tugboat propeller would not be best suited for a high-speed passenger liner.

Propellers have an optimum operational speed. Some propellers are optimized for thrust in one direction over another. A common small ROV thruster on the market today uses such a propeller for forward and downward thrusting. The advantage to this propeller arrangement is better thrust performance in the forward direction while sacrificing turning reversal and upward thrust performance. Other propellers have equal thrusting capabilities in both directions. Both have their strengths and weaknesses. All ROVs are slow systems and should make use of propellers' maximizing power at slow speeds in order to counter the combined tether/vehicle drag. Remember, an ROV is a tugboat and not a speedboat.

Propellers also produce both cavitations and propeller-tip vortices, causing substantial amounts of drag. As the spinning propeller moves water across the blade, the centrifugal force (instead of moving aft to produce a forward thrust vector) throws the water toward the tips of the blade, spilling over the end of the blade in turbulent flow. Kort nozzles form a basic hub around the propeller to substantially reduce the instance of tip vortices. The kort nozzle then maintains the water volume within the thruster unit, allowing for more efficient movement of the water mass in the desired vectored direction. Propeller cavitations are a lesser problem due to the speed at which the small ROV thruster propeller turns and are inconsequential to this analysis.

Thrust to drag and bollard pull

The following factors come into play when calculating vehicle speed and ability to operate in current:

- *Bollard pull* is a direct measurement of the ability of the vehicle to pull on a cable. Values provided by manufacturers can vary due to lack of standards for testing: 'Actual bollard pull can only be measured in full scale, and is performed during so-called bollard pull trials. Unfortunately the test results are not only dependent on the performance of the [vehicle] itself, but also on test method and set-up, on trial site and on environmental conditions …' (Jukola and Skogman, 2002).
- *Hydrodynamics* is another aspect of ROV design that must be considered holistically. Although a vehicle shape and size may make it very hydrodynamic, i.e. certain smaller enclosed systems, there is often a trade-off in stability. Some manufacturers seem to spend considerable effort making their ROVs more hydrodynamic in the horizontal plane, but in deep sea operations diving to depth may consume considerable time.

It is bollard pull, vehicle hydrodynamics, and tether drag together that determine most limitations on vehicle performance. The smaller the tether cable diameter, the better – in all respects (except, of course, power delivery). Stiffer tethers can be difficult to handle, but they typically provide less drag in the water than their more flexible counterparts. Flexible tethers are much nicer for storage and handling, but they tend to get tangled or hang up more often than those that are slightly stiffer.

The use of ROVs in current is an issue that is constantly debated among users, designers, and manufacturers. This is not a topic that can be settled by comparing specifications of one vehicle to another. One of the most common misconceptions is that maximum speed equates to an ability to deal with current. When operating at depth (versus at the surface), the greatest influence of current is on the tether cable. It is the ability of the vehicle to pull this cable that allows it to operate in stronger currents. A vehicle with more power, but not necessarily more speed, will be better able to handle the tether (an example of which would be bollard pull of a tugboat versus that of a speedboat). The most effective way to determine a vehicle's ability

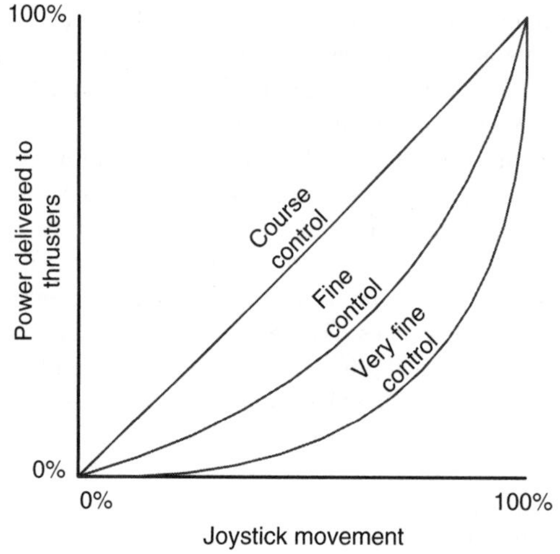

Figure 2.31 *Joystick control matrix variation based upon varying power delivery versus joystick position.*

to operate in current is to test the vehicle in current. Experience of the operator can have a significant effect on how the vehicle performs in higher current situations. Realistically, no small surfaced-powered ROV can be considered effective in any current over three knots.

2.3.6 Vehicle Control

Vehicle manufacturers use a variety of techniques for gross versus fine control of vehicle movement to conform to the operating environment. One small vehicle manufacturer makes use of a horizontal and vertical gain setting to allow for varying power versus joystick position combinations. Another vehicle manufacturer allows for variable power delivery scaling via software on the controller. The reason this power scaling is necessary is that when towing the tether and vehicle combination to the work site, the full power complement is needed for the muscling operation. Once at the work site, finer adjustments are needed to ease the ROV into and out of tight places. If the power were set to full gain in a confined area, a quarter joystick movement could over-ramp the power so quickly that the vehicle could ram into a wall, damaging the equipment and causing some embarrassing personnel reviews.

The joystick control matrix can also significantly affect the ease of control over the vehicle (Figures 2.31 and 2.32). For example, if during a small turning adjustment, such as 20°, the thrusters may ramp up power so quickly that the operator cannot stop the turn until after reaching the 90° rotation point. Effective control of the vehicle would be lost.

2.3.6.1 *Control versus speed*

Unlike underwater vehicles built for high speed (examples of high-speed underwater vehicles are a torpedo or a nuclear attack submarine), most observation-class ROV

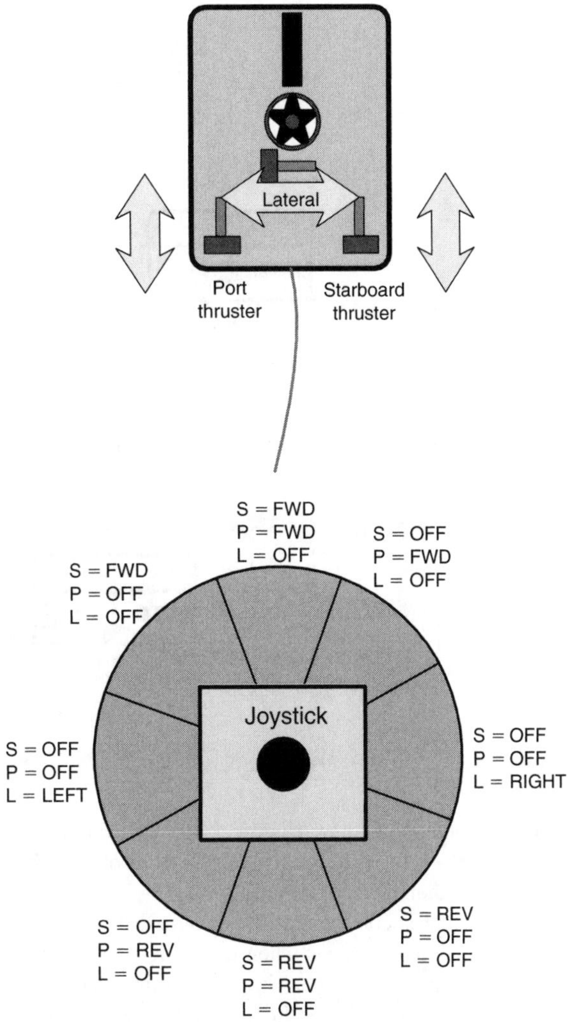

Figure 2.32 *Joystick position versus thruster activation on a four-thruster configuration.*

submersibles are designed for speeds no greater than 3 knots. In fact, somewhere in the speed range of 6–8 knots for underwater vehicles, interesting hydrodynamic forces act upon the system, which require strong design and engineering considerations that address drag and control issues. At higher speeds, small imperfections in vehicle ballasting and trim propagate to larger forces that simple thruster input may not overcome. As an anecdote to unexpected consequences for high-speed underwater travel, during trials for the USS *Albacore* (AGSS 569) it was noted with some surprise (especially by the Commanding Officer) that the submarine snap rolled in the direction of the turn during high-speed maneuvering!

For the ROV at higher speeds, any thruster that thrusts on a plane perpendicular to the relative water flow will have the net vector of the thrust reaction move in the

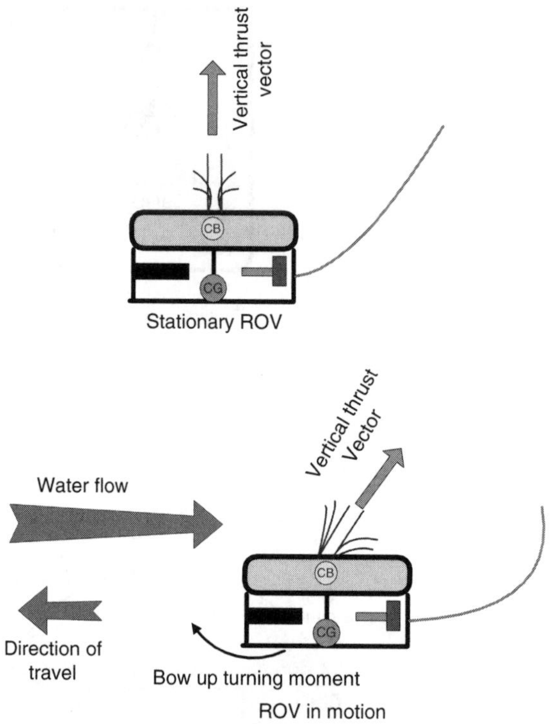

Figure 2.33 *Apparent thrust vector change due to water flow across vehicle.*

direction of the water flow (Figure 2.33). Also, one current vehicle manufacturer makes use of vectored thrust for vehicle control, which mitigates the thrust vector problem at higher speeds.

At what point is control over the vehicle lost? The answer to that is quite simple. The loss of control happens when the vehicle's thrusters can no longer counter the forces acting upon the vehicle while performing a given task. Once the hydrodynamic forces exceed the thruster's ability to counter these forces (on any given plane), control is lost. One of the variables must be changed in order to regain control.

2.3.6.2 *Auto stabilization*

With sensor feedback fed into the vehicle control module, any number of parameters may be used in vehicle control through a system of closed-loop control routines. Just as dogs follow a scent to its source, ROVs can use sensor input for positive navigation. Advances are currently being made for tracking chemical plumes from environmental hazards or chemical spills. A much simpler version of this technique is the rudimentary auto depth/altitude/heading.

Auto depth is easily maintained through input from the vehicle's pressure-sensitive depth transducer. Auto altitude is equally simple, but the vehicle manufacturer is seldom the same company as the sensor manufacturer (causing some issues with communication standards and protocols between sensor and vehicle). The most

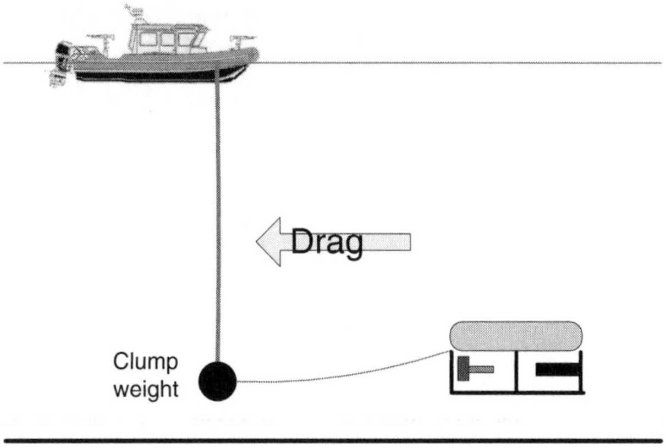

Figure 2.34 *Clump weight deployed ROV.*

common compass modules used in observation-class ROV systems are the inexpensive flux gate-type. These flux gate-type compasses have a sampling rate (while accurate) slower than the yaw swing rate of most small vehicles, which cause the vehicle to 'chase the heading'. Flux gate auto heading is better than no auto heading, but several manufacturers of small systems have countered this 'heading chase' problem by using a gyro.

Gyros for small ROVs come in two basic types, the slaved gyro and the rate gyro. The slaved gyro samples the magnetic compass to slave the gyro periodically to correspond with its magnetic counterpart. Since the auto heading function of an ROV is simply a heading hold function, some manufacturers have gotten away with using a simple rate gyro for 'heading stabilization'. When the heading hold function is slaved to a gyro only, sensing a turn away from the initial setting and a rate at which the turn is progressing (i.e. the rate gyro has no reference to any magnetic heading), the vehicle is then only referenced to a given direction. Hence the term 'heading stabilization' due to the lack of any reference to a specific compass direction.

2.3.7 Deployment Techniques

Deployment methods vary, but there are a few common methods that have proven successful. The deployment methods can be divided into two main categories: Direct deployed and TMS deployed.

2.3.7.1 *Directly deployed*

For station-keeping operations with smaller ROVs, the vehicle can be directly deployed from the deck of the boat. Larger vehicles can be directly deployed, but the risk of damage increases as the weight of the vehicle increases. This is due to the vehicle's momentum building through vessel sway while the vehicle is suspended in air (i.e. the vehicle becomes a 'wrecker's ball'). Directly deployed vehicles are more vulnerable to any currents prevalent from the surface to the operating depth.

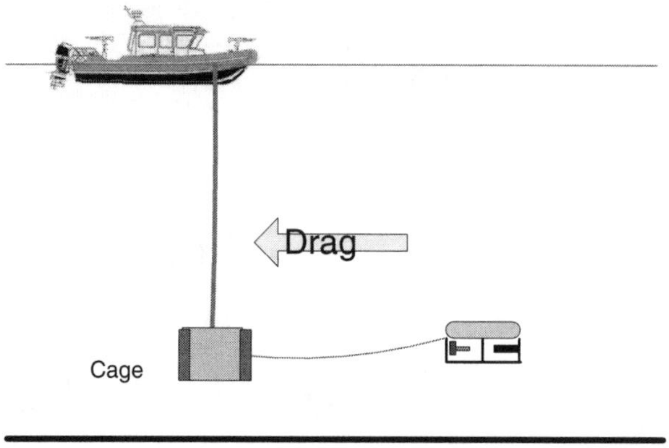

Figure 2.35 *Cage deployed ROV.*

2.3.7.2 *Tether management system (TMS)*

The tether management system (TMS) can be part of a cage deployment system (Figure 2.35) or can simply be attached to the so-called clump weight (Figure 2.34). The main function for the TMS is to manage a soft tether cable – the link from the TMS to the ROV for electrical power and sensors, including video and telemetry. The tether cable allows the ROV to make excursions at depth for a distance of 500 feet or more from the point of the clump weight. Some refer to the TMS as the entire system of cage or 'top hat' deployment, tether management, vehicle protection, and junction point for the surface/vehicle link. Technically, the TMS is the tether handling machinery only.

Clump deployed

The use of a clump weight has become prevalent in the observation-class category. If working on or near the sea bottom, clump weights enable the ROV operator to easily manage the tether 'lay' from the insertion point next to the vessel all the way to the clump weight location next to the bottom.

This allows the weight to absorb the cross-section drag of the current, relieving the submersible of the tether drag from the surface to the working depth. The vessel can be maneuvered to a point directly above the work site, thus locating the center point of an operational circle at the clump weight. In short, the vehicle only needs to drag the tether length between the clump weight and the vehicle for operations on the bottom.

Cage deployed

Cages are used for larger ROVs to protect the vehicle against abrasions and deployment damage due to the instability of most vessels of opportunity while underway.

Cages also function as a negatively buoyant anchor to overcome the drag imposed from the cross-section of the cable presented to the current (between the platform and the cage) at shallower depths (Figure 2.35). This allows the weight of the cage to fight the current instead of the vehicle fighting the current. The cage further provides room for a tether management system to meter the softer tether in small amounts, thus lowering the risk of tether entanglement. The cage umbilical is normally made of durable material (steel, Kevlar, etc.) with the conductors for the vehicle buried within the core of the umbilical. For deeper diving submersibles, the umbilical encases fiber-optic data links to the cage, requiring digital modem feeds from the cage to the control unit at the surface.

Chapter 3
ROV Components

This chapter discusses the major components (Figure 3.1) of a typical ROV system along with everyday underwater tasks ROVs perform.

As discussed in the prior chapter on ROV design, the design team must consider the overall system. To reinforce the importance of this point, a few additional comments about the design process are warranted.

An ROV is essentially a robot. What differentiates a robot from its immovable counterparts is its ability to move under its own power. Along with that power of locomotion comes the ability to navigate the robot, with ever increasing levels of autonomy to achieve some set goal. While the ROV system, by its nature, is one of the simplest robotic designs, complex assignments can be accomplished with a variety of closed-loop aids to navigation. Some ROV manufacturers are aggressively embracing the open source computer-based control models, allowing users to design their own navigation and control matrix. This is an exciting development in the field of subsea robotics and will allow development of new techniques, which will only be limited by the user's imagination. This concept takes the control of the development of navigation capabilities (which is the mission) from the hands of the design engineer (who may or may not understand the user's needs) into the hands of the end user (who does understand the needs). Designing efficient and cost-effective systems with the user in mind is critical to the success of the product and ultimately the mission.

Do not over-design the system. The old saying goes that a chain is only as strong as its weakest link. Accordingly, all components of an ROV system should be rated to the maximum operating depth of the underwater environment anticipated, including

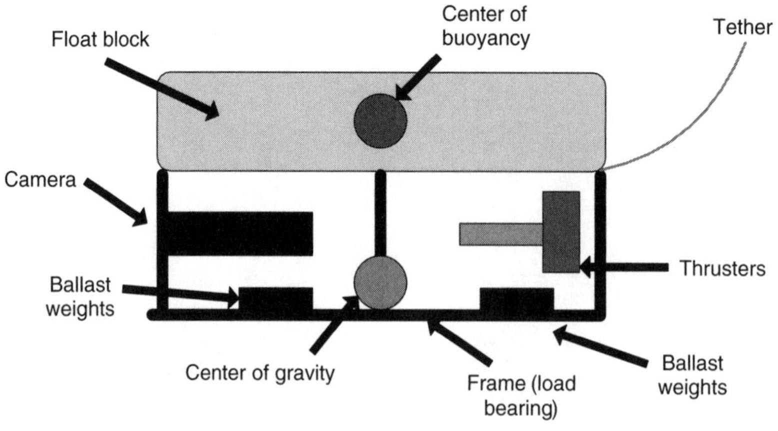

Figure 3.1 *ROV submersible components.*

safety factors. However, they should not be over-designed. As the operating depth proceeds into deeper water, larger component wall thicknesses will be required for the air-filled spaces (pressure-resistant housings) on the vehicle. This increased wall thickness results in an increased vehicle weight, which requires a larger floatation system to counter the additional weight. This causes an increase in drag due to a larger cross-section, which requires more power. More power drives the cable to become larger, which increases drag, etc. It quickly becomes a vicious design spiral.

Careful consideration should be given during the design phase of any ROV system to avoid over-engineering the vehicle. By saving weight and cost during the design process, the user will receive an ROV that has the capability of providing a cost-effective operation. This is easier said than done, as 'bells and whistles' are often added during the process, or the 'latest and greatest' components are chosen without regard to the impact on the overall system. Keep these ideas in mind as the various component choices are presented in the remainder of this chapter.

3.1 MECHANICAL AND ELECTRO/MECHANICAL SYSTEMS

Since weight is one of the most critical design factors, the components/subsystems having a significant impact in this area are discussed first.

3.1.1 Frame

The frame of the ROV provides a firm platform for mounting, or attaching, the necessary mechanical, electrical, and propulsion components. This includes special tooling/instruments such as sonar, cameras, lighting, manipulator, scientific sensor, and sampling equipment. ROV frames have been made of materials ranging from plastic composites to aluminum tubing. In general, the materials used are chosen to give the maximum strength with the minimum weight. Since weight has to be offset with buoyancy, this is critical.

The ROV frame must also comply with regulations concerning load and lift path strength. The frame can range in size from 6 in × 6 in to 20 ft × 20 ft. The size of the frame is dependent upon the following criteria:

- Weight of the complete ROV unit in air
- Volume of the onboard equipment
- Volume of the sensors and tooling
- Volume of the buoyancy
- Load-bearing criteria of the frame.

3.1.2 Buoyancy

Archimedes' principle states: An object immersed in a fluid experiences a buoyant force that is equal in magnitude to the force of gravity on the displaced fluid. Thus, the objective of underwater vehicle flotation systems is to counteract the negative buoyancy effect of heavier than water materials on the submersible (frame, pressure housings, etc.) with lighter than water materials; A near neutrally buoyant state is the goal. The flotation foam should maintain its form and resistance to water pressure at the anticipated operating depth. The most common underwater vehicle flotation materials encompass two broad categories: Rigid polyurethane foam and syntactic foam.

The term 'rigid polyurethane foam' comprises two polymer types: Polyisocyanurate formulations and polyurethane formulas. There are distinct differences between the two, both in the manner in which they are produced and in their ultimate performance.

Polyisocyanurate foams (or 'trimer foams') are generally low-density, insulation-grade foams, usually made in large blocks via a continuous extrusion process. These blocks are then put through cutting machines to make sheets and other shapes. ROV manufacturers generally cut, shape, and sand these inexpensive foams, then coat them with either a fiberglass covering or a thick layer of paint to help with abrasion and water intrusion resistance. These resilient foam blocks have been tested to depths of 1000 feet of seawater (fsw) and have proven to be an inexpensive and effective flotation system for shallow water applications (Figure 3.2).

Polyisocyanurate foams have excellent insulating value, good compressive-strength properties, and temperature resistance up to 300°F. They are made in high volumes at densities between 1.8 and 6 lb per cubic foot, and are reasonably inexpensive. Their stiff, brittle consistency and their propensity to shed dust (friability) when abraded can serve to identify these foams.

For deep-water applications, syntactic foam has been the foam of choice. Syntactic foam is simply an air/microballoon structure encased within a resin body. The amount of trapped air within the resin structure will determine the density as well as the durability of the foam at deeper depths. The technology, however, is quite costly and is normally saved for the larger deep-diving ROV systems.

3.1.3 Propulsion and Thrust

The propulsion system significantly impacts the vehicle design. The type of thrusters, their configuration, and the power source to drive them usually take priority over many of the other components.

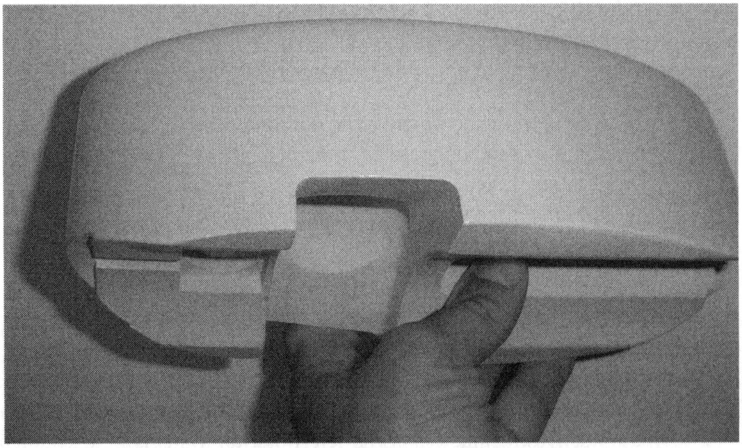

(a)

Figure 3.2 *Polyurethane fiberglass encased and simple painted float blocks.*

3.1.3.1 *Propulsion systems*

ROV propulsion systems come in three different types: Electrical, hydraulic, and ducted jet propulsion. These different types have been developed to suit the size of vehicle and anticipated type of work. In some cases, the actual location of the work task

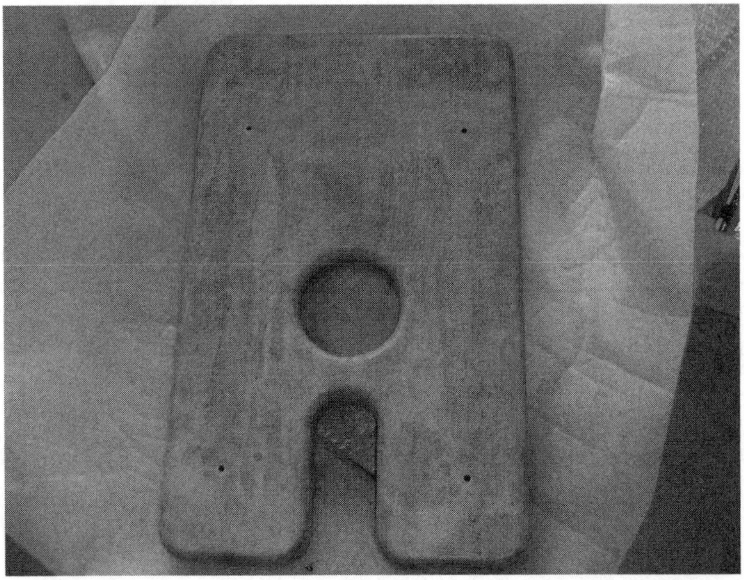

(b)

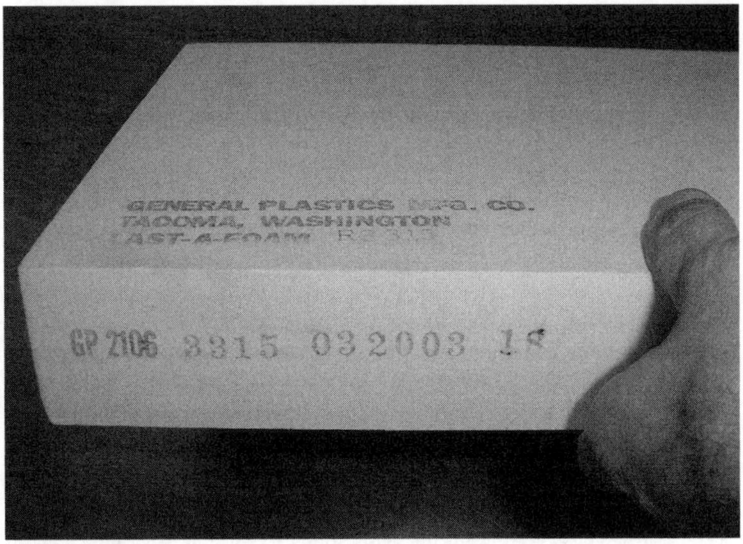

(c)

Figure 3.2 *(Continued)*

has dictated the type of propulsion used. For example, if the vehicle is operated in the vicinity of loosely consolidated debris, which could be pulled into rotating thrusters, ducted jet thruster systems could be used. If the vehicle requires heavy duty tooling for intervention, the vehicle could be operated with hydraulics (including thruster power). Hydraulic pump systems are driven by an electrical motor on the vehicle, requiring a change in energy from electrical to mechanical to hydraulic – a process that is quite energy inefficient. A definite need for high mechanical force is required to justify such an energy loss and corresponding costs.

The main goal for the design of ROV propulsion systems is to have high thrust-to-physical size/drag and power-input ratios. The driving force in the area of propulsion systems is the desire of ROV operators to extend the equipment's operating envelope. The more powerful the propulsion of the ROV, the stronger the sea current in which the vehicle can operate. Consequently, this extends the system's performance envelope.

Another concern is the reliability of the propulsion system and its associated sub-components. In the early development of the ROV, a general practice was to replace and refit electric motor units every 50–100 hours of operation. This increased the inventory of parts required and the possibilities of errors by the technicians in reassembling the motors. Thus, investing in a reliable design from the beginning can save both time and money.

The propulsion system has to be a trade-off between what the ROV requires for the performance of a work task and the practical dimensions of the ROV. Typically, the more thruster power required, the heavier the equipment on the ROV. All parts of the ROV system will grow exponentially larger with the power requirement continuing to increase. Thus, observation ROVs are normally restricted to a few minor work tasks without major modifications that would move them to the next heavier class.

3.1.3.2 *Thruster basics*

The ROV's propulsion system is made up of two or more thrusters that propel the vehicle in a manner that allows navigation to the work site. Thrusters must be positioned on the vehicle so that the moment arm of their thrust force, relative to the central mass of the vehicle, allows a proper amount of maneuverability and controllability.

Thrust vectoring is the only means of locomotion for an ROV. There are numerous placement options for thrusters to allow varying degrees of maneuverability. Maneuvering is achieved through asymmetrical thrusting based upon thruster placement as well as varying thruster output.

The three-thruster arrangement (Figure 3.3) allows only fore/aft/yaw, while the fourth thruster also allows lateral translation. The five-thruster variation allows all four horizontal thrusters to thrust in any horizontal direction simultaneously.

Also, placing the thruster off the longitudinal axis of the vehicle (Figure 3.4) will allow a better turning moment, while still providing the vehicle with strong longitudinal stability.

One problem with multiple horizontal thrusters along the same axis, without counter-rotating propellers, is the torque steer issue (Figure 3.5). With two or more thrusters operating on the same plane of motion, a counter-reaction to this turning moment will result. Just as the propeller of a helicopter must be countered by the tail rotor or a counter-rotating main rotor, the ROV must have counter-rotating thruster propellers in order to avoid the torque of the thrusters rolling the vehicle counter to

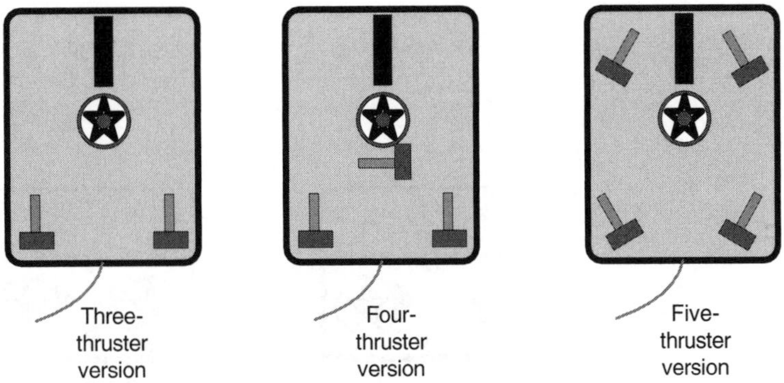

Figure 3.3 *Thruster arrangement.*

Figure 3.4 *Thruster aligned off the longitudinal axis.*

the direction of propeller rotation. If this roll does occur, the resulting asymmetrical thrust and drag loading could give rise to course deviations – the effect of which is known as 'torque steering'.

3.1.3.3 *Thruster design*

Underwater electrical thrusters are composed of the following major components:

- Power source
- Electric motor
- Motor controller (this may be part of the thruster or may be part of a separate driver board)
- Thruster housing and attachment to vehicle frame
- Gearing mechanism (if thruster is geared)
- Drive shafts, seals, and couplings
- Propeller
- Kort nozzle and stators.

The most critical of these components will be discussed in more detail in the following sections.

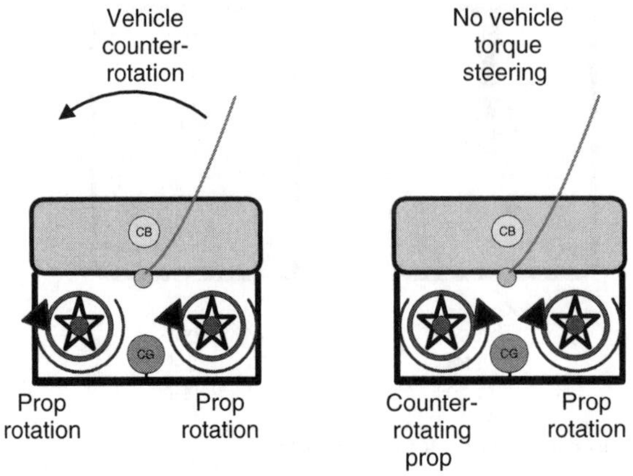

Figure 3.5 *Thruster rotational effect upon vehicle.*

Power source

On a surface-powered ROV system, power arrives to the vehicle from a surface power source. The power can be in any form from basic shore power (e.g. 110 VAC 60 Hz or 220 VAC 50 Hz – which is standard for most consumer electrical power delivery worldwide) to a DC battery source. For an observation-class ROV system running on DC power, the AC source is first rectified to DC on the surface, then sent to the submersible for distribution to the thrusters. The driver and distribution system location will vary between manufacturers and may be anywhere from on the surface control station, within the electronics bottle of the submersible, to within the actual thruster unit. The purpose of this power source is the delivery of sufficient power to drive the thruster through its work task.

Electric motor

Electric motors come in many shapes, sizes, and technologies, each designed for different functions. By far the most common thruster motor on observation-class ROV systems is the DC motor, due to its power, availability, variety, reliability, and ease of interface. The DC motor, however, has some difficult design and operational characteristics. Factors that make it less than perfect for this application include:

- The optimum motor speed is much higher than the normal in-water propeller rotation speed, thus requiring gearing to gain the most efficient speed of operation
- DC motors consume a high amount of current
- They require a rather complex pulse width modulation (PWM) motor control scheme to obtain precise operations.

Permanent magnet DC motors Per Clark and Owings (2003) the permanent magnet DC motor has, within the mechanism, two permanent magnets that provide a magnetic field within which the armature rotates. The rotating center portion of the motor (the armature) has an odd number of poles – each of which has its own

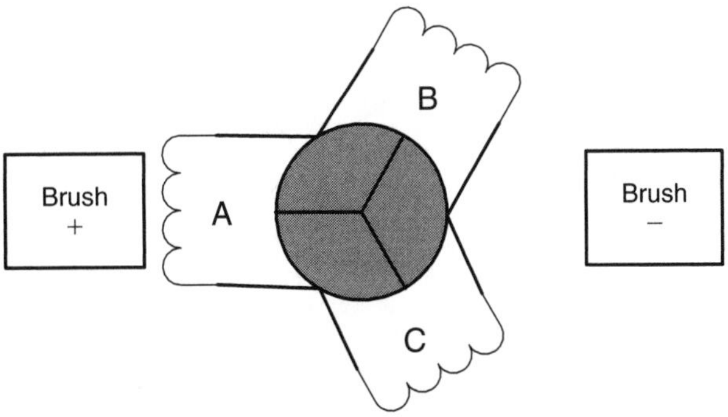

Figure 3.6 *Commutator and brushes.*

winding (Figure 3.6). The winding is connected to a contact pad on the center shaft called the commutator. Brushes attached to the (+) and (−) wires of the motor provide power to the windings in such a fashion that one pole will be repelled from the permanent magnet nearest it and another winding will be attracted to another pole. As the armature rotates, the commutator changes, determining which winding gets which polarity of the magnetic field. An armature always has an odd number of poles, and this ensures that the poles of the armature can never line up with their opposite magnet in the motor, which would stop all motion.

Near the center shaft of the armature are three plates attached to their respective windings (A, B, and C) around the poles. The brushes that feed power to the motor will be exactly opposite each other, which enables the magnetic fields in the armature to forever trail the static magnetic fields of the magnets. This causes the motor to turn. The more current that flows in the windings, the stronger the magnetic field in the armature, and the faster the motor turns.

Even as the current flowing in the windings creates an electromagnetic field that causes the motor to turn, the act of the windings moving through the static magnetic field of the motor causes a current in the windings. This current is opposite in polarity to the current the motor is drawing from the power source. The end result of this current, and the countercurrent (CEMF – counter electromotive force), is that as the motor turns faster it actually draws less current.

This is important because the armature will eventually reach a point where the CEMF and the draw current balance out at the load placed on the motor and the motor attains a steady state. The point where the motor has no load is the point where it is most efficient. It is also the point where the motor is the weakest in its working range. The point where the motor is strongest is when there is no CEMF; All current flowing is causing the motor to try to move. This state is when the armature is not turning at all. This is called the stall or startup current and is when the motor's torque will be the strongest. This point is at the opposite end of the motor speed range from the steady-state velocity (Figure 3.7).

While the maximum efficiency of an electric motor is at the no-load point, the purpose of having the motor in the first place is to do work (i.e. produce mechanical rotary motion that may be converted into linear motion or some other type of work). The degree of turning force delivered to the drive shaft is known as motor torque.

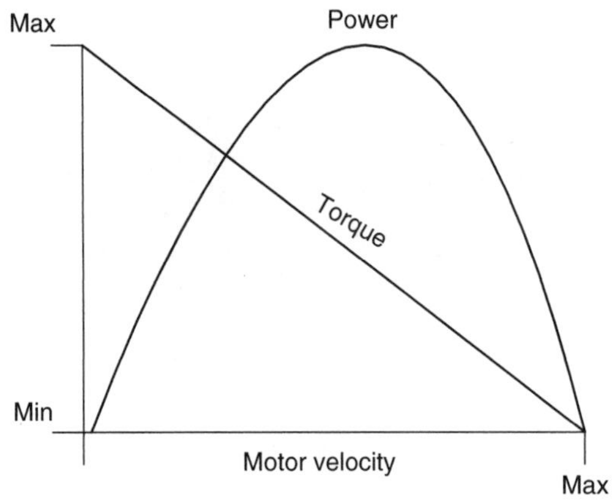

Figure 3.7 *Motor velocity versus torque and power.*

Motor torque is defined as the angular force the motor can deliver at a given distance from the shaft. If a motor can lift 1 kilogram from a pulley with a radius of 1 m it would have a torque of 1 newton-meter. One newton equals 1 kilogram-meter/second^2, which is equal to 0.225 pounds (1 inch equals 2.54 centimeters, 100 centimeters are in a meter, and there are 2π radians in one revolution).

The formula for mechanical power in watts is equal to torque times the angular velocity in radians/second. This formula is used to describe the power of a motor at any point in its working range. A DC motor's maximum power is at half its maximum torque and half its maximum rotational velocity (also known as the no-load velocity). This is simple to visualize from the discussion of the motor working range: Where the maximum angular velocity (highest revolutions per minute (RPM)) has the lowest torque, and where the torque is the highest, the angular velocity is zero (i.e. motor stall or start torque).

Note that as motor velocity (RPM) increases, the torque decreases; At some point the power stops rising and starts to fall, which is the point of maximum power.

When sizing a DC motor for an ROV, the motor should be running near its highest efficiency speed, rather than its highest power, in order to get the longest running time. In most DC motors this will be at about 10 percent of its stall torque, which will be less than its torque at maximum power. So, if the maximum power needed for the operation is determined, the motor can be properly sized. A measure of efficiency and operational life can then be obtained by oversizing the motor for the task at hand.

Brushless DC motors For brushless DC motors, a sensor is used to determine the armature position. The input from the sensor triggers an external circuit that reverses the feed current polarity appropriately. Brushless motors have a number of advantages that include longer service life, less operating noise (from an electrical standpoint), and, in some cases, greater efficiency.

Gearing

DC motors can run from 8000 to 20 000 RPM and higher. Clearly, this is far too fast for ROV applications if vehicle control is to be maintained. Thus, to match the efficient operational speed of the motor with the efficient speed of the thruster's propeller, the motor will require gearing. Gearing allows two distinctive benefits – the power delivered to the propeller is both slower and more powerful. Further, with the proper selection of a gearbox with a proper reduction ratio, the maximum efficiency speed of the motor can match the maximum efficiency of the thruster's propeller/kort nozzle combination.

Drive shafts, seals, and couplings

The shafts, seals, and couplings for an ROV thruster are much like those for a motorboat. The shaft is designed to provide torque to the propeller while the seal maintains a watertight barrier that prevents water ingress into the motor mechanism.

Drive shafts and couplings vary with the type of propeller driving mechanism. Direct drive shafts, magnetic couplings, and mechanical (i.e. geared) couplings are all used to drive the propeller. Technology advances are being exploited in attempts to miniaturize thrusters. In one case, a new type of thruster housing places the drive mechanism on the hub of the propeller instead of at the drive shaft, allowing better torque and more efficient propeller tip flow management. Others are developing miniature electric ring thrusters, where the propeller, which can be hubless, is driven by an external 'ring' motor built into the surrounding nozzle. Such a design eliminates the need for shafts, and sometimes seals, altogether.

There are various methods for sealing underwater thrusters. Some manufactures use fluid-filled thruster housings to lower the difference in pressure between the sea water and the internal thruster housing pressure by simply matching the two pressures (internal and external). Still others use a lubricant bath between the air-filled spaces and the outside water (Figure 3.8). A common and highly reliable technique is the use of a magnetically coupled shaft, which allows the air-filled housing to remain sealed (Figure 3.9).

Propeller design

The propeller is a turning lifting body designed to move and vector water opposite to the direction of motion. Many thruster propellers are designed so their efficiency is much higher in one direction (most often in the forward and the down directions) than in the other. Propellers have a nominal speed of maximum efficiency, which is hopefully near the vehicle's normal operating speed. Some propellers are designed for speed and others are designed for power. When selecting a propeller, choose one with the ROV's operating envelope in mind. The desired operating objectives should be achieved through efficient propeller output below the normal operating envelope.

Kort nozzle

A kort nozzle is common on most underwater thruster models. The efficacy of a kort nozzle is the mechanism's help in reducing the amount of propeller vortices generated as the propeller turns at high speeds. The nozzle, which surrounds the propeller blades, also helps with reducing the incidence of foreign object ingestion into the thruster propeller. Also, stators help reduce the tendency of rotating propellers' swirling discharge, which tends to lower propeller efficiency and cause unwanted thruster torque acting upon the entire vehicle.

(a)

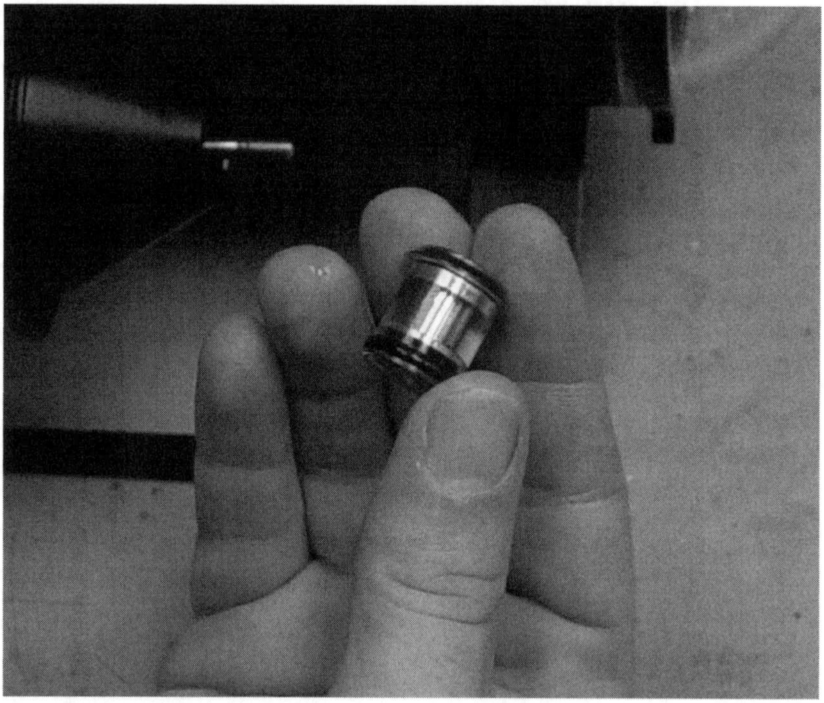

(b)

Figure 3.8 *Fluid sealing of direct dive thruster coupling.*

(c)

(d)

Figure 3.8 *(Continued)*

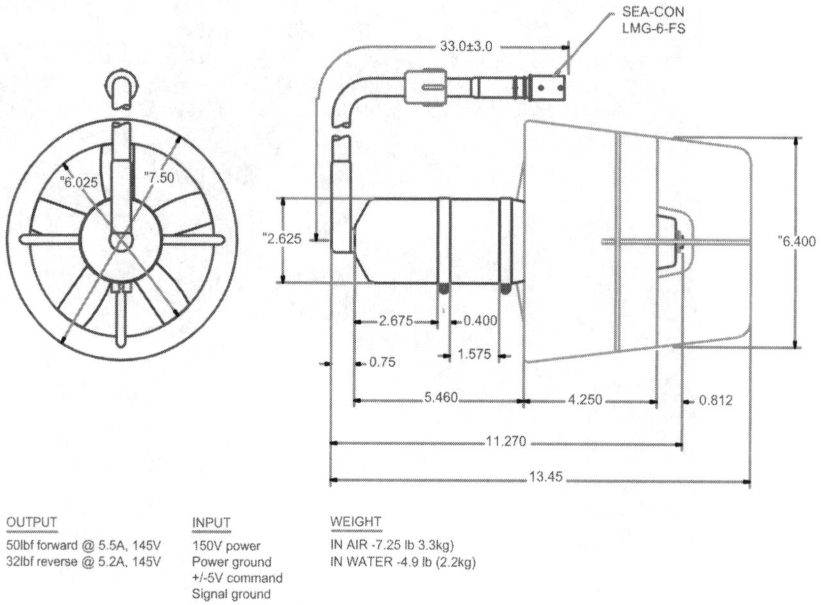

Figure 3.9 *Magnetically coupled thruster diagram.*

3.2 PRIMARY SUBSYSTEMS

The ability to sense the environment, either visually or through other means, and perform work at the desired location, is the mission of the ROV. The subsystems necessary for this task are discussed in the following sections.

3.2.1 Lighting

This explanation of lighting comes courtesy of Ronan Gray of Deep Sea Power & Light. The need for underwater lighting becomes apparent below a few feet from the surface. Ambient visible light is quickly attenuated by a combination of scattering and absorption, thus requiring artificial lighting to view items underwater with any degree of clarity. We see things in color because objects reflect wavelengths of light that represent the colors of the visible spectrum. Artificial lighting is therefore necessary near the illuminated object to view it in true color with intensity. Underwater lamps provide this capability.

Lamps convert electrical energy into light. The main types or classes of artificial lamps/light sources used in underwater lighting are incandescent, fluorescent, high-intensity gas discharge, and light-emitting diode (LED) – each with its strengths and weaknesses. All types of light are meant to augment the natural light present in the environment. Table 3.1 shows the major types of artificial lighting systems, as well as their respective characteristics.

- *Incandescent* – The incandescent lamp was the first artificial light bulb invented. Electricity is passed through a thin metal element, heating it to a high enough

Table 3.1 *Light source characteristics (table and lighting description courtesy of Deep Sea Power and Light).*

Source	*Lumens/watt*	*Life (h)*	*Color*	*Size*	*Ballast*
Incandescent	15–25	50–2500	Reddish	M–L	No
Tungsten–halogen	18–33	25–4000	Reddish	S–M	No
Fluorescent	40–90	10 000	Varies	L	Yes
Green fluor.	125	10 000	Green	L	Yes
Mercury	20–58	20 000	Bluish	M	Yes
Metal halide	70–125	10 000	Varies	M	Yes
High-Press. sodium	65–140	24 000	Pink	M	Yes/I
Xenon arc	20–40	400–2000	Daylight	VS	Yes/I
HMI/CID	70–100	200–2000	Daylight	S	Yes/I
Low-press. sodium	100–185	18 000	Yellow	L	Yes
Xenon flash	30–60	NA	Daylight	M	NA

V, very; S, small; M, medium; L, large; I, ignitor required; NA, not applicable.

temperature to glow (thus producing light). It is inefficient as a lighting source with approximately 90 percent of the energy wasted as heat. Halogen bulbs are an improved incandescent. Light energy output is about 15 percent of energy input, instead of 10 percent, allowing them to produce about 50 percent more light from the same amount of electrical power. However, the halogen bulb capsule is under high pressure instead of a vacuum or low-pressure noble gas (as with regular incandescent lamps) and, although much smaller, its hotter filament temperature causes the bulbs to have a very hot surface. This means that such glass bulbs can explode if broken, or if operated with residue (such as fingerprints) on them. The risk of burns or fire is also greater than other bulbs, leading to their prohibition in some underwater applications. Halogen capsules can be put inside regular bulbs or dichroic reflectors, either for aesthetics or for safety. Good halogen bulbs produce a sunshine-like white light, while regular incandescent bulbs produce a light between sunlight and candlelight.

- *Fluorescent* – A fluorescent lamp is a type of lamp that uses electricity to excite mercury vapor in argon or neon gas, producing short-wave ultraviolet light. This light then causes a phosphor coating on the light tube to fluoresce, producing visible light. Fluorescent bulbs are about 40 percent efficient, meaning that for the same amount of light they use one-fourth the power and produce one-sixth the heat of a regular incandescent. Fluorescents typically do not have the luminescent output capacity per unit volume of other types of lighting, making them (in many underwater applications) a poor choice for underwater artificial light sources.
- *High-intensity discharge* – High-intensity discharge (HID) lamps include the following types of electrical lights: Mercury vapor, metal halide, high-pressure sodium and, less common, xenon short-arc lamps. The light-producing element of these lamp types is a well-stabilized arc discharge contained within a refractory envelope (arc tube) with wall loading (power intensity per unit area of the arc tube) in excess of $3\,W/cm^2$ ($19.4\,W/in^2$). Compared to fluorescent

and incandescent lamps, HID lamps produce a large quantity of light in a small package, making them well suited for mounting on underwater vehicles. The most common HID lights used in underwater work are of the metal halide type.

- *LED* – A light-emitting diode (LED) is a semiconductor device that emits incoherent narrow-spectrum light when electrically biased in the forward direction. This effect is a form of electroluminescence. The color of the emitted light depends on the chemical composition of the semiconducting material used, and can be near-ultraviolet, visible, or infrared. LED technology is useful for underwater lighting because of its low power consumption, low heat generation, instantaneous on/off control, continuity of color throughout the life of the diode, extremely long life, and relatively low cost of manufacture. LED lighting is a rapidly evolving technology – look for more usage of LEDs in the underwater lighting field soon.

Most observation-class ROV systems use the smaller lighting systems, including halogen and metal halide HID lighting.

The efficiency metric for lamps is efficacy, which is defined as light output in lumens divided by energy input in watts, with units of lumens per watt (LPW). Lamp efficacy refers to the lamp's rated light output per nominal lamp watts. System efficacy refers to the lamp's rated light output per system watts, which include the ballast losses (if applicable). Efficacy may be expressed as 'initial efficacy', using rated initial lumens at the beginning of lamp life. Alternatively, efficacy may be expressed as 'mean efficacy', using rated mean lumens over the lamp's lifetime; Mean lumens are usually given at 40 percent of the lamp's rated life and indicate the degree of lumen depreciation as the lamp ages.

An efficient reflector will not only maximize the light output that falls on the target, but will also direct heat forward and away from the lamp. The shape of the reflector will be the main determinant in how the light output is directed. Most are parabolic, but ellipsoidal reflectors are often used in underwater applications to focus light through a small opening in a pressure housing. The surface condition of a reflector will determine how the light output will be dispersed and diffused. The majority of reflectors are made of pure, highly polished aluminum that will reflect light back at roughly the same angle to the normal at which it was incident. By adding dimples or peens to the surface, the reflected light is dispersed or spread out. When a plain white surface is used, the reflected light is diffused in all directions.

3.2.2 Cameras

Currently, most small ROV systems use inexpensive charge-coupled device (CCD) cameras as their main viewing device. These camera systems are mounted on small circuit boards and produce a video signal transmitted in a format sent up the tether to the video capture device on the surface. The actual protocol of the signal emanating from the camera and control box (after transmission through the tether) is manufacturer-specific, but usually falls under either composite or RF (radio frequency) video. The protocol of the video signal will determine the receiving adapter on the viewing device. Refer to the manufacturer's instructions for the specific protocol and/or adapter for the system.

Production of an ROV camera assembly can be accomplished on any electronics bench with rudimentary equipment. A simple chip camera system sold through any surveillance camera manufacturing company is mounted on a block along with a motor and gearing system for panning and/or tilting, plus focusing (if manual focusing is desired). Once the camera is mounted (Figure 3.10), simple wiring and switching will accomplish both control of the individual camera as well as switching between various camera systems aboard the vehicle.

(a)

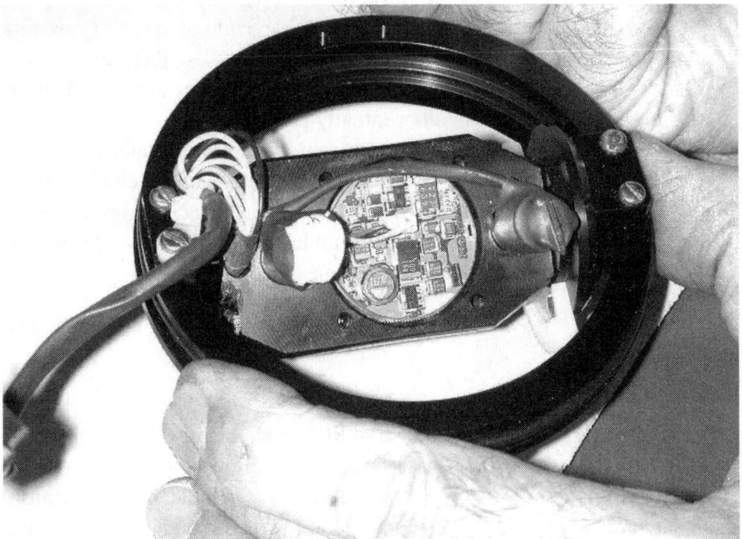

(b)

Figure 3.10 *Typical arrangement – CCD camera system mounted to ring with tilting mechanism.*

(c)

Figure 3.10 *(Continued)*

Various regions of the world use different video formats. In the USA, as well as a few other countries, NTSC (National Television Standards Committee) is the standard format, while most of Europe, Africa, and Asia use PAL (Phased Array Lines) format. The SECAM format used predominately in France has been declining in recent years and will in all likelihood eventually be eliminated.

Camera technology is evolving rapidly. The High Definition format is quickly being adapted to ROVs as it trickles down to the smaller vehicles due to decreasing size and lower cost structure. Digital still camera technology is also being adapted for high-resolution image capture of underwater items. Look for major improvements in size, functionality, and cost in the near future.

3.2.3 Sensors

As stated earlier, most industrial ROV systems provide the capability to transmit data from the submersible to the surface. This allows the ROV system to deliver a suite of instruments to the work site, powered by the vehicle, with data transmitted through the tether to the surface. Any combination of sensor and instrument (heading/gyro/depth, etc.) is available as payload to the modern ROV system, assuming proper data protocol transmissions and power delivery are available. Figure 3.11 shows (left to right) a sonar, hydrology sensor, and manipulator configured for attachment to an ROV.

Major issues regarding the integration of sensors involves the data transmission protocol and the method through which this transmission takes place. The ROV

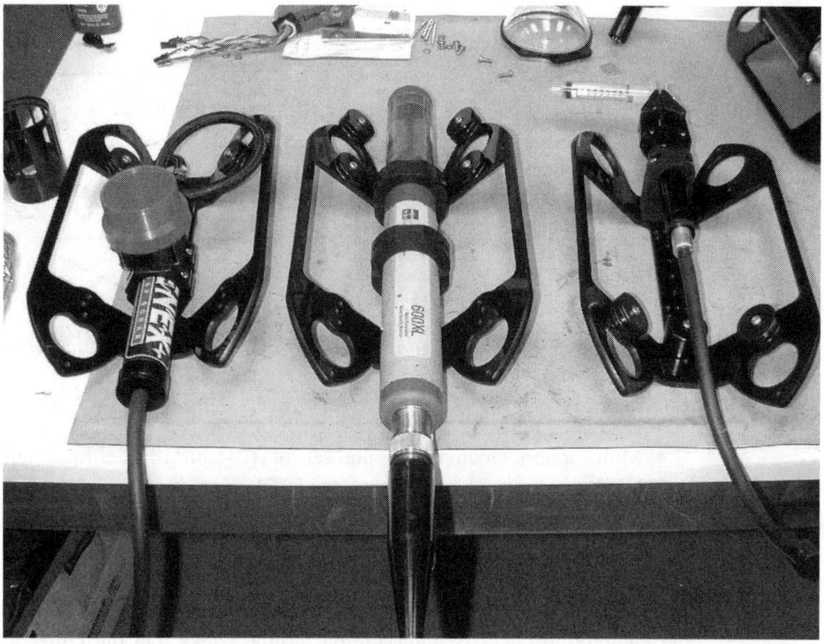

Figure 3.11 *Various sensors for mounting to ROV.*

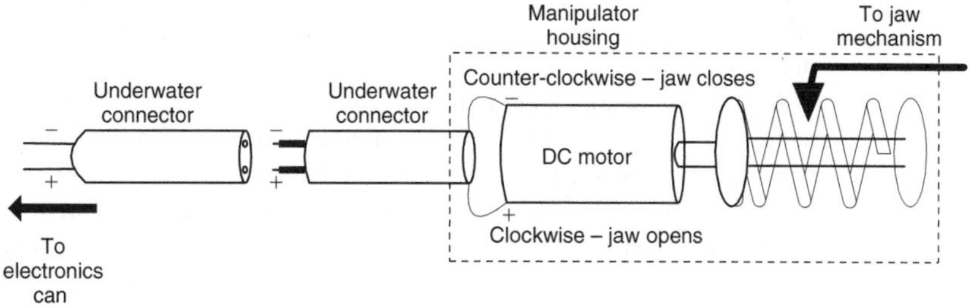

Figure 3.12 *Typical setup for small ROV manipulator.*

manufacturer must provide a throughput within the ROV system to allow for sensor integration. Some examples of common sensors packages placed aboard ROV systems in industrial applications include:

- Radiation sensors
- CTD (conductivity/temperature/depth) sensors
- Pressure-sensitive depth transducer
- Magnetic flux gate compass module
- Slaved or rate gyro for heading stabilization
- Ultrasonic thickness gauges for measuring metal thickness and quality
- Imaging sonar
- Acoustic positioning

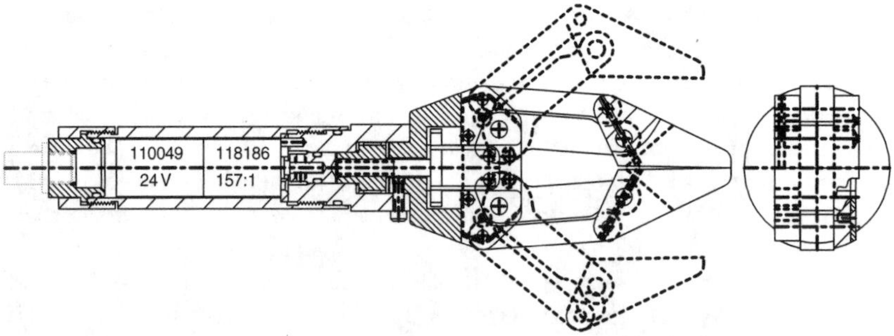

Figure 3.13 *Small manipulator CAD drawing (courtesy of Inuktun Services Ltd.).*

- Digital cameras
- Multi-parameter environmental sensors (e.g. turbidity, chlorophyll, DO, pH, and ORP sensors, which are discussed in Chapter 6).

3.2.4 Manipulator and Tool Pack

Most professional ROV systems allow for a simple A/B power source to provide the locomotion needs of intervention tooling packs. On the larger hydraulic ROV systems, a simple independent A and B connector is provided to power turning or cutting equipment for subsea work. On observation-class systems, a simple 12 or 24 VDC source is provided to run manipulators or other small tools needed for specific jobs. Power can also be redirected from main thruster power for more demanding mechanical work.

A basic single-function manipulator package, common on many small ROV systems, consists of a 24 VDC electric motor running a worm gear to open and close small grabber arms for light intervention duties (Figures 3.12 and 3.13). A common problem with small manipulators without a limit switch at the end of the travel of the worm gear is that the ROV operator will continue activating the manipulator. When the worm gear reaches the end of travel, the motor can become stalled at the full close or full open point of the jaw. If the jaw is left in this position (i.e. with the worm gear stalled at the end of its travel and torqued against the end of the worm gear stop) and allowed to sit for a length of time, oxidation can seize the screw/teeth, preventing it from moving the worm gear away from the stop. If this happens, the jaw must be either squeezed/pulled to take the pressure off the worm gear, and the motor then activated, or the mechanism must be disassembled to unfreeze the worm gear.

3.3 ELECTRICAL CONSIDERATIONS

The following sections discuss specific issues and relationships regarding the tether, power, data, and the connectors that bring it all together.

3.3.1 The Tether

The tether and the umbilical are essentially the same item. The cable linking the surface to the cage or tether management system (TMS) is termed the 'Umbilical', while the

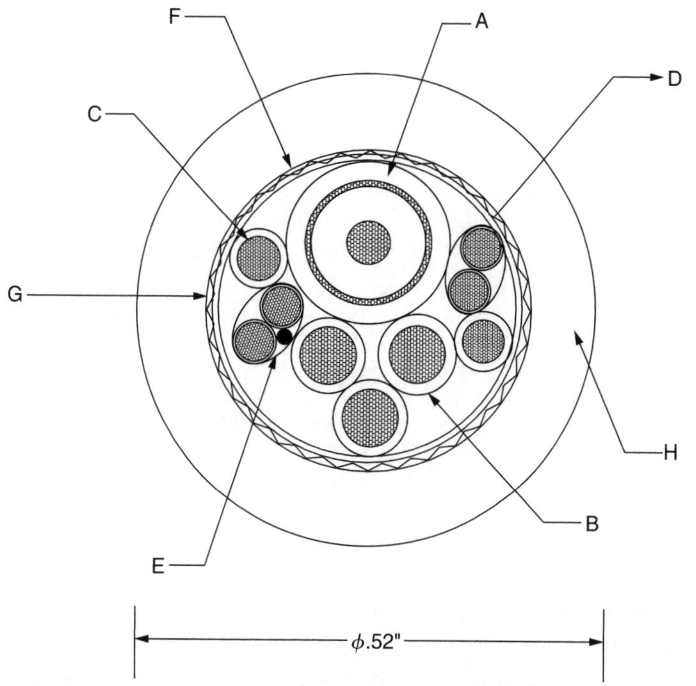

A – 75-ohm mini coax. Cap. 16.6 pf/ft. 22-7 TC, XLPO-Foam, Alum,1 spiral T.C. shield
B – 3×3 #18 (19/30) T.C. (OR/BL/PUR)
C – 2×3 #22 (19/34) T.C. (WH)
D – 1 3× #24 AWG TP, (19/36) T.C.
E – 1×3 #24 AWG TSP, (19/36) T.C. polypropylene insulation 0.010 wall
F – MYLAR tape
G – KEVLAR weave
H – 0.045 Green cellular foam polyurethane

Figure 3.14 *Cross-section of neutrally buoyant tether.*

cable from the TMS to the submersible is termed the 'Tether'. Any combination of electrical junctions is possible in order to achieve power transmission and/or data relay. For instance, AC power may be transmitted from the surface through the umbilical to the cage, where it is then changed to DC to power the submersible's thrusters and electronics. Further, video and data may be transmitted from the surface to the cage via fiber-optics (to lessen the noise due to AC power transmission), then changed to copper for the portion from the cage to the submersible, thus eliminating the AC noise problem. Figure 3.14 is an example of the neutrally buoyant tether for the Outland 1000 observation-class ROV system (courtesy of Outland Technology).

The umbilical/tether can be made up of a number of components:

- Conductors for transmitting power from the surface to the submersible
- Control throughput for telemetry (conducting metal or fiber optic)
- Video/data transmission throughput (conducting metal or fiber optic)
- Strength member allowing for higher tensile strength of cable structure
- Lighter-than-water filler that helps the cable assembly achieve neutral buoyancy
- Protective outer jacket for tear and abrasion resistance.

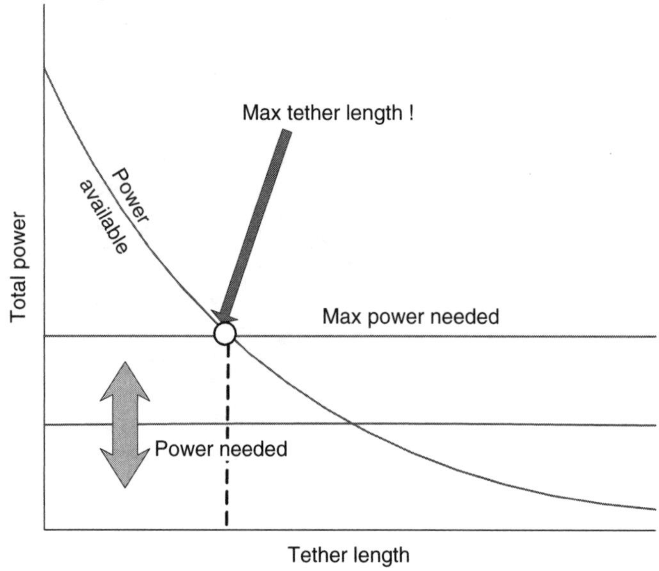

Figure 3.15 *Diagram depicting the power budget for power through the tether.*

Most observation-class ROV systems use direct current power for transmission along the tether to power the submersible. The tether length is critical in determining the power available for use at the vehicle. The power available to the vehicle must be sufficient to operate all of the electrical equipment on the submersible. The electrical resistance of the conductors within the tether, especially over longer lengths, could reduce the vehicle power sufficiently during high-load conditions to effect operations.

The maximum tether length for a given power requirement is a function of the size of the conductor, the voltage, and the resistance. For example, using a water pipe analogy, there is only a certain amount of water that will flow through a pipeline at a given pressure. The longer the pipe, the higher the internal resistance to movement of the water. As long as the water requirements at the receiving end do not exceed the delivery capacity of the pipe (at a given pressure), the system delivery of water will be adequate. If there were to be a sudden increase in the water requirement (a fire requiring water, everyone watering their lawn simultaneously, etc.), the only way to get adequate water to the delivery end would be to increase the pressure or to decrease the resistance (i.e. shorten the pipe length or increase the diameter) of the pipe. The same holds true in electrical terms between tether length, total power required, voltage, and resistance (Figure 3.15).

Ohm's law deals with the relationship between voltage and current in an ideal conductor. This relationship states that the potential difference (voltage) across an ideal conductor is proportional to the current through it. So, the voltage (V, or universally as E) is equal to the current (I) times the resistance (R). This is stated mathematically as $V = IR$. Further, power (measured in watts) delivered to a circuit is a product of the voltage and the current.

Thus, based on Ohm's law, the voltage drop over a length of cable can be calculated by using the formula, $V = IR$, where V is the voltage drop, I is the current draw of

Table 3.2 *Standard copper wire gauge resistance over nominal lengths (example and table courtesy of Deep Sea Power and Light).*

Wire gauge	*Ohms/1000 ft (approx.)*
20	10
18	6
16	4
14	2.5
12	1.5

the vehicle in amps, and R is the total electrical resistance of the power conductor within the tether in ohms. The current draw of a particular component (light, thruster, camera, etc.) can be calculated if the wattage and voltage of the component are known. The current draw is equal to the component wattage divided by the component voltage (or amps = watts/volts).

For example, referring to the table of electrical resistances for various wire gauges (Table 3.2), the voltage required to operate a 24-volt/300-watt light at 24 volts over 250 feet of 16-gauge cable can be calculated as follows: The current draw, I, of a 24-volt/300-watt lamp operating at 24 volts is 300 watts/24 volts = 12.5 amps. The resistance of 16-gauge wire is approximately 4 ohms/1000 feet (Table 3.2). Since the total path of the circuit is from the power supply to the light and back to the power supply, the total resistance of the cable is twice the length of the cable times the linear resistance, or for this example, $R = (2 \times 250\,\text{ft}) \times (4\,\text{ohms}/1000\,\text{ft}) = 2.0$ ohms. Since $V = IR$, the voltage drop, V, is equal to 12.5 amps $\times$ 2.0 ohms = 25 volts. This means that 25 volts is lost due to resistance, so the power supply will need to provide at least 49 volts (the 24 volts necessary to operate the light plus the additional voltage loss of 25 volts) to power this 24-volt/300-watt light over a 250-foot cable.

3.3.2 Power Source

The ROV system is made up of a series of compromises. The type of power delivered to the submersible is a trade-off of cost, safety, and needed performance. Direct current (DC) allows for lower cost and weight of tether components; Since inductance noise is minimal, it allows for less shielding of conductors in close proximity to the power line. Alternating current (AC) allows longer transmission distances than that available to DC while using smaller conductors.

Most operators of ROV systems specify a power source independent of the vessel of opportunity. The reason for this separation of supply is that the time the vessel is in most need of its power is normally the time when the submersible is most in need of its power. Submersible systems attempting to escape a hazardous bottom condition have been known to lose power at critical moments while the vessel is making power-draining repositioning thrusts on its engines. This can cause entanglement of the vehicle. With a separate power source, submersible maneuvering power is separated from the power needs of the vessel.

With the advent of the lightweight micro-generators for use with small ROVs, the portability of the ROV system is significantly enhanced. Some operators prefer usage of the battery/inverter combination for systems requiring AC power. Also, some smaller systems use only DC as their power source. Either method should have the power source capable of supplying uninterrupted power to the system at its maximum sustained current draw for the length of the anticipated operation.

3.3.3 AC Versus DC Considerations

Electrical power transmission techniques are an important factor in ROV system design due to their effect upon component weights, electrical noise propagation and safety considerations. The DC method of power transmission predominates the observation-class ROV systems due to the lack of need for shielding of components, weight considerations for portability, and the expense of power transmission devices. On larger ROV systems, AC power is used for the umbilical due to its long power transmission distances, which are not seen by the smaller systems. AC power in close proximity to video conductors could cause electrical noise to propagate due to EMF (electromotive force) conditions. The shielding necessary to lower this EMF effect could cause the otherwise neutrally buoyant tether to become negatively buoyant, resulting in vehicle control problems. And the heavy and bulky transformers are a nuisance during travel to a job site or as checked baggage aboard aircraft.

Larger work-class systems normally use AC power transmission from the surface down the umbilical to the cage (the umbilical normally uses fiber-optic transmission, lowering the EMF noise through the video) since the umbilical does not require neutral buoyancy. At the cage, the AC power is then rectified to DC to run the submersible through the neutrally buoyant tether that runs between the cage and vehicle.

3.3.4 Data Throughput

The wider the data pipeline from the submersible to the surface, the greater the ability for the vehicle to deliver to the operator the necessary job-specific data as well as sensory feedback needed to properly control the vehicle. With the cost of broadband fiber-optic transmission equipment dropping into the range of most small ROV equipment manufacturers' budgets, more applications and sensors should soon become available to the ROV marketplace. The ROV is simply a delivery platform for transporting the sensor package to the work location. The only limitation to full sensor feedback to the operator will remain one of lack of funding and imagination. The Human-Robot Interface (the intuitive interaction protocol between the human operator and the robotic vehicle) is still in its infancy; However, sensors are still outstretching the human's ability to interpret this data fast enough to react to the feedback in a timely fashion. This subject is probably the most exciting field of development for the future of robotics and will be of considerable interest to the next generation of ROV pilots.

3.3.5 Data Transmission and Protocol

Most small ROV manufacturers simply provide a spare twisted pair of conductors for hard-wire communication of sensors from the vehicle to the surface. The strength of this method is that the sensor vendor does not need engineering support from the ROV manufacturer in order to design these sensor interfaces. The weakness is that unless the sensor manufacturers collude to form a set of transmission standards, each sensor connected to the system 'hogs' the data transmission line to the detriment of other sensors needed for the task. A specific example of this problem is the need for concurrent use of an imaging sonar system and an acoustic positioning system. Unless the manufacturers of each sensor package agree upon a transmission protocol to share the single data line, only one instrument may use the line at a time. A few manufacturers have adapted industry standard protocols for such transmissions, including TCP/IP, RS-485, and other standard protocols. The most common protocol, RS-232, while useful and seemingly ubiquitous in the computer industry, is distance limited through conductors, thus causing transmission problems over longer lengths of tether.

The move toward open source PC-based sensor data processing has led to the production of data protocol converters for use in ROV sensor interpretation. Most small ROV sensor manufacturers transmit data with the RS-485 protocol, requiring a converter at the surface to both isolate the signal and to convert it to USB (or RS-232) protocol for easy processing with a standard laptop computer. Standards for these protocol converters are slow in evolving (due to the size of the customer base). Thus, the ROV system integrator must become familiar with the wiring and pin arrangement for these converters to assure data transmission from the sensor, through the vehicle and tether to the software at the surface, is achieved.

3.3.6 Underwater Connectors

The underwater connector is said to be the bane of the ROV business. Salt water is highly conductive, causing any exposed electrical component submerged in salt water to short to ground. The result is the 'Ubiquitous ground fault'. The purpose of an underwater connector is to conduct needed electrical currents through the connector while at the same time squeezing the water path and sealing the connection to lower the risk of electrical leakage to ground.

The underwater connector is lined with synthetic rubber that blocks the ingress path of water while allowing a positive electrical connection. Connectors sometimes experience cathodic delamination, causing rubber peeling and flaking from the connector walls. Connector maintenance (Figure 3.16) should include:

- Use small amounts of silicone grease to lubricate the connector, thus allowing easier slide on and off. Using too much grease, a widespread problem, can interfere with sealing.
- Always pull the connector by its body instead of its tail (cable), since the wire splice is located in the connection. Pulling on the tail could part the solder joint and ruin the electrical continuity within the connector.
- Keep the connectors as clean as possible through regularly scheduled maintenance tasks that include cleaning the contacts and lubricating the rubber lining.

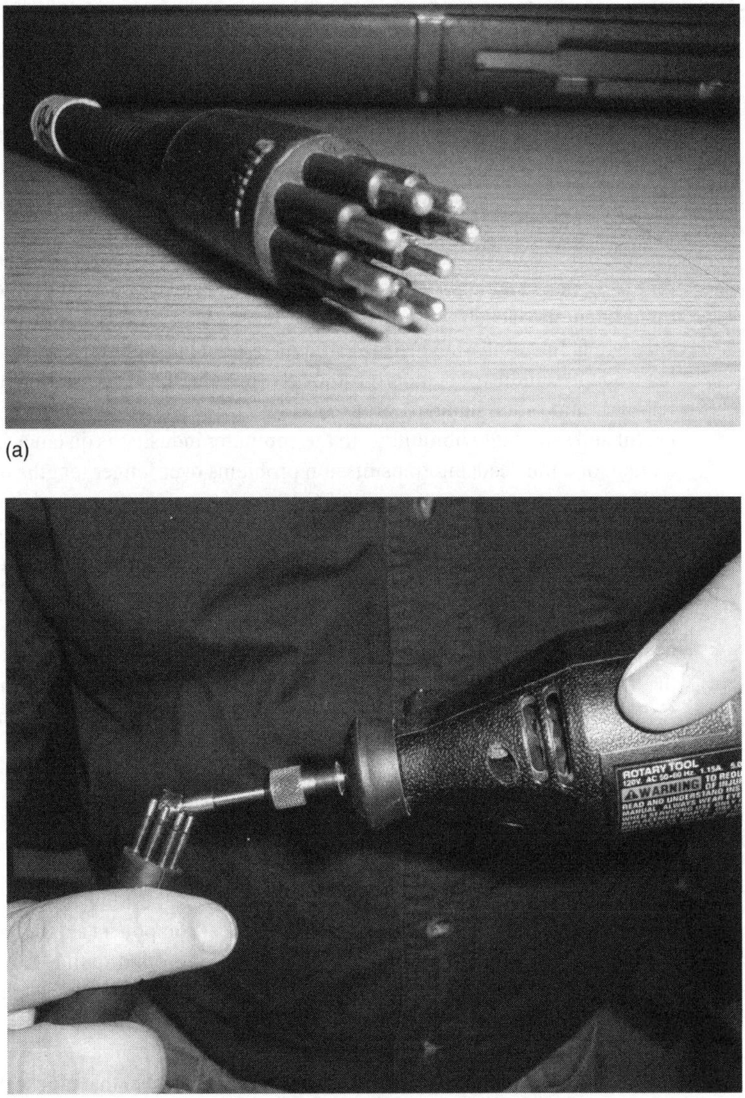

(a)

(b)

Figure 3.16 *Underwater connectors must be serviced to assure proper electrical connectivity.*

- Spray the connector body with silicone spray to keep the housing from drying out, which could result in flaking and rubber degradation.

Even when the contacts are right and the connector has good design features, the connector must be appropriate for the intended use and environment. The connector materials must be able to withstand the environmental conditions without degradation. For example, extended exposure to sunlight (ultraviolet energy) will cause damage

Figure 3.17 *Internal workings of male/female underwater connectors.*

to neoprene, and many steels will corrode in sea water. Check that the connector will fully withstand the environment.

The connector must not adversely affect the application. For example, all ferrous materials (steel, etc.) should be avoided in cases where the connector's magnetic signature might affect the system. In extreme cases, even the nickel used under gold plating could have an effect and should be reviewed.

The physical size of the connector, its weight, ease of use (and appropriateness for the application), durability, submergence (depth) rating, field repairability, etc. should all be assessed. The use of oil-filled cables or connectors should be considered.

Ease of installation and use is especially important, so realistically appraise the technical ability of those personnel who will actually install or use the equipment. If they are inexperienced, a more 'user-friendly' connector may be a better choice. And, if possible, train operators in the basics of proper connector use: Use only a little lubricant, avoid over-tightening, note acceptable cable bending radii, provide grounding wires for steel connectors in aluminum bulkheads, etc.

Splicing and repairing underwater cables and connectors, while quite simple, requires some basic precautions to avoid water ingress into the electrical spaces, thus grounding the connection. Examples are shown in Figures 3.17–3.22.

3.4 CONTROL SYSTEMS

The control system controls the different functions of the ROV, from controlling the propulsion system to switching of the light(s) and video camera(s). From simple relay control systems in the past to today's digital fiber optics, these systems are equipped

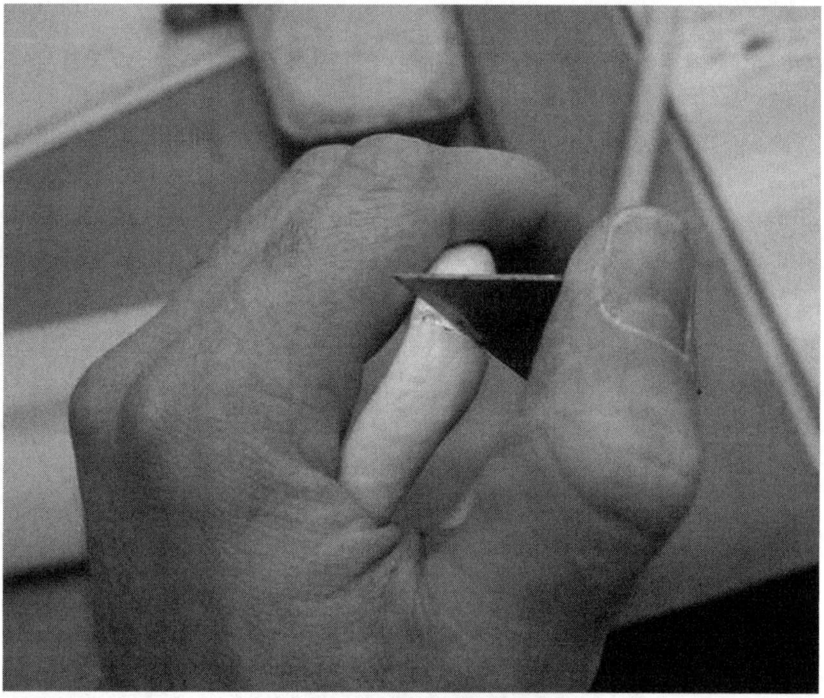

Figure 3.18 *Slicing is conducted by peeling back the conductors of the tether.*

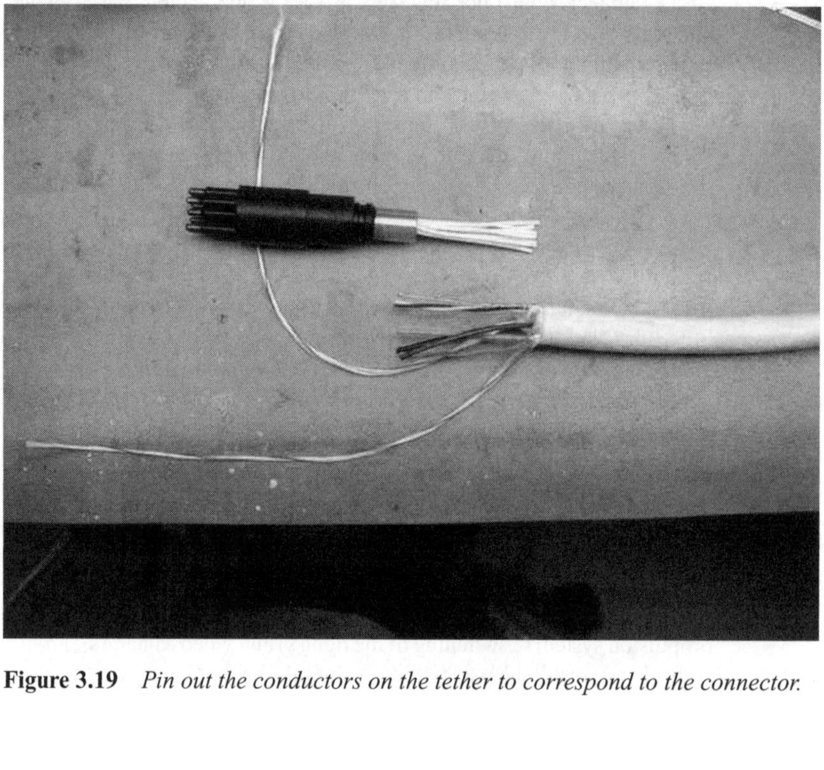

Figure 3.19 *Pin out the conductors on the tether to correspond to the connector.*

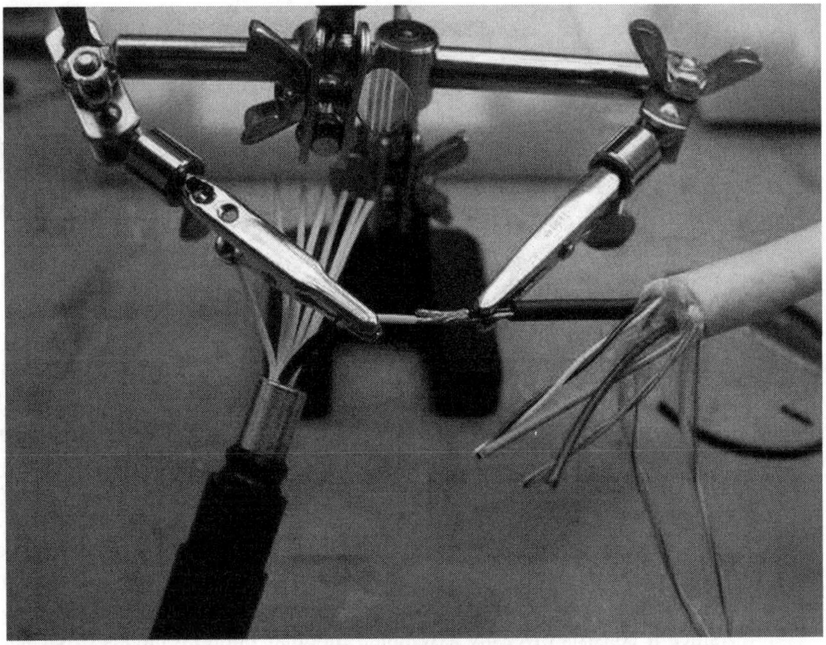

Figure 3.20 *The electrical connection is made through standard bench techniques.*

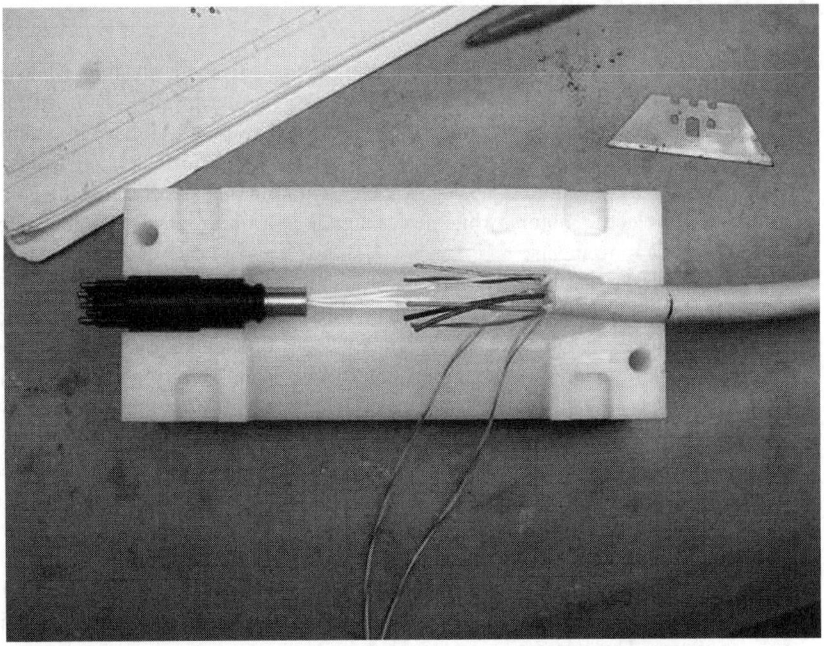

Figure 3.21 *The connected conductors are laid into a potting mold then sealed with potting compound.*

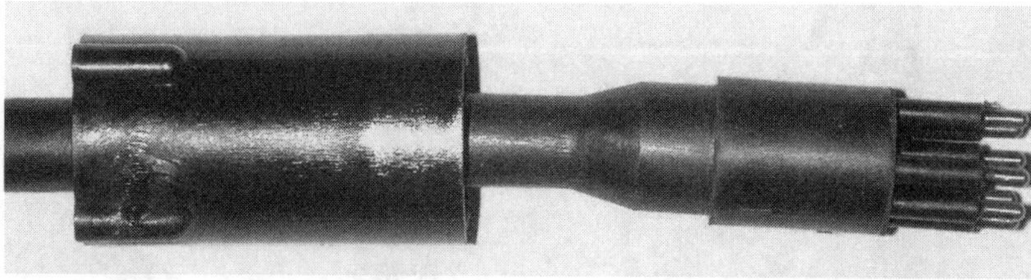

Figure 3.22 *Finished connector along with plastic guard.*

with a computer and subsystem control interface. The control system has to manage the input from the operator at the surface and convert it into actions subsea. The data required by the operator on the surface to accurately determine the position in the water is collected by sensors (sonar and acoustic positioning) and transmitted to the operator.

Over the last 10–15 years, computers utilized for these purposes have been designer computers with sophisticated computer programs and control sequences. Today, one can find standard computers in the heart of these systems. There has been a shift back to simpler control systems recently with the commercial advent of the PLC (Programmable Logic Computer). This is used in numerous manufacturing processes since it consists of easily assembled modular building blocks of switches, analog in/outputs, and digital in/outputs.

3.4.1 The Control Station

Control stations vary from large containers, with their spacious enclosed working area for work class systems, to simple PC gaming joysticks with PHDs (personal head-mounted displays) for some micro-ROV systems. All have in common a video display and some form of controlling mechanism (normally a joystick, such as at Figure 3.23). On older analog systems, a simple rheostat controls the variable power to the electric motors, while newer digital controls are necessary for more advanced vehicle movements.

With the rise of robotics as a sub-discipline within electronics, further focus highlighted the need to control robotic systems based upon intuitive interaction through emulation of human sensory inputs. Under older analog systems, a command of 'look left/go left' was a complex control command requiring the operation of several rheostats to gain vector thrusting to achieve the desired motion. As digital control systems arose, more complex control matrices could be implemented much more easily through allowing the circuit to proportionally control a thruster based upon the simple position of a joystick control. The advent of the modern industrial joystick coupled with programmable logic circuits has allowed easier control of the vehicle while operating through a much simpler and more intuitive interface. The more sensors available to the 'human' that allow intuitive interaction with the 'robot', the easier it is for the operator to figuratively operate the vehicle from the vehicle's point of view. This interaction protocol between operator and vehicle has become known as the Human–Robot Interface and is the subject of intensive current research. Look for major developments within this area of robotics over both the short and long terms.

(a)

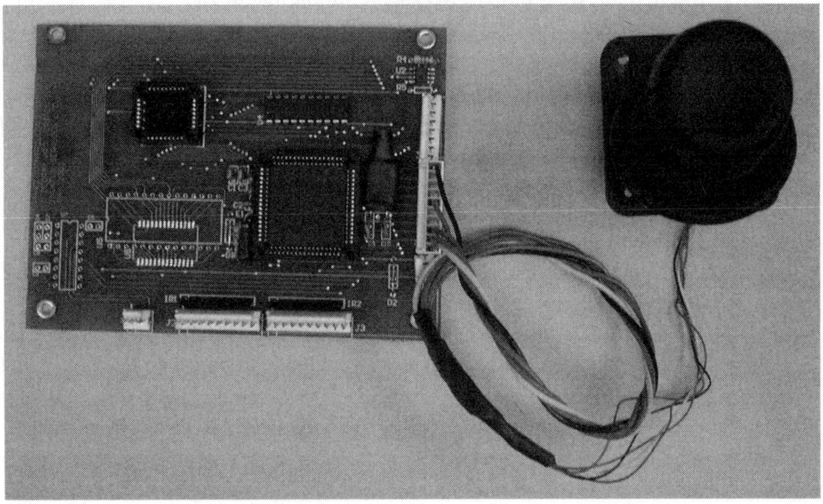

(b)

Figure 3.23 *Industrial joystick and circuit board used in vehicle control.*

3.4.2 Motor Control Electronics

Since observation-class ROV systems use mainly electronic motors for thruster-based locomotion, a study of basic motor control is in order.

A basic control of direction and proportional scaling of electrical motors is necessary to finely control the motion of the submersible. If only 'On' or 'Off' were the choices of motor control via switches, the operator would quickly lose control of the

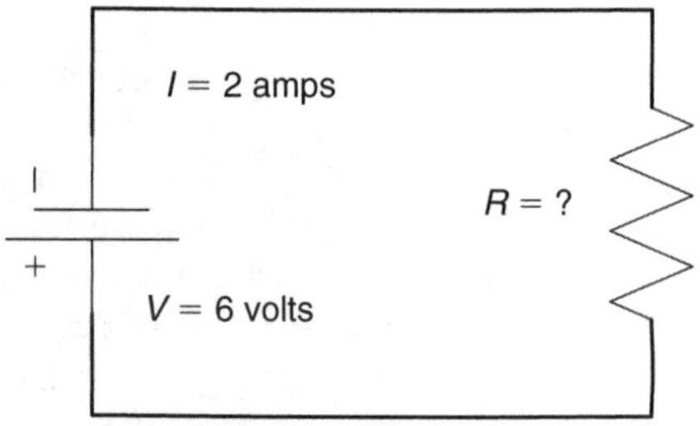

Question:	Answer
$I = 2$ amps	$V = IR$
$V = 6$ volts	$R = V/I$
$R = ?$ ohms	$R = 3.0$

Figure 3.24 *Simple electrical circuit.*

vehicle due to the inability to make the fine corrections needed for accurate navigation. In the early days of ROVs, the simple analog rheostat was used for motor control. It was quite a difficult task to control a vehicle with the operation of three or four independent rheostat knobs while attempting to fly a straight line. Later came digital control of electric motors and the finer science of robotics took a great leap forward.

The basic electronic circuit that made the control of electronic motors used in robotics and industrial components so incredibly useful is known as the 'H-bridge'. An understanding of the H-bridge (discussed later in this section) and the digital control of that H-bridge will help significantly with the understanding of robotic locomotion.

Consider the analysis of a simple electric circuit (Figure 3.24). As discussed earlier in this chapter, Ohm's law gives a relationship between voltage (V), current (I) and resistance (R) that is stated as $V = IR$. In Figure 3.24, the current and voltage are known, thus the resistance can be calculated to be 3.0 ohms.

3.4.2.1 *Inductors*

An inductor is an energy storage device that can be as simple as a single loop of wire or consist of many turns of wire wound around a core. Energy is stored in the form of a magnetic field in or around the inductor. Whenever current flows through a wire, it creates a magnetic field around the wire. By placing multiple turns of wire around a loop, the magnetic field is concentrated into a smaller space, where it can be more useful. When applying a voltage across an inductor, current starts to flow. It does not instantly rise to some level, but rather increases gradually over time (Figure 3.25). The relationship of voltage to current vs. time gives rise to what is known as inductance. The higher the inductance, the longer it takes for a given voltage to produce a given current – sort of a 'shock absorber' for electronics. Whenever there is a moving or changing magnetic field in the presence of an inductor, that change attempts to generate a current in the inductor. An externally applied current

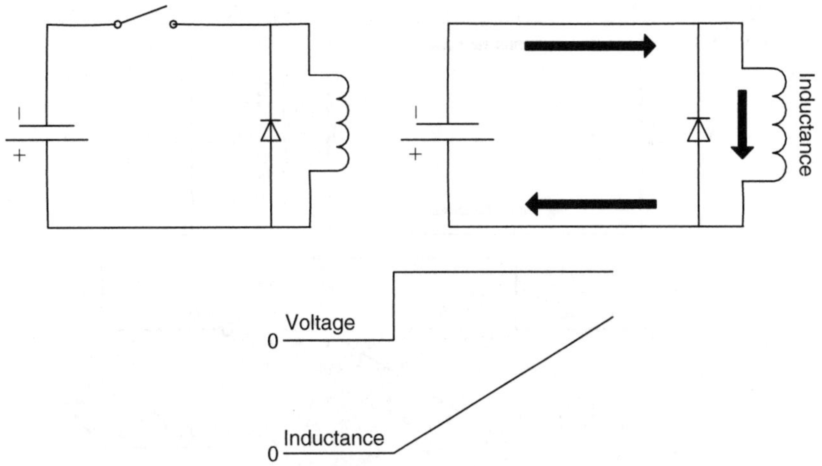

Figure 3.25 *Inductor/voltage interaction without resistance.*

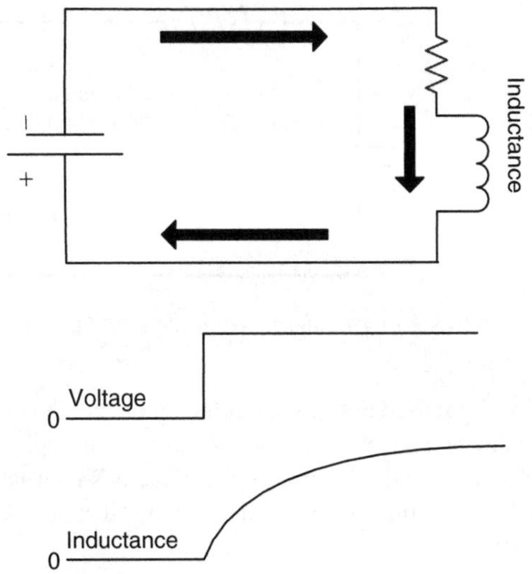

Figure 3.26 *Inductor/voltage interaction with resistance.*

produces an increasing magnetic field, which in turn produces a current opposing that applied externally, hence the inability to create an instantaneous current change in an inductor. This property makes inductors useful as filters in power supplies. The basic mathematical expression for inductance (L) is $V = L \times \mathrm{d}I/\mathrm{d}T$.

The ideal world and the real world depart, since real inductors have resistance. In this world, the current eventually levels out, leaving the strength of the magnetic field plus the level of the stored energy as proportional to the current (Figure 3.26).

So, what happens when the switch is opened? The current dissipates quickly in the arc (Figure 3.27).

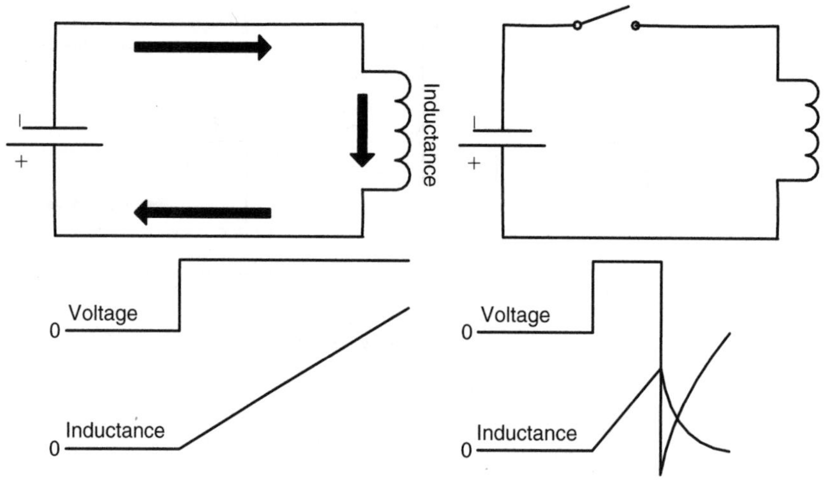

Figure 3.27 *Inductor/voltage interaction as circuit is opened.*

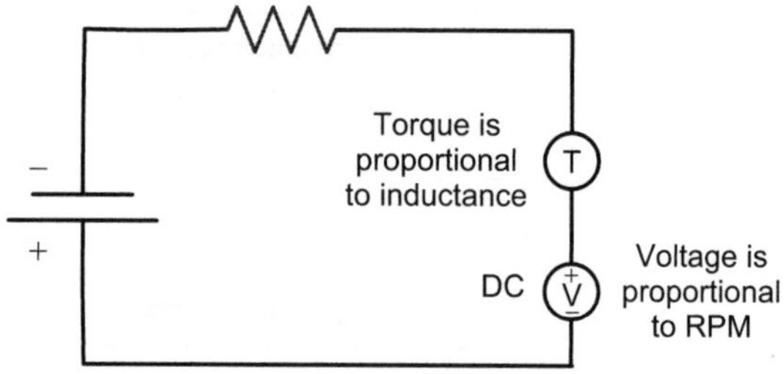

Figure 3.28 *Diagram depicting inductance/torque interaction.*

Diodes are used to suppress arcing, allowing the recirculation currents to dissipate more slowly. The current continues to flow through the inductor, but power is dissipated across the diode through the inductor's internal resistance. This is the basis of robotic locomotion electronic dampening (sort of a shock absorber for your thruster's drive train).

Permanent magnet DC motors can be modeled as an inductor, a voltage source, and a resistor. In this case, the torque of the motor is proportional to the current and the internal voltage source is proportional to the RPM (Figure 3.28) or back EMF (when the current is released and the motor turns into a generator as the motor spools down).

The stall point of an electric motor is at the point of highest torque as well as the point of highest current, and is proportional to the internal resistance of the motor.

That leads to the concern over back EMF within thruster control electronics design – once the DC motor is disengaged, the turning motor mass continues to rotate the coils within the armature. This rotating mass converts the electric motor into a generator, rapidly reversing and spiking the voltage in the reverse direction. Unless there is some circuit protection within the driver board circuit, damage to the control electronics will often result (Figure 3.29).

Figure 3.29 *Damaged driver board due to back EMF.*

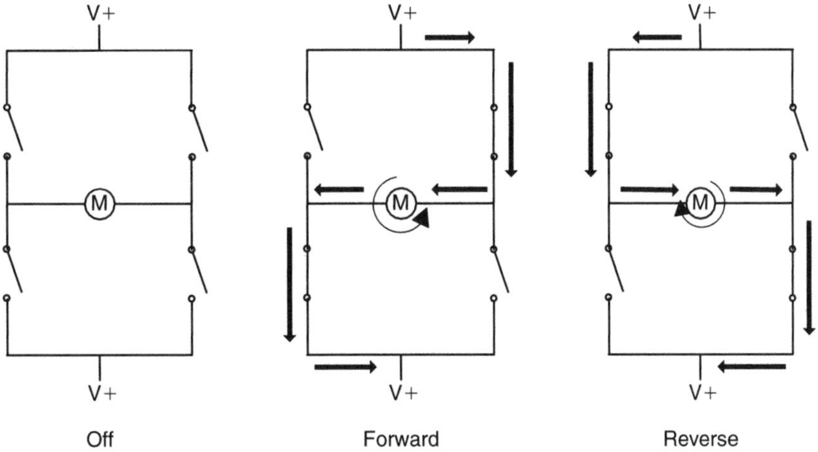

Figure 3.30 *H-bridge diagram depicting basic operation.*

3.4.2.2 *The H-bridge*

The circuit that controls the electrical motor is known as the 'H-bridge' due to its resemblance to the letter 'H' (Figure 3.30). Through variation of the switching, as well as inductance filtering (described earlier), the direction and the ramp-up speed are controlled through this circuit in an elegant and simplistic fashion.

3.4.2.3 *PWM control*

Pulse width modulation (PWM) is a modulation technique that generates variable-width pulses to represent the amplitude of an analog input signal. The output switching transistor is on more of the time for a high-amplitude signal and off more of the time for a low-amplitude signal. The digital nature (fully on or off) of the PWM circuit is less costly to fabricate than an analog circuit that does not drift over time.

PWM is widely used in ROV applications to control the speed of a DC motor and/or the brightness of a light bulb. For example, if the line were closed for 1 μs, opened for 1 μs, and continuously repeated, the target would receive an average of 50 percent of the voltage and run at half speed or the bulb at half brightness. If the line were closed for 1 μs and open for 3 μs, the target would receive an average of 25 percent.

There are other methods by which analog signals are modulated for motor control, but observation-class ROV systems predominate with the PWM mode due to cost and simplicity of design.

Chapter 4
Underwater Acoustics and Positioning

Mathematics is the language of science. All of the properties of underwater acoustics can be expressed in mathematical terms. Later in this chapter, as well as in Chapter 5, two of the practical applications of underwater acoustics (acoustic positioning and imaging/profiling sonar) will be explored. In the initial sections of this chapter, the theoretical basis of underwater acoustics is presented to segue into the acoustic positioning and sonar applications.

This section is fairly technical and is math-based for those who would like to explore the theoretical aspects of underwater acoustics to gain a better insight into positioning and sonar. For those less interested in the technical background, and more concerned with the practical applications of these technologies, proceed to Section 4.2 of this chapter for acoustic positioning as well as Chapter 5 for sonar.

Section 4.1 on underwater acoustics is based on a publication by Kongsberg Maritime entitled *Introduction to Underwater Acoustics.* A special thanks goes to Arndt Olsen with Kongsberg Maritime for obtaining permissions to include this material.

4.1 UNDERWATER ACOUSTICS

4.1.1 Introduction

Acoustic sound transmission represents the basic techniques for underwater navigation, telemetry, echo sounder, and sonar systems. Common for these systems are the use of underwater pressure wave signals that propagate with a speed of approximately 1500 m/s through the water (Figure 4.1).

When the pressure wave hits the sea bottom, or another object, a reflected signal is transmitted back and is detected by the system. The reflected signal contains information about the nature of the reflected object.

For a navigation and telemetry system, the communication is based upon an active exchange of acoustic signals between two or more intelligent units.

Transmission of underwater signals is influenced by a number of physical limitations, which together limit the range, accuracy, and reliability of a navigation or telemetry system.

The factors described in this section include:

- Transmitted power
- Transmission loss
- Transducer configuration
- Directivity and bandwidth of receiver
- Environmental noise

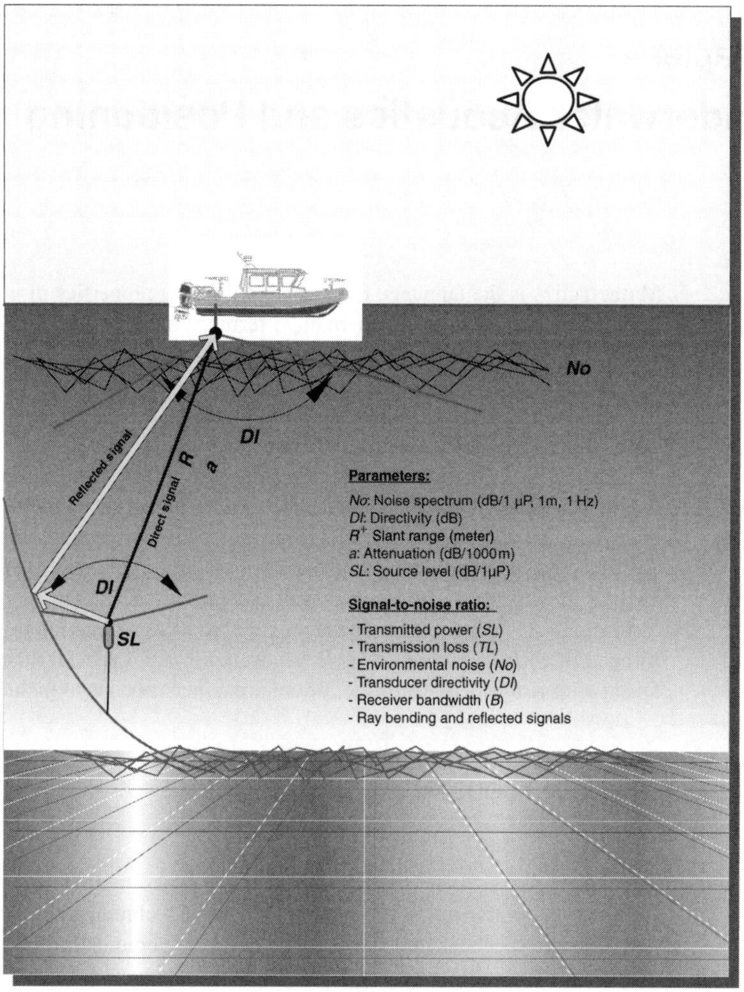

Figure 4.1 *Parameters of the sonar equation.*

- Requirements of positive signal-to-noise ratio for reliable signal detection
- Ray bending and reflected signals.

The signal-to-noise ratio obtained can be calculated by the sonar equation.

4.1.2 Sound Propagation

4.1.2.1 *Pressure*

A basic unit in underwater acoustics is pressure, measured in μPa (micropascal) or μbar. The Pa (Pascal) is now the international standard. It belongs to the MKS system, where 1 μPa = 10^6 newton/m^2. The μbar belongs to the CGS system.

$$1\,\mu\text{bar} = 10^5\,\mu\text{Pa}$$
$$0\,\text{dB re }1\,\mu\text{bar} = 100\,\text{dB re }1\,\mu\text{Pa}.$$

The μbar is a very small unit so negative decibels will rarely occur, if ever. To convert from μbar to μPa, simply add 100 dB.

4.1.2.2 *Intensity*

The sound intensity is defined as the energy passing through a unit area per second. The intensity is related to pressure by:

$$I = p^2/\rho c,$$

where

I = intensity,
p = pressure,
ρ = water density, and
c = speed of sound in water.

4.1.2.3 *Decibel*

The decibel is widely used in acoustic calculations. It provides a convenient way of handling large numbers and large changes in variables. It also permits quantities to be multiplied together simply by adding their decibel equivalents. The decibel notation of intensity I is:

$$10 \log I/I_o,$$

where I_o is a reference intensity.

The decibel notation of the corresponding pressure is:

$$10 \log (p^2/\rho c)/(p_o^2/\rho c) = 20 \log p^2/p_o,$$

where p_o is the reference pressure corresponding to I_o.

Normally, p_o is taken to 1 μPa, and I_o will then be the intensity of a plane wave with pressure 1 μPa.

Example

A pressure $p = 100\,\mu$Pa
In decibels: 20 log 100/1 = 40 dB re 1 μPa
The intensity will also be 40 dB re 'the intensity of a plane wave with pressure 1 μPa'.

As shown in the example, the decibel number is the same for pressure and intensity. It is therefore common practice to speak of sound level rather than pressure and intensity. The reference level is in both cases a plane wave with pressure 1 μPa.

4.1.2.4 *Transmission loss*

Geometrical spreading

When sound is radiated from a source and propagated in the water, it will be spread in different directions. The wave front covers a larger and larger area. For this reason the sound intensity decreases (with increasing distance from the source). When the distance from the source has become much larger than the source dimensions, the source can be regarded as a point source, and the wave front takes the form as a part of an expanding sphere. The area increases with the square of the distance (Figure 4.2) from the source, making the sound intensity decrease with the square of the distance.

Let I and I_o be the sound intensities in the distances r and r_o. Then:

$$I_o/I = (r/r_o)^2.$$

Expressed in decibels, the geometrical spread loss is:

$$TL_1 = 10\log I_o/I = 20\log r/r_o.$$

Usually a reference point is taken 1 meter in front of the source. Setting $r_o = 1$ meter we get:

$$TL_1 = 20\log r,$$

where r is measured in meters.

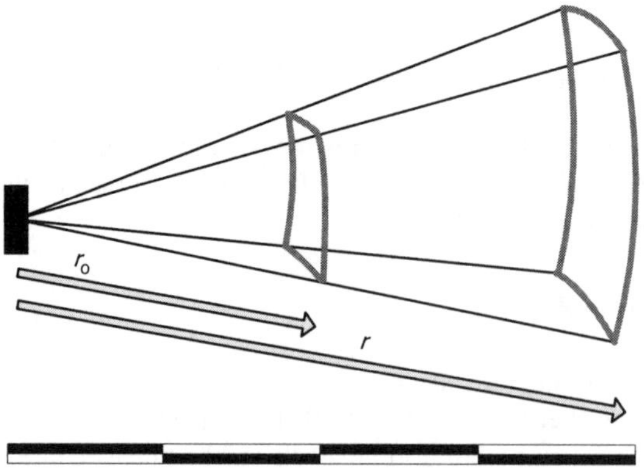

Figure 4.2 *Transmission loss due to geometrical spreading.*

Absorption loss

When the sound propagates through the water, part of the energy is absorbed by the water and converted to heat. For each meter a certain fraction of the energy is lost:

$$dI = -A \cdot I dr,$$

where A is a loss factor. This formula is a differential equation with the solution:

$$I(r) = [I(r_o)/(e - Ar_o)] \cdot e^{-Ar}.$$

$I(r_o)$ is the intensity at the distance r_o:

$$TL_2 = 10 \log I(r_o)/I(r) = \alpha (r - r_o),$$

where $\alpha = 10 \mathrm{A} \log (e)$.

Expressed in decibels, the absorption loss is proportional to the distance traveled. For each meter travelled a certain number of decibels is lost.

If r_o is the reference distance 1 meter, and if the range r is much larger than 1 meter, the absorption loss will approximately be:

$$TL_2 = \alpha r,$$

where α is named the absorption coefficient. Figure 4.3 shows absorption loss coefficient as a function of frequency. The value of α depends strongly on the frequency. It also depends on salinity, temperature, and pressure.

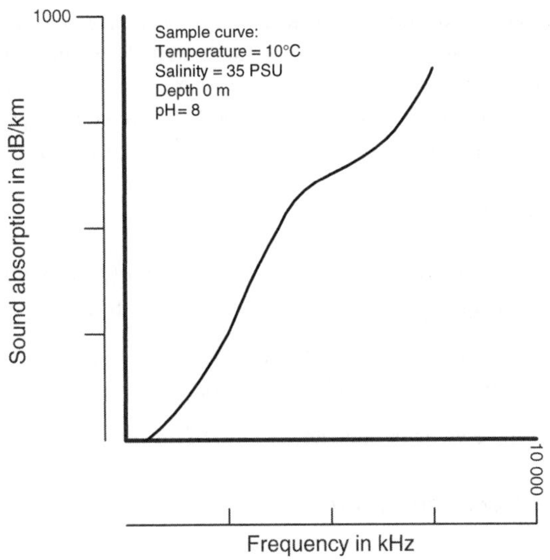

Figure 4.3 *Absorption loss.*

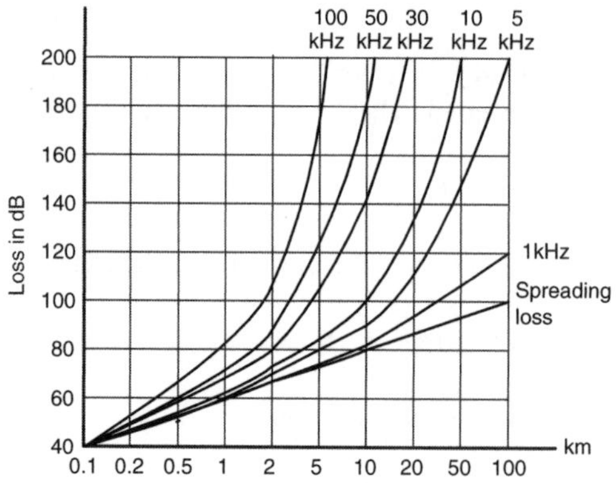

Figure 4.4 *Propagation loss versus range (one-way transmission loss* TL = *20 log* R + αR*).*

One-way transmission loss

The total transmission loss, which the sound suffers when it travels from the transducer to the target (Figure 4.4), is the sum of the spreading loss and the absorption loss:

$$TL = 20 \log r + \alpha r,$$

where r is measured in meters and α is measured in dB/meter.

4.1.3 Transducers

4.1.3.1 *Construction*

A modern transducer is based on piezoelectric ceramic properties, which change physical shape when an electrical current is introduced (Figure 4.5). The change in shape, or vibration, causes a pressure wave, and when the transducer receives a pressure wave, the material transforms the wave into an electrical current. Thus, the transducer may act as both sound source and receiver.

4.1.3.2 *Efficiency*

When the transducer converts electrical energy to sound energy or vice versa, parts of the energy is lost in friction and dielectric loss. Typical transducer efficiency is:

- 50 percent for a ceramic transducer
- 25 percent for a nickel transducer.

The efficiency is defined as the ratio of power out to power in.

4.1.3.3 *Transducer bandwidth*

Normally a transducer is resonant. This means that they offer maximum sensitivity at the frequency they are designed for. Outside this frequency the sensitivity drops.

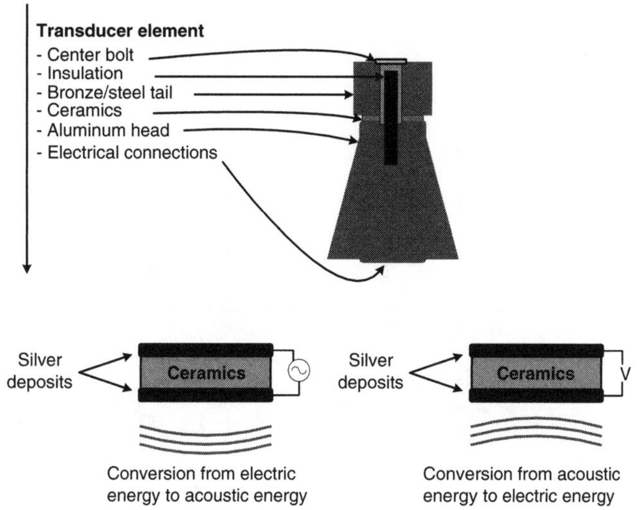

Principle

Transmit:

Electrical signal (converts to) Mechanical ceramic vibrations (converts to) Acoustic signal

Receive:

Acoustic signal (converts to) Mechanical ceramic vibrations (converts to) Electric signal

Figure 4.5 *Cross-section of a typical ceramic transducer.*

Typically, the Q-value is between 5 and 10, where:

$$Q = \text{center frequency/bandwidth (between 3 dB points).}$$

4.1.3.4 *Beam pattern*

The beam pattern shows the transducer sensitivity in different directions. It has a main lobe, normally perpendicular to the transducer face. The direction in which the sensitivity is maximum is called the *beam axis*. It also has unwanted side lobes and unwanted back radiation.

An important parameter is the beam width, defined as the angle between the two 3 dB points. As a rough rule of thumb, the beam width is connected with the size of the transducer by:

$$\beta = \lambda / L,$$

where:

β = beam width in radians
λ = wavelength
L = linear dimension of the active transducer area (side for a rectangular area, diameter for a circular).

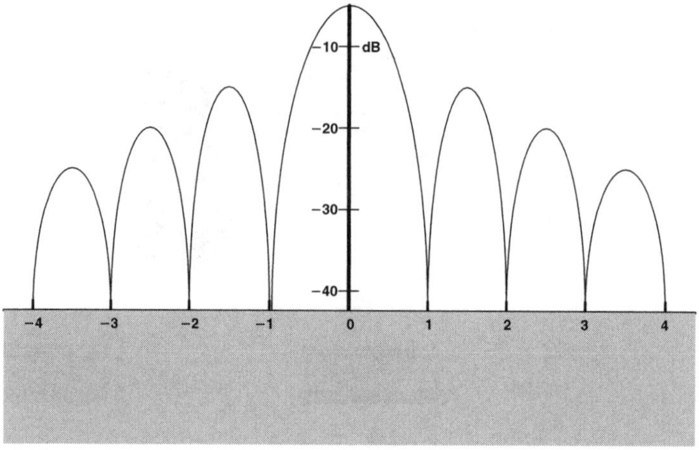

Figure 4.6 *Beam pattern of a continuous line.*

This rule is not valid for very small transducers, i.e. $L < \lambda$.

The theoretical beam pattern of a continuous line transducer is a sin x/x function, namely:

$$b(\theta) = (\sin(\pi L/\lambda \cdot \sin\theta)/\pi L/\lambda \cdot \sin\theta)^2,$$

where:

θ = angle from the beam axis
L = line length
b = transducer power response.

Figure 4.6 shows this pattern in a Cartesian plot. Note that the side lobes are gradually decreasing. The first side lobe is 13 dB below the point of maximum response.

A transducer having a rectangular active area vibrating uniformly as a piston will have this beam pattern in the two planes parallel to the sides.

In many transducers the side lobes are reduced by a technique called tapering.

4.1.3.5 *Directivity index (DI)*

With reference to transmission, the directivity index of a transducer can be defined by:

$$DI = 10\log(I_o/I_m)$$

where:

I_o = the radiated intensity
I_m = the mean intensity for all directions (including back radiation).

Both intensities are measured in the same distance from the transducer.

The mean intensity I is equal to the intensity we would get from an omnidirectional source, if this was given the same power and had the same efficiency as the transducer.

Table 4.1 *Directivity index examples.*

Omnidirectional source	$DI = 0$ dB
Transducer with equal radiation everywhere in one half-plane and zero back radiation	$DI = 3$ dB
Typical echo sounder transducer	$DI = 25$ dB
Wide beam transducer	$DI = 4$ dB
Medium beam transducer	$DI = 9$ dB
Narrow beam transducer	$DI = 15$ dB
HiPAP (USBL surface station)	$DI = 25$ dB

We could therefore also define the directivity index as the ratio between the radiated intensity at the beam axis and the intensity an omnidirectional source would have given at the same point.

A transducer, which has a rectangular active area vibrating uniformly as a piston, will have this beam pattern in the two planes parallel to the sides.

As shown in Table 4.1, the narrower the beam, the higher the DI.

The directivity index for a transducer with beam pattern $b(\theta, \Phi)$ and the mean intensity is found by integration over all directions, with solid angle element $d\Omega$, and division by the total solid angle 4π:

$$I_m = (1/4\pi) \int_{4\pi} I_o \cdot b(\theta, \Phi)\, d\Omega.$$

According to the definition of DI:

$$DI = 10 \log (4/\int_{4\pi} b(\theta, \Phi) d\Omega).$$

Calculation of DI after this formula is, however, no easy job, not even for the simplest transducer.

If the transducer side or diameter is larger than λ, the directivity index is approximately:

$$DI = 10 \log (4\pi A/\lambda^2),$$

where A is the active transducer area, $A = L^2$.

When the beam width is known another approximate formula can be used. For a rectangular transducer the beam pattern in the two planes parallel to the sides are sin x/x function as mentioned previously.

The response is 3 dB down at:

$$(L/\lambda) \sin \theta_{3\,\mathrm{dB}} = 0.443.$$

Inserting this in the formula above gives:

$$DI = 10 \log (2.47/[\sin (\beta_1/2) \cdot \sin (\beta_2/2)]).$$

4.1.3.6 *Transmitting response*

The transmitting power response (S) of a transducer is the pressure produced at the beam axis 1 meter from the transducer by a unit electrical input. The electrical input unit may be volt, ampere, or watt. A typical value for the transmitting response for a ceramic transducer is:

$$S = 193 \text{ dB re } 1\,\mu\text{Pa per watt.}$$

4.1.3.7 *Source level*

The source level SL of sonar or an echo sounder is the sound pressure in the transmitted pulse at the beam axis 1 meter from the transducer. If the transmitting response (S) is known, then the source level is:

$$SL = S + 10 \log P,$$

where:

P = transmitter power
S = transmitting power response.

A widely used formula is:

$$SL = 170.9 + 10 \log P + E + DI,$$

where:

SL = source level in dB re 1 μPa
P = transmitter power in watts
E = 10 log η
η = transducer efficiency
DI = directivity index.

The constant 170.9 incorporates conversion from watts to pascals, and can be derived as follows:

$$SL = 10 \log (1/4\pi) + 10 \log P + E + DI.$$

The factor $1/4\pi$ represents the source level from an omnidirectional source, supplied with 1 watt electrical power and with 100 percent efficiency. When 1 watt sound power is distributed over a sphere with radius 1 meter and surface 4π meter^2 the sound intensity will be:

$$(1/4\pi) \text{ watt/m}^2.$$

The connection between pressure and intensity is:

$$p = \sqrt{I \rho c},$$

where

I = intensity
ρ = density of water
p = pressure
c = sound velocity (in water).

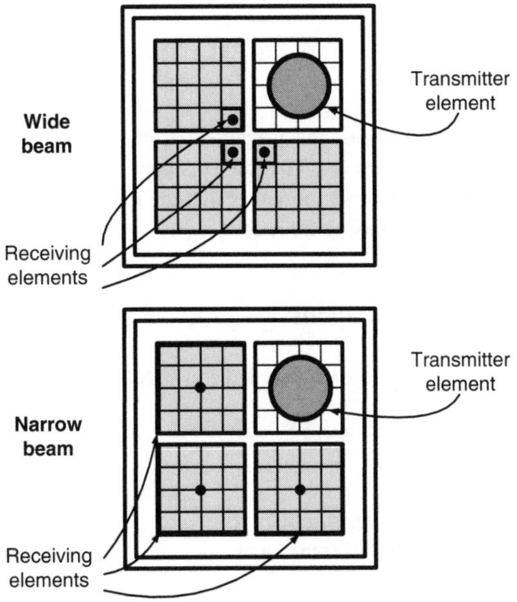

Figure 4.7 *Multi-element transducers.*

Seawater at temperature 10°C with salinity 35 (PPT) at the sea surface has the following values:

$\rho = 1027\,\text{kg/m}^3$
$c = 1490\,\text{m/s}$

A sound density of $1/4\,\text{watt/m}^2$ will in this environment correspond to a sound pressure of:

$$p = 349 \times 10^6\,\mu\text{Pa} = 170.9\,\text{dB re }1\,\mu\text{Pa}.$$

4.1.3.8 *Medium beam/narrow beam TD*

The transducer used in USBL (ultrashort baseline – Europeans use the term SSBL or 'super short baseline') mode (acoustic positioning) normally consists of three different groups of elements. This is to be able to calculate a three-dimensional bearing to the transponders.

- The beam width of a transducer can be changed during operation. This is achieved by combining more transducer elements in series/parallel.
- A more narrow beam gives higher directivity (higher gain and higher noise suppression from outside the beam), but will give a smaller signal 'footprint'.
- For the *narrow beam transducer* in Figure 4.7, the 3 dB point in wide beam mode is 160°, while in narrow beam mode at the 3 dB point it is 30°.
- When using the transducers in SBL (short baseline) or LBL (long baseline) mode, no angle measurements are done and only the R-group (reference) is used. Dedicated SBL/LBL transducers containing only one element or one group can be used.

4.1.4 Acoustic Noise

4.1.4.1 *Environmental*

Noise from thrusters and propellers from surface vessels is the dominating environmental noise source. This noise is approximately 40 dB above normal sea noise. Common for all noise sources is that the noise level drops approximately 10 dB per decade with increasing frequency.

4.1.4.2 *Noise level calculations*

The noise level at the system detector is calculated by the following equation:

$$N = (N_o - 10 \log (B) - DI),$$

where:

B = detector bandwidth
DI = directivity of transducer.

4.1.4.3 *Thruster noise*

The noise from the thruster is changing depending on the thruster. On pitch-controlled thrusters (fixed RPM), the noise level is actually higher when running idle (0 percent pitch) than running with load. In addition, the impact of the thruster noise is determined by the direction of the (azimuth) thruster.

Running a thruster on low RPM and high pitch normally generates less noise than a thruster on high RPM and low pitch. In general, thrusters with variable RPM/fixed pitch generate less noise than thrusters with fixed RPM/variable pitch.

4.1.4.4 *Sound paths*

The velocity of sound is an increasing function of water temperature, pressure, and salinity. Variations of these parameters produce velocity changes, which in turn cause a sound wave to refract or change its direction of propagation. If the velocity gradient increases, the ray curvature is concave upwards (Figure 4.8). If the velocity gradient is negative, the ray curvature is concave downwards.

The refraction of the sound paths represents the major limitations of a reliable underwater navigation and telemetry system. The multi-path conditions can vary significantly depending upon ocean depth, type of bottom, and transducer–transducer configuration and their respective beam patterns. The multipath transmissions result in a time and frequency smearing of the received signal as illustrated.

There are several ways of attacking this problem. The obvious solution is to eliminate the multiple arrivals by combining careful signal detection design with the use of a directional transducer beam. A directional receiving beam discriminates against energy outside of the arrival direction and directional transmit beam project the energy, so that a minimum number of propagation paths are excited.

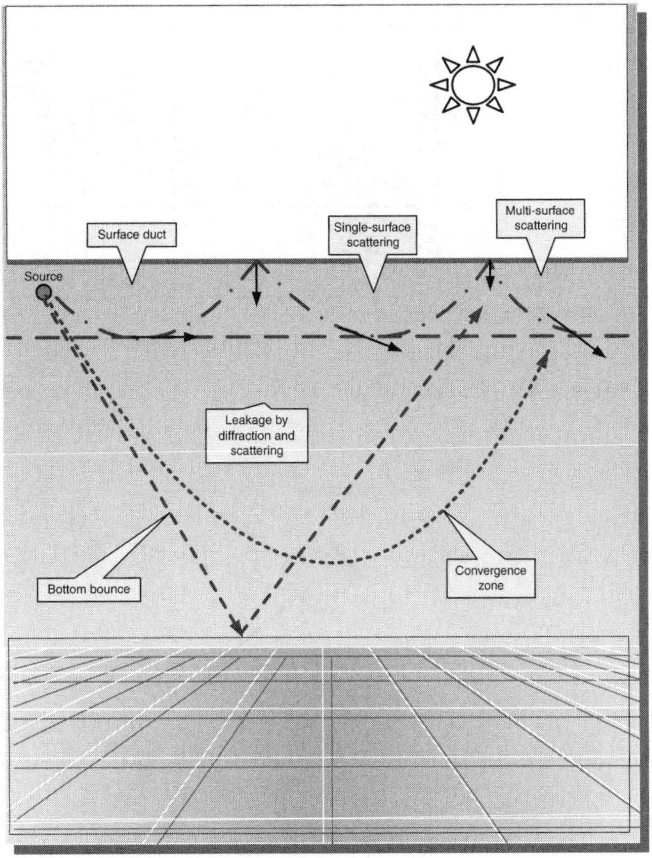

Figure 4.8 *Ray diagrams for deep-water propagation.*

4.1.4.5 *Sound velocity*

From Bowditch (2002), the speed of sound in seawater is a function of its density, compressibility, and, to a minor extent, the ratio of specific heat at constant pressure to that at constant volume (Figure 4.9). As these properties depend upon the temperature, salinity, and pressure (depth) of seawater, it is customary to relate the speed of sound directly to the water temperature, salinity, and pressure. An increase in any of these three properties causes an increase in the sound speed; The converse is also true.

The speed of sound changes by 3–5 meters per second per °C temperature change, by about 1.3 meters per second per PPT (PSU) salinity change, and by about 1.7 meters per second per 100 m depth change. A simplified formula adapted from Wilson's (1960) equation for the computation of the sound speed in seawater is:

$$U = 1449 + 4.6T - 0.055T^2 + 0.0003T^3 + 1.39(S - 35) + 0.017D,$$

where U is the speed (m/s), T is the temperature (°C), S is the salinity (PSU), and D is depth (m).

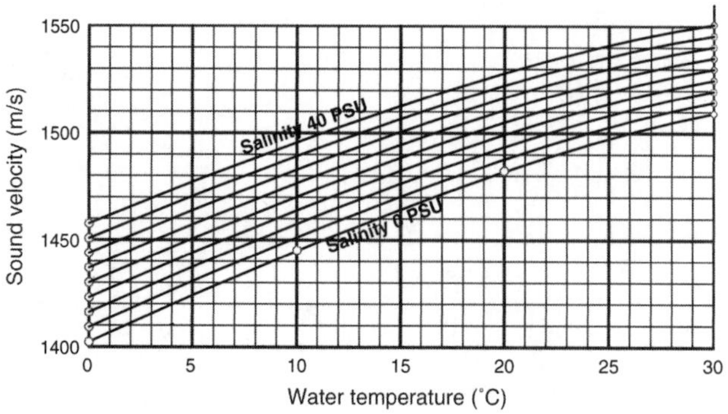

Figure 4.9 *Example of velocity change with changing temperature and salinity.*

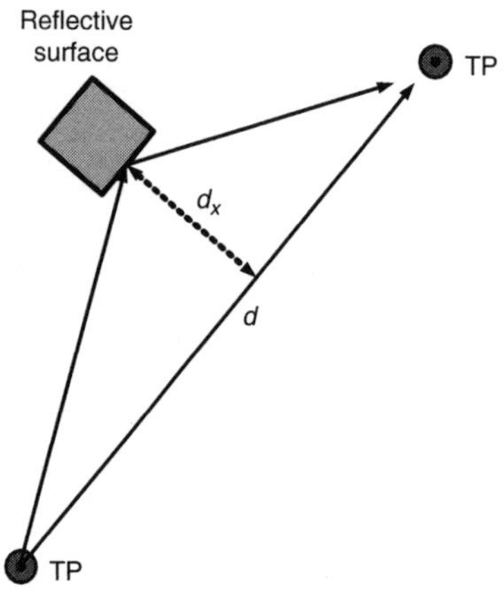

Figure 4.10 *Two LBL transponders measuring baseline.*

4.1.4.6 *Reflections*

Reflections can be caused when the signal bounces off a subsea structure, seabed, riser, ship's hull, or surface (Figure 4.10). Normally the reflection is not perfect, meaning the reflected pulse has less energy than the direct pulse and should not cause problems. Sometimes the pulse is so strong it might cause problems for the pulse detection in the receiver.

When two pulses (sine waves) are added, the resultant can be stronger or weaker than a single pulse. Adding two signals that are 180° out of phase and of equal strength will create no signal at all (Figure 4.11).

If the signal path of the reflection is 0.5λ (or multiples of this) longer, the above situation might occur.

Example (refer to Figure 4.11):

$$f = 30\,\text{kHz} \approx \lambda = 0.05\,\text{m}$$
$$d = 100\,m$$

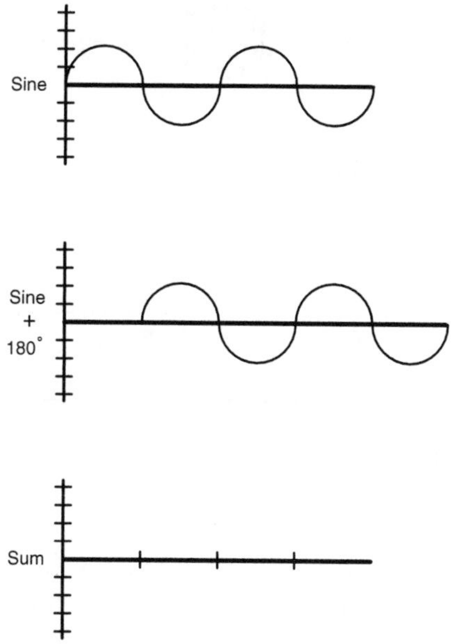

Figure 4.11 *Sine wave interference.*

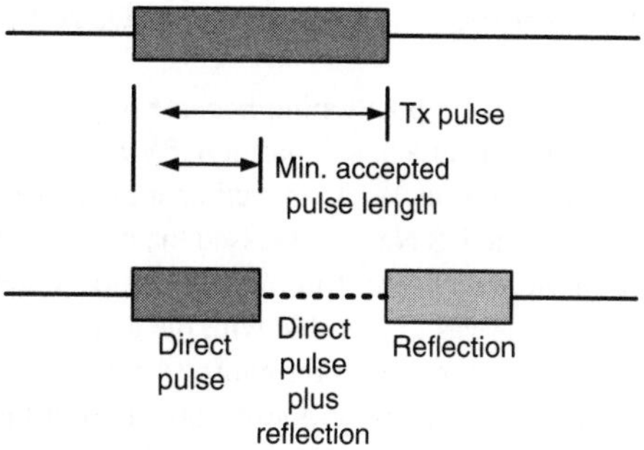

Figure 4.12 *Received pulse.*

180° phase shift (1/2λ):

$$d_x = \surd((100.025/2)^2 - (100/2)^2) = 1.12\,\text{m}$$

180° phase shift (2 + 1/2λ):

$$d_x = \surd((100.125/2)^2 - (100/2)^2) = 2.50\,\text{m}$$

This shows that a surface with 100 percent reflection will create no resulting signal at the receiver given the distances. The same problem can be caused by ray bending or reflections from risers or a ship's hull.

Even reflections that are slightly delayed might cause problems. The receiver has certain criteria in order to accept the signal as a pulse. One of them is the pulse length. As Figures 4.12 and 4.13 illustrate, a worst-case condition will be that the direct pulse

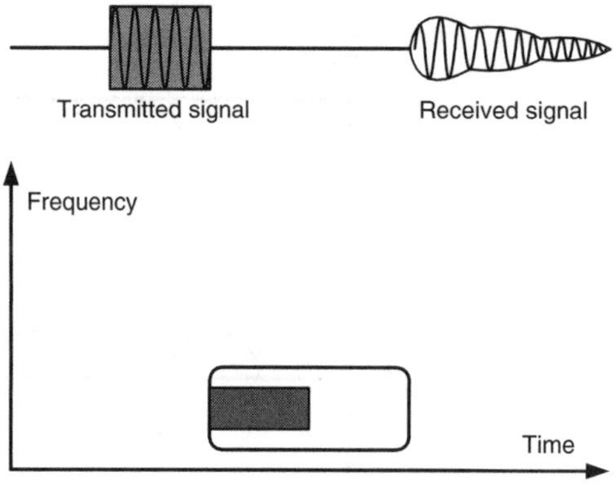

Figure 4.13 *Multi-path transmission: Time and frequency smearing of acoustic signals.*

length is too short to be acknowledged as a pulse before the reflection cancels it. Then the remainder of the reflection is just long enough to be accepted as a pulse.

4.2 ACOUSTIC POSITIONING

4.2.1 Acoustic Positioning – A Technological Development

The need for acoustic positioning became apparent with the loss, then difficulty in locating, the atomic attack submarine USS *Thresher*, which sank in 8400 feet (2560 meters) of sea water in 1963, as well as a nuclear bomb lost at sea off the coast of Spain in 1966. The US Navy possessed the manned submersible capability to dive to the depth of the wreck site. But precision underwater navigation, through any other means but visual, was impossible with the available technology.

In the 1970s, as the search for hydrocarbons migrated into deeper water, the need for repeatable high-accuracy bottom positioning became necessary to place drill-string into the exact position referenced earlier through seismic instrumentation. Since radio frequency waves penetrate just a few wavelengths into water, some other form of precision navigation technology needed development. Thus was born acoustic positioning technology.

Today, water conditions in ports and harbors, as well as littoral areas around the world, are such that visual navigation below the surface is either difficult or impossible due to low visibility conditions. The need for underwater acoustic positioning remains high.

4.2.2 What is Positioning?

According to Bowditch (2002), positioning is defined as 'The process of determining, at a particular point in time, the precise physical location of the craft, vehicle, person or site.' The position determination can vary in quality (degree of certainty as to its accuracy), in relativity (positioning relative to any number of reference frames), and point reference (versus a line of position that is a mathematical position referenced along a given line, circle, or sphere).

A position can be derived from any number of means, including deduced (also termed dead reckoning), resolved (resolving bearing referenced to known fixed or moving objects – also termed 'geodetic' when resolved relative to known earth-fixed objects), estimated (also termed SWAG – normally claimed by helmsman immediately before striking submerged rocks/objects), and claimed (or announced – 'I claim this island in the name of King George').

4.2.3 Theory of Positioning

All positioning is a simple matter of referencing a position relative to some other known position. From the earliest mariners, navigation was performed through 'line of sight' with the coast. As explorers ventured further from sight of land, navigation with reference to the stars became common. Navigation with reference to the North Star for determination of latitude was the earliest version of celestial navigation. Accurate determination of one's latitude could be gained by measuring the angular height of the North Star above the visible horizon.

The determination of position upon a known line of latitude formed a rudimentary 'line of position' in that a known position is resolved. In order to gain a higher positional resolution, a second line of position and then a third (and so forth) will be required to intersect the lines of position and resolve for two- and three-dimensional accuracy. This theory works for celestial navigation, GPS positioning, and (of course) acoustic positioning.

4.2.4 Basics of Acoustic Positioning

The basic underwater 'speaker' is a transducer. This device changes electrical energy into mechanical energy to generate a sound pulse in water. For transducers used in underwater positioning, the typical transducer produces an omnidirectional sound beam capable of being picked up by other transducers in all directions from the signal source.

Acoustic positioning is a basic sound propagation and triangulation problem. The technology itself is simple, but the inherent physical errors require understanding and consideration in order to gain an accurate positional resolution.

As discussed above, water density is affected by water temperature, pressure, and salinity. This density also directly affects the speed of sound transmission in water. If an accurate round-trip time/speed can be calculated, the distance to a vehicle from a reference point can be ascertained. Therefore, the simple formula $R \times T = D$ (rate $\times$ time $=$ distance) can be used. The time function is easily measurable. The rate question is dependent upon the medium through which the sound travels. The speed of sound (or 'sonic' speed) through various media is listed in Table 4.2.

As shown in Table 4.2, pure water and sea water have different sound propagation speeds. For underwater port security tasks, varying degrees of water temperature and salinity conditions will be experienced. The industry-accepted default value for sound speed in water is 4921 feet/second (or 1500 meters per second). If the extreme speed of pure water (4724 ft/s or 1440 m/s) to the median (4921 ft/s or 1500 m/s) is

Table 4.2 *Speed of sound in various materials at 68°F/20°C at 1 atmosphere.*

Material	*Speed (ft/s)*	*Speed (m/s)*
Air	1125	343
Air (32°F/0°C)	1086	331
Helium	3297	1005
Hydrogen	4265	1300
Pure water	4724	1440
Sea water	5118	1560
Iron and steel	16 404	5000
Glass	14 764	4500
Aluminum	16 732	5100
Hard wood	13 123	4000

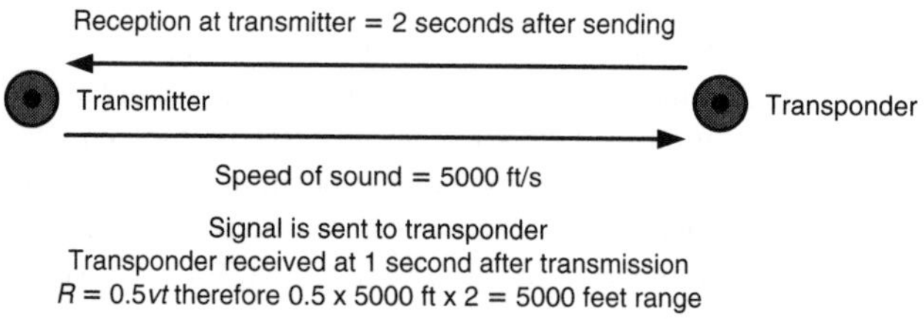
Reception at transmitter = 2 seconds after sending
Transmitter
Transponder
Speed of sound = 5000 ft/s
Signal is sent to transponder
Transponder received at 1 second after transmission
R = 0.5vt therefore 0.5 x 5000 ft x 2 = 5000 feet range

Figure 4.14 *Sample transmitting time calculation.*

experienced, the difference of 197 feet/second (60 meters/second) is approximately a 4 percent error (or 4 feet over a 100-foot distance). If this level of maximum error is acceptable, use the default sonic speed setting. Otherwise, consult the temperature/salinity tables for your specific conditions. Make speed adjustments within the software of the positioning system based upon those conditions.

The range to an acoustic beacon/transponder is a simple calculation:

$$R = \tfrac{1}{2} vt,$$

where R, range, is half the round trip time, t, multiplied by the velocity, v.

So, if there is any latency time for a transponder/responder to process the signal, subtract that out of the time equation to produce a clean range to the beacon.

Using round numbers, if the speed of sound in water is 5000 feet per second and it takes 2 seconds (disregarding any latency time) from sending the signal to reception back at the transmitter, the beacon is 5000 feet away (Figure 4.14).

All underwater acoustical range is 'slant range' for computing raw offset to the vehicle (more to point, from transducer to transducer). When tracking underwater vehicles from the surface, the depth of the vehicle is easily resolved from the vehicle's depth gauge. With the slant range and the depth producing two sides of a right-angled triangle, simple trig will resolve the third side (Figure 4.15).

For determining the distance from a transducer, the resolved equidistant line of offset is known as a 'line of position' (LOP). This is defined as a line of known distance from a point (i.e. the transponder). The easy part of this equation is resolving the range

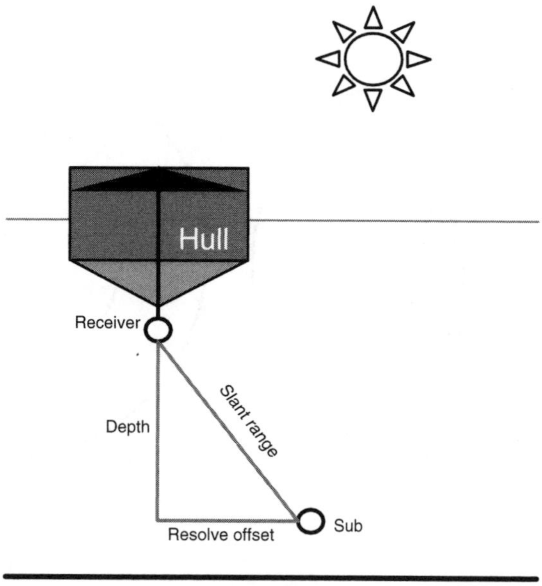

Figure 4.15 *Computation of vehicle offset.*

question. The remainder of this chapter (and the difficult part of this technology) now focuses on resolving the bearing.

With two transducers, the lines of position will correctly resolve to two different locations (Figure 4.16). This has been experienced in the field, giving a resolution to an in-water position that showed the vehicle to be on land! The distance from each transducer is known, but the LOPs cannot resolve the different locations. This is termed 'baseline ambiguity'. A third transducer is needed in order to resolve the exact point. The software that comes with some of the positioning systems currently on the market allows the operator to select/assume a side of the baseline for operations. This allows disregarding readings on one side of the baseline, thus allowing a two-transducer arrangement.

In the scenarios shown in Figures 4.16 and 4.17, certain assumptions can be made. If it is determined that the submersible will be operated only on one side of the baseline (as shown in Figure 4.17 – it is impossible to operate the system on land), all positional resolutions on one side of the baseline can be disregarded, producing accurate navigation with only two surface stations.

With underwater acoustic navigation and positioning, there are various types of underwater markers functioning as transmitters, receivers or both. The six main classifications (Milne, 1983) are:

1. *Interrogator* – A transmitter/receiver that sends out an interrogation signal on one frequency and receives a reply on a second frequency. The channel spacing for these transmit/receive signals is often 500 Hz (0.5 kHz) apart.
2. *Transponder* – A receiver/transmitter installed on the seabed or a submersible (relay), which, on receipt of an interrogation signal from the *interrogator* (command) on one frequency, sends out a reply signal on a second frequency.

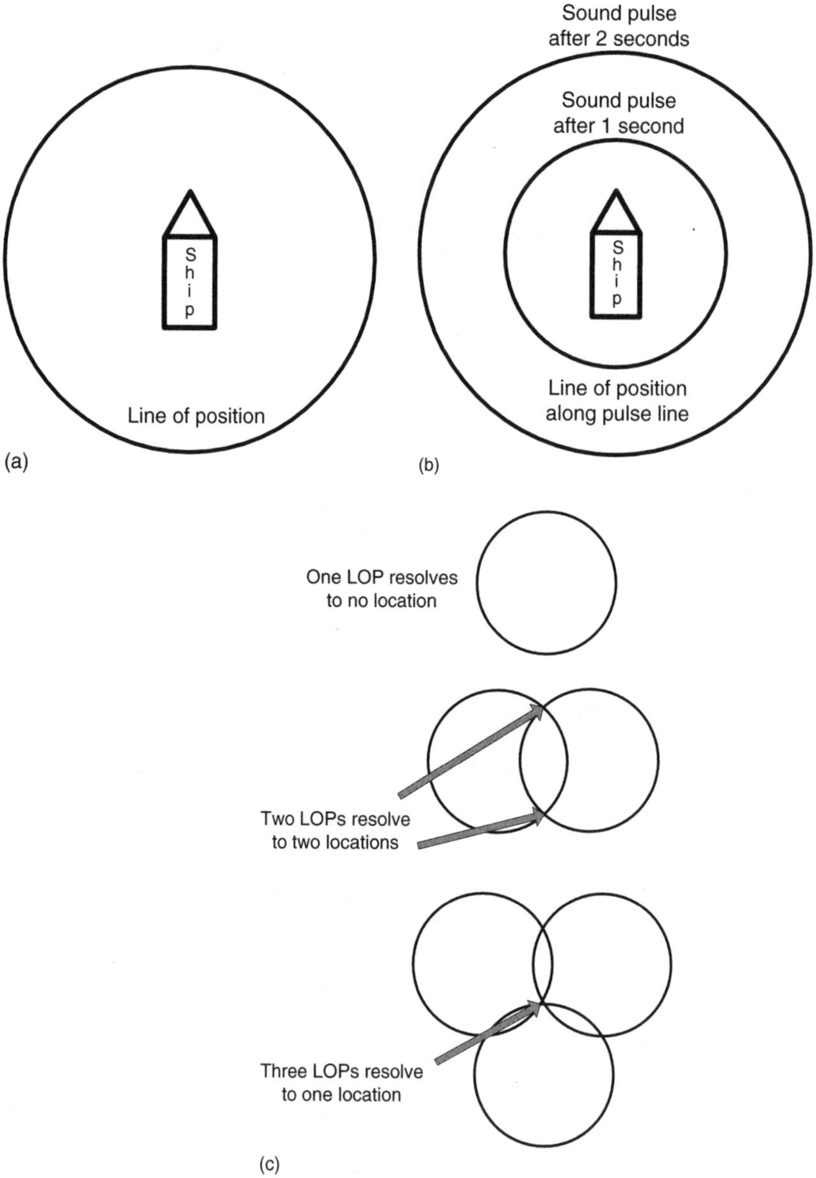

Figure 4.16 *An omnidirectional pulse propagating outward at sonic speed.*

3. *Beacon/pinger* – A transmitter attached to the seabed or a submersible, which continually sends out a pulse on a particular frequency, i.e. free running.
4. *Hydrophone* – An underwater 'microphone' used to receive acoustic signals. This term also sometimes refers to a directional or omnidirectional receiver system (*hydrophone* plus receiver electronics), hull mounted, which is capable of receiving a reply from either a *transponder* or a beacon/pinger.

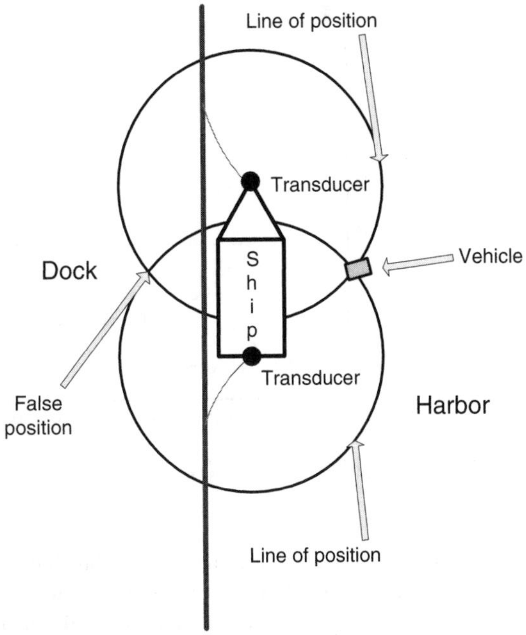

Figure 4.17 *False position resolution.*

5. *Transducer* – A sonar *transducer* is the 'antenna' of a *transponder* or *interrogator*. Either connected by a cable to the electronics package or hard mounted to it, a *transducer* can both transmit and receive acoustic signals. This is in contrast to a *hydrophone*, which is specified for receive operation only.
6. *Responder* – A transmitter fitted to a submersible or on the seabed, which can be triggered by a hard-wired external control signal (command) to transmit an interrogation signal for receipt by a receiver (*hydrophone*).

The particular arrangement of these items as well as their signal sequence determines the physical parameters of the overall positional accuracy.

4.2.5 Sound Propagation, Threshold, and Multi-Path

Sound propagated in water as well as in air possesses many of the same characteristics. A shout in a canyon or in a theater can be echoed several times to the receiver (in this case, your ear). A concert played in an enclosed auditorium sounds much different than when played in an open-air location. Picking out the sound of one instrument in a full symphony is often a difficult task. Thus, trying to pick up a positioning beacon within the cacophony of underwater sounds within a busy harbor is also a challenge.

Range measurements are made by measuring the time it takes an acoustic signal (a 'ping') to travel between the end-points of the range of interest. In order for the range measurement (and hence the position determination) to be successful, the acoustic signal must be detected. A signal is 'detected' if a pressure wave, of the proper

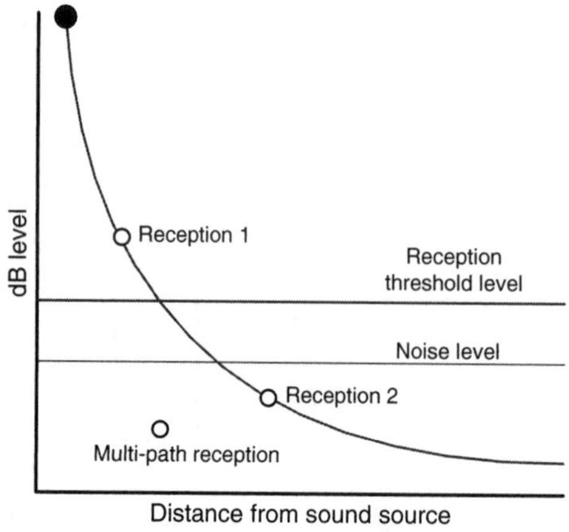

Figure 4.18 *Reception threshold level.*

frequency, has amplitude greater than a set threshold. All signal detection really means is hearing the 'ping'. The 'ping' must be of the proper frequency and loud enough to hear (i.e. a signal-to-noise ratio of greater than 1).

Figure 4.18 shows an example of the reception of a signal from above and below the noise and reception threshold level; The signal is accepted in '1' and rejected in '2'.

The performance of an acoustic positioning system can be predicted using a sonar equation that expresses the relationships between signal received and surrounding in-band noise. If a signal pulse results in a negative signal-to-noise ratio (or less than 5 dB), then most acoustic positioning systems will fail to detect the incoming signal (Wernli, 1998).

$$\text{Signal-to-noise ratio (dB)} = (E - N),$$

where:

$E = (SL - TL)$
$N = 20 \log_{10} (NT)$
$NT = \surd(NA^2 + NS^2 + NR^2)$,

with:

E = received signal sound pressure in dB
N = total received 'in-band' noise sound pressure level
SL = source level in dB
TL = one-way transmission loss in dB
NT = total noise pressure level in μPa
NA = ambient noise pressure level (noise in the environment, both natural and man-made) in μPa
NS = self noise pressure level (noise generated by the acoustic receiver itself) in μPa
NR = reverberation pressure level (reverberations or echoes remaining from previous pings) in μPa.

The source level (*SL*) is the acoustic power, measured in decibels, transmitted into the water by the equipment.

There are two reasons why an in-band signal will not be detected: The signal is either too weak (too quiet) or there is too much noise.

4.2.5.1 *Weak signals*

There are many possible causes for a signal being too weak. The most common cause by far is signal blockage. The acoustic signal cannot pass through certain objects, most notably those containing air (boat hull, fish, sea grass, etc.). Blockage will make it appear that the signal is too quiet and it may not be detected. Blockage can also be by air bubbles, kelp, rock outcroppings, mud, etc. The best way to combat signal blockage is to avoid it.

Other causes of a weak signal include damaged equipment (broken station or transducer), using the system at too great a range, the presence of thermoclines, surface effects, using a system in a liquid with a greater sound adsorption, etc. There are a large number of possible causes for a weaker than expected signal and often it will be difficult to determine the cause.

4.2.5.2 *Noise*

Noise is essentially the detection of an unwanted signal, thus drowning out the wanted signal. In order for a signal to be detected, there must be a method of separating the signal from any noise present. Essentially, the signal must be louder than the noise. Most hardware performs signal extraction through *filtering* and *thresholding*. Filtering is the operation of passing a signal at specific frequencies. Thresholding is the operation of classifying a signal based on amplitude. Basically, the system listens on a specific channel (frequency) for a signal above a certain amplitude level (threshold). If a signal of the appropriate frequency is received and it has amplitude above the threshold it is a valid signal, not noise, and will be allowed to pass.

Sound propagation in water is subject to a number of challenging environmental factors. Over the shorter distances covered by observation-class ROV positioning systems, many of the physical factors affecting sonic speed (i.e. speed of sound in water) are insignificant. The significant physical factors over these shorter distances are sound threshold considerations and the multi-path phenomenon.

4.2.5.3 *Multi-path*

In Figure 4.19, all of the sounds shown originated from the same source (e.g. an acoustic beacon) but arrive at the receiver at different times. If each of these receptions appeared as a beacon on a submersible, it would show three different distances – none of which would be correct, since none of these are line of sight. Sounds '1' and '2' have multiple reflections to get to the receiver, while sound '3' has just one reflection. If the sound source is a measured output level (units measured in dB or decibels), the receiver can be set to reject all reception below a certain dB level, thereby rejecting all multi-path signals.

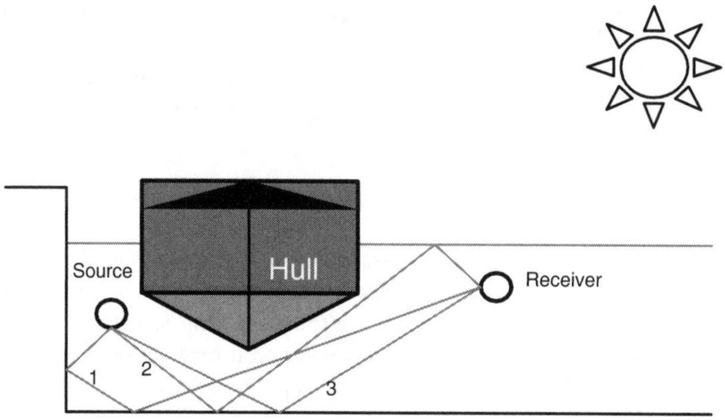

Figure 4.19 *Multi-path signals.*

Acoustic positioning systems update at regular intervals. The update rate is limited by the source and reception offset. Once an acoustic sound signal has been generated, it must go to the end of its reception range and come back to the transmitter/receiver (thereby generating a range through timing difference) before the next pulse can be sent. This is the reason acoustic positioning cannot maintain real-time positioning feedback.

4.2.6 Types of Positioning Technology

Any discussion of positioning must first begin with an explanation of the term 'frame of reference'.

4.2.6.1 *Frame of reference*

Whenever a position is given for anything anywhere in the universe, that position must be made within a coordinate reference frame. If a geo-referenced position is given, there are many unanswered questions simply to specify which mathematical model is used to generate that position. Some frame of reference examples are:

- 'Two feet from my front door' (house reference frame)
- 'The left seat of my car' (car reference frame)
- 'Twenty feet off the port beam' (ship reference frame)
- 'Control Yoke is located at station 45', i.e. 45 inches aft of the originating datum on the engineering drawings of an example aircraft (aircraft reference frame)
- 'Latitude 30 North/Longitude 90 West (WGS 84)' (earth reference frame).

In acoustic positioning, the raw positioning data is resolved from the acoustic beacon/tracking device to the transducer array (arranged in a known pattern). Your raw position will be given in an *x*/*y*/*z* offset from that referenced transducer array. The imaginary line forming the sides of the reference triangle(s) is known as a 'baseline'. The frames of reference can be fixed to the transducer array itself, a physical object

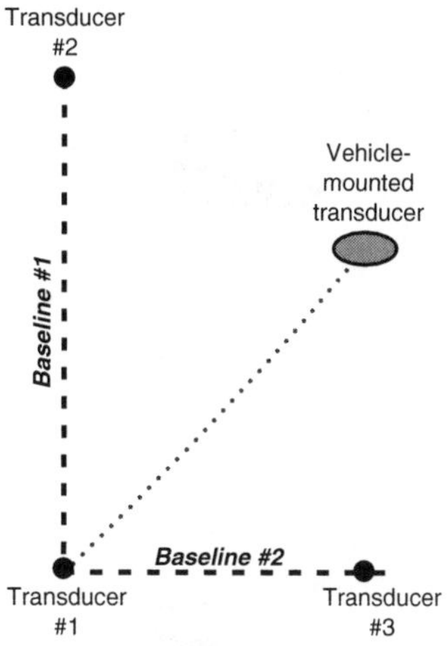

Figure 4.20 *Basic arrangement of transducers into baselines for angular measurements.*

(such as a dam, a ship or a pier), or to the earth. All offset coordinates may then be given with reference to that 'frame of reference'.

The arrangement of the various acoustic sensors as well as their relative placement will determine both the accuracy of the positional resolution as well as the ease of placement and system cost. All arrangements use some form of active send/receive in order to positively measure the time of transmission and reception over the array of transducers. The known spacing and angular offset of the transducer array forms what is known as a 'baseline' from which to measure angular distance (Figure 4.20). A comparison of different types of acoustic positioning arrangements is depicted in Figure 4.21.

4.2.6.2 *Short baseline (SBL)*

An SBL system (Figure 4.22) is normally fitted to a vessel or fixed object with the vessel or fixed object used as the frame of reference. A number of (at least three but typically four) acoustic transducers are fitted in a triangle or a rectangle on the lower part of the vessel or fixed object. The distances between the transducers (the 'baselines') are made as large as practical given physical space limitations; Typically they are at least 30 feet long. The position of each transducer within a coordinate frame fixed to the vessel or fixed object is determined by conventional survey techniques or from the 'as built' survey.

The term 'short' is used as a comparison with 'long base line' (LBL) techniques. If the distances from the transducers to an acoustic beacon are measured as described for LBL, then the position of the beacon, within the coordinate reference frame, can be computed. Moreover, if redundant measurements are made, a best estimate can be determined that is more accurate than the basic position calculation by averaging several fixes.

SBL systems transmit from one but receive on all transducers. The result is one distance (or range) measurement and a number of range (or time) differences.

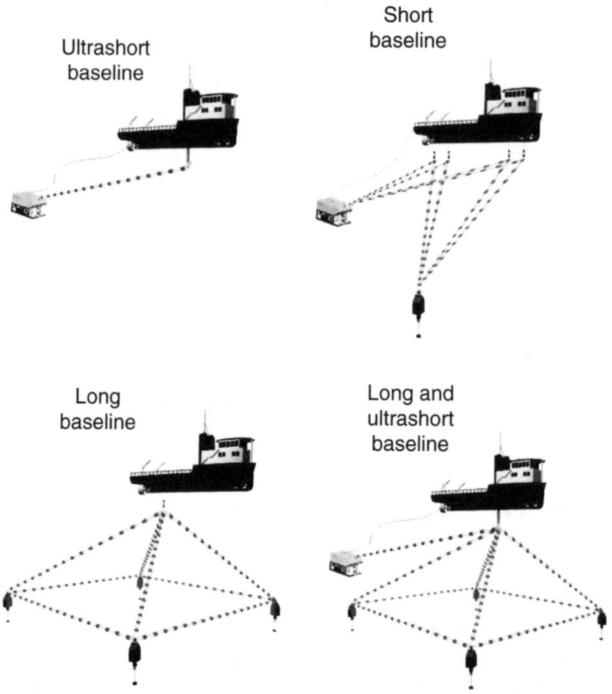

Figure 4.21 *Comparative positioning system arrangements (graphic by Sonardyne).*

With an SBL system, the coordinate frame is fixed to the vessel – which is subject to roll (change in list), pitch (change in trim), and yaw (change in heading) motion. This 'disadvantage' can be overcome by using additional equipment such as a VRU (vertical reference unit) to measure roll and pitch and a gyro-compass to measure heading. The coordinates of the beacon can then be transformed mathematically to remove the effect of these rotational motions.

At the shorter transducer spacing distances (i.e. less than 30 feet), SBL produces a higher error level than does LBL due to the greater impact of multi-path and other range measurement errors. Also, any errors in transducer spacing measurement, heading indicator errors, GPS inaccuracies, and vessel instability (for instance, due to vessel motion) are propagated through to position inaccuracies. SBL is not normally considered survey quality.

4.2.6.3 *Ultrashort baseline (USBL)*

USBL principles (Figure 4.23) are very similar to SBL principles (in which an array of acoustic transducers is deployed on the surface vessel) except that the transducers are all built into a single transceiver assembly – or the array of transducers is replaced by an array of transducer elements in a single transceiver assembly.

The distances or ranges are measured as they are in an SBL system but the time differences have decreased. Systems using sinusoidal signals measure the 'time phase'

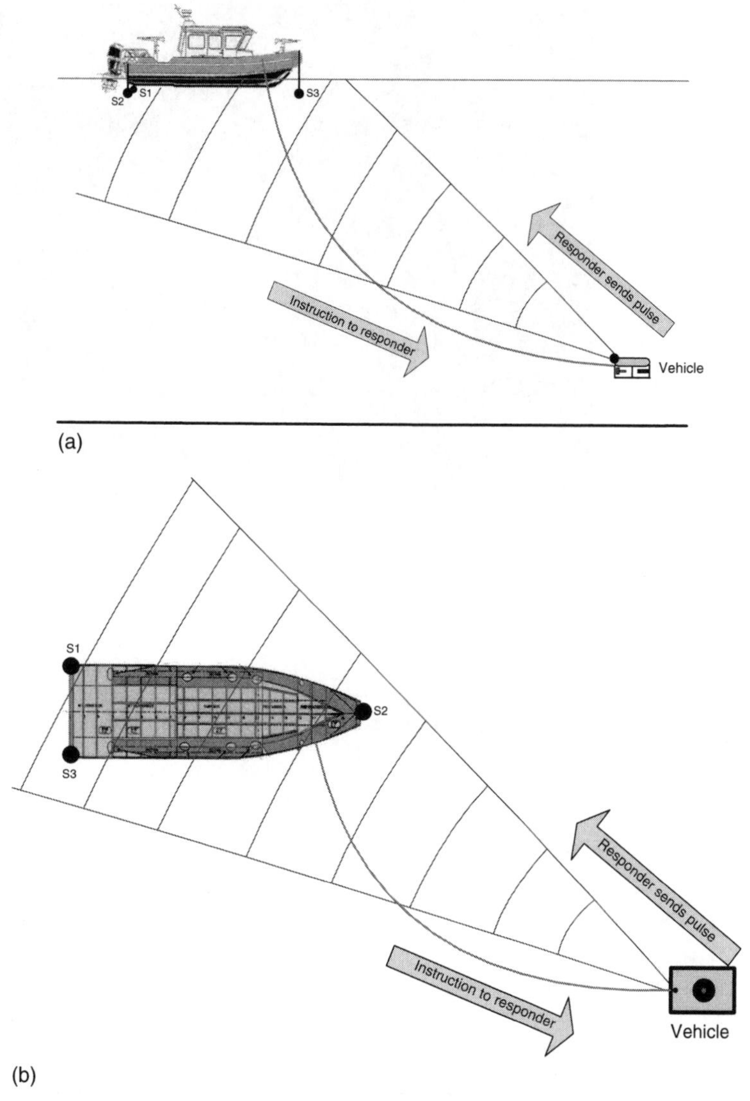

Figure 4.22 *Typical short baseline system.*

of the signal in each element with respect to a reference in the receiver. The 'time-phase differences' between transducer elements are computed by subtraction and then the system is equivalent to an SBL system.

Another practical difference is that the transducer elements are in a transceiver assembly that is placed somewhere in the vessel frame. The attitude of the assembly in the vessel frame must be measured during installation. It should be remembered that, intrinsically, a USBL system positions a beacon in a frame fixed to the transceiver assembly, not directly in a vessel-fixed frame as in the SBL case.

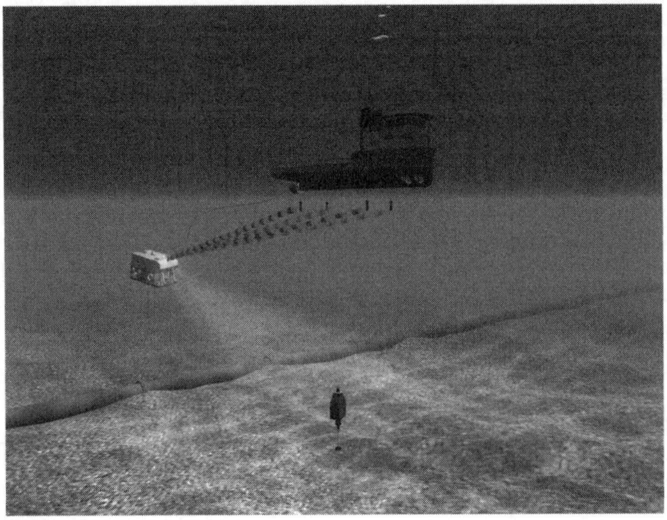

(c)

Figure 4.22 *(Continued)*

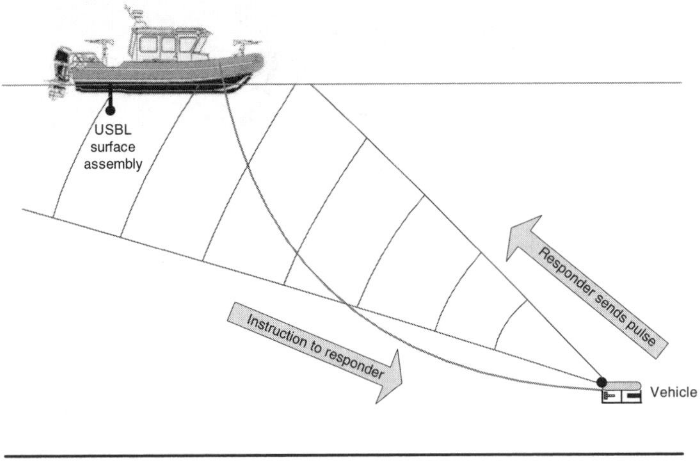

(a)

Figure 4.23 *Typical USBL arrangement.*

4.2.6.4 *Long baseline (LBL)*

An LBL system consists of a number of acoustic transponder beacons moored in fixed locations on the seabed or mounted on fixed locations of objects such as ships or oil platforms to be surveyed (Figures 4.24–4.26). The positions of the beacons are described in a coordinate frame fixed to the seabed or the referenced object (an example of a non-sea floor LBL system is the vessel-referenced ship hull inspection

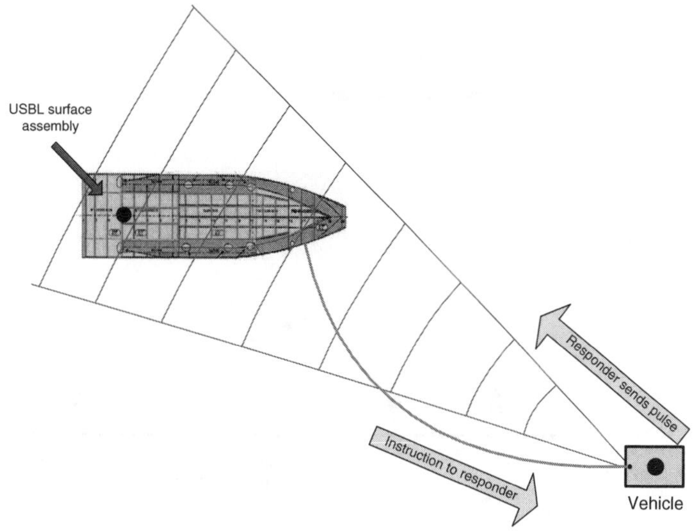

(b)

(c)

Figure 4.23 *(Continued)*

system). The distances between the transponders form the 'baselines' used by the system.

These transponders are interrogated by the interrogator, which is installed on the ROV, the diver, the submersible, or the tow fish to be positioned. The distance from the transducer/interrogator to a transponder beacon can be measured by causing the transducer/interrogator to transmit a short acoustic signal, which the transponder detects, causing it to transmit an acoustic signal in response on a discrete channel.

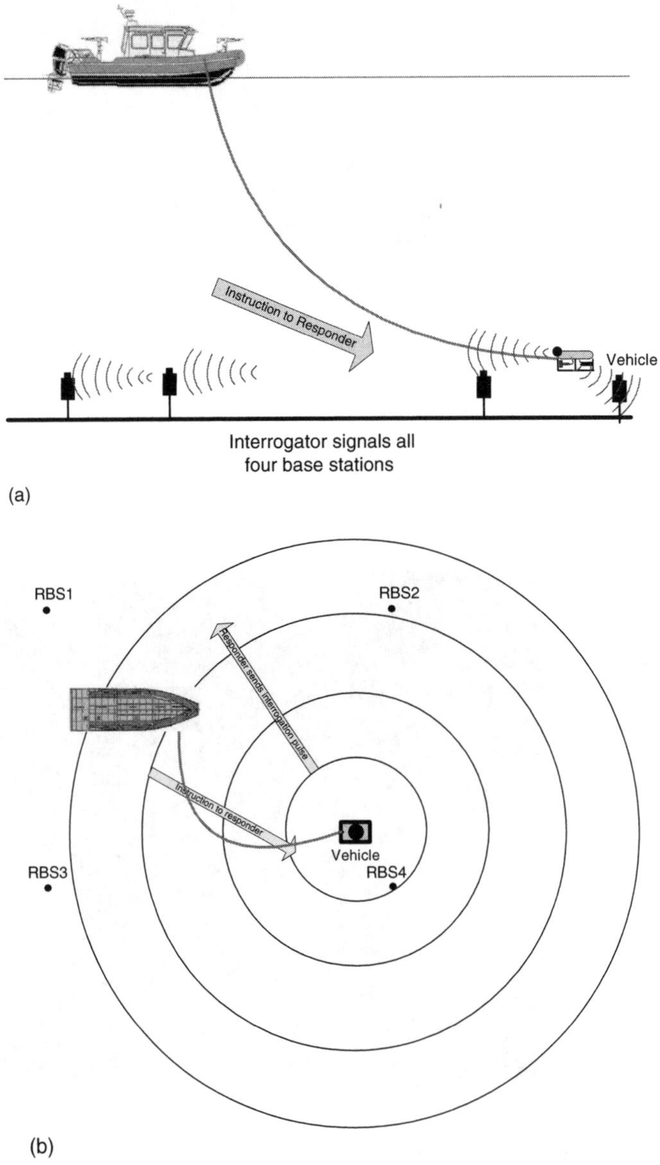

Figure 4.24 *Long baseline system.*

The time from the transmission of the first signal to the reception of the second is measured. As sound travels through the water at a known speed, the distance between the transducer and the beacon can be estimated. The process is repeated for the remaining beacons and the position of the vessel relative to the array of beacons is then calculated or estimated.

In principle, navigation can be achieved using just two transponder beacons, but in that case there is a possible ambiguity as to which side of the baseline (a line drawn

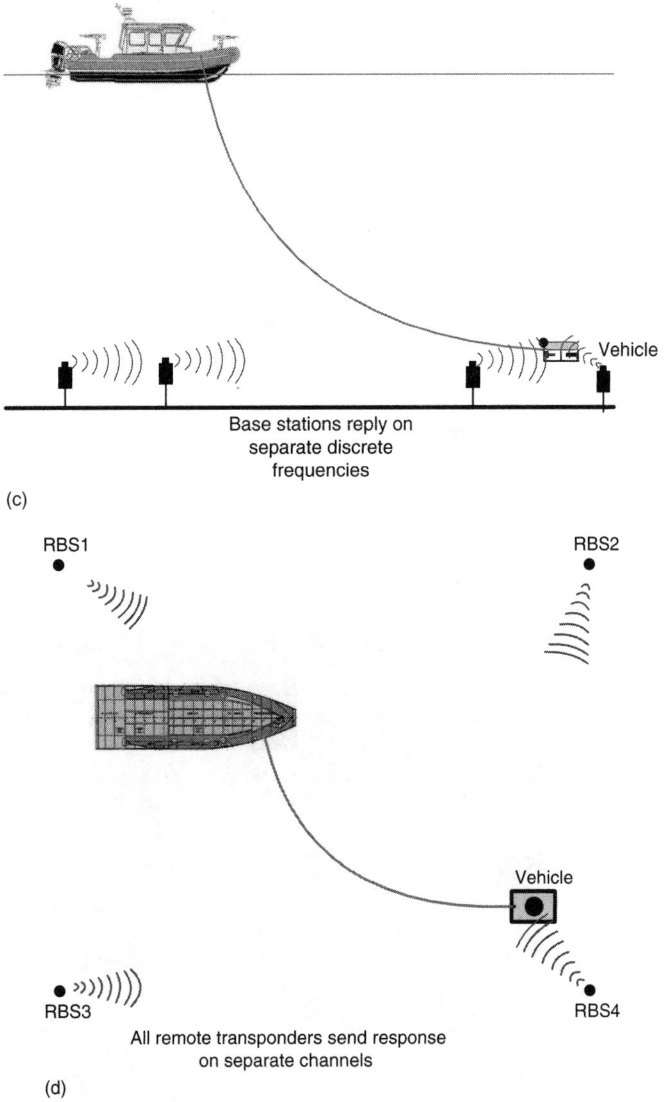

Figure 4.24 *(Continued)*

between the beacons) the vessel may be on (the so-called 'baseline ambiguity' – see previous section of this chapter). In addition, the depth or height of the transducer has to be assumed, unless there is an embedded depth transducer that encodes depth measurements with the transponder response. Three transponder beacons is the minimum required for unambiguous navigation in three dimensions; Four is the minimum required for some degree of redundancy. This is useful for checks on the quality of navigation.

The term 'long-baseline' is used because, in general, the baseline distances are much greater for LBL than for SBL (short base line) and certainly for USBL

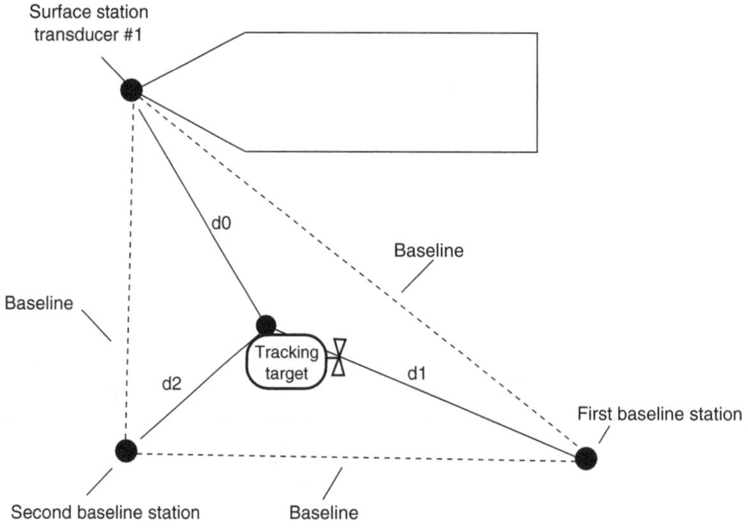

Figure 4.25 *Typical arrangement for an LBL system with transponders mounted on the sea floor (graphic by Desert Star LLC).*

Figure 4.26 *LBL setup on vessel (graphic by Sonardyne).*

(ultrashort baseline). Because the baselines are much larger, an LBL system is very accurate and position fixes are very robust compared with the SBL and USBL versions. In addition, the transponder beacons are mounted fixed in the desired reference frame, such as on the sea floor for sea floor surveys or on a ship's hull for ship hull surveys. This removes most of the problems associated with vessel motion.

The vessel-referenced ship hull positioning system is useful in underwater port security needs (Figures 4.27 and 4.28). This system simply applies the LBL system and

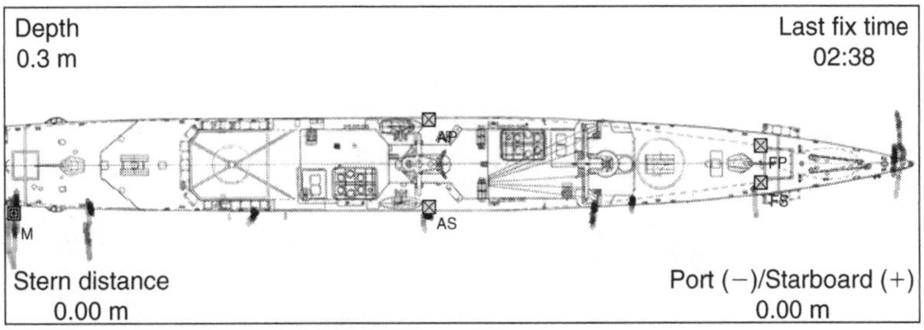

Figure 4.27 *Depiction of vessel with ship hull inspection system transponders registered onto vessel drawing (M = Mobile Station, AP = Aft Port Station, AS = Aft Starboard Station, FP = Forward Port Station, FS = Forward Starboard Station).*

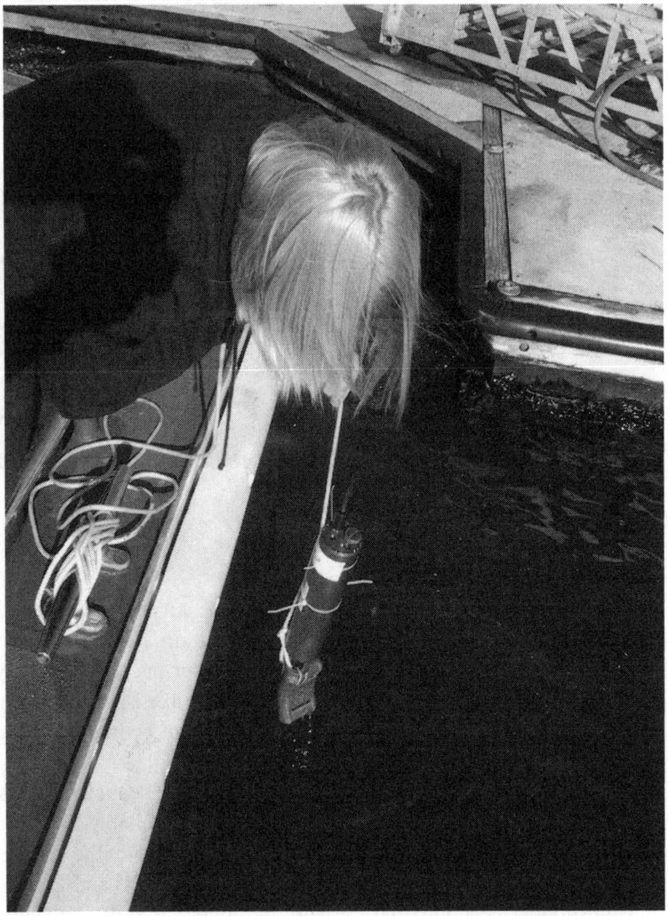

(a)

Figure 4.28 *Ship hull inspection system being deployed.*

(b)

Figure 4.28 *(Continued)*

uses as a frame of reference the vessel to be surveyed – depicted by scaled drawings of that vessel. In ship hull inspections, certain assumptions can be made allowing positive navigation with only one transponder in 'sight'. If only one transponder is communicating, a deduction that the submersible is co-located in the quadrant of the ship that contains the requisite transponder can be made as long as the submersible is in visual contact with the hull.

For geo-referenced arrangements, the array of seabed beacons/transponders needs to be calibrated. There are several techniques available for achieving this. The most appropriate technique depends on the requirements of the task and the available hardware. With the continuing integration of LBL, SBL and USBL systems, intelligent transponder beacons (that measure baselines directly), and satellite navigation systems, the calibration of seabed arrays is becoming a quick and simple operation. The operator will be free to choose the techniques appropriate to the requirements and the task based upon manufacturer-supplied specifications.

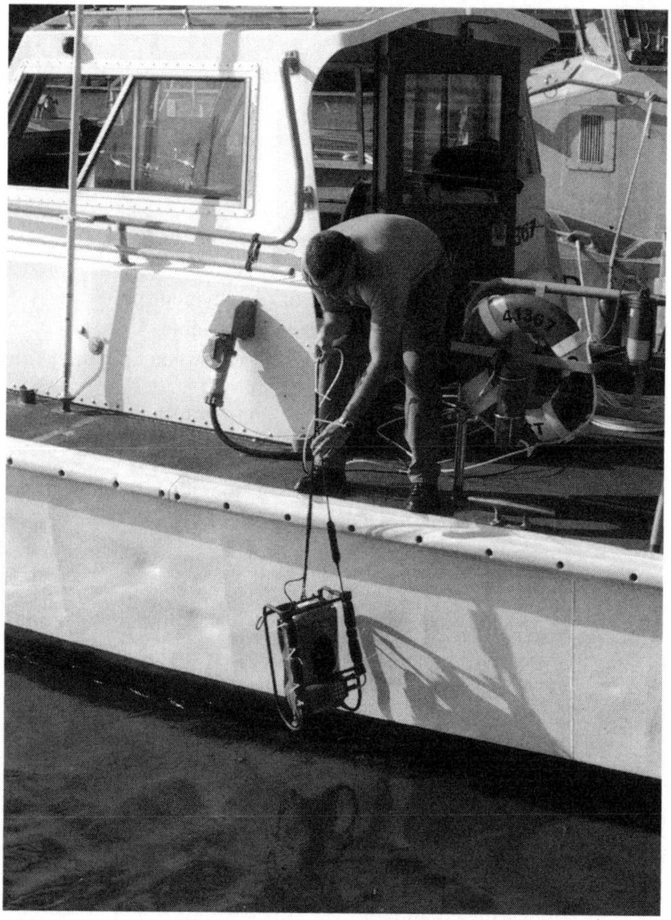

(c)

Figure 4.28 *(Continued)*

4.2.7 Advantages and Disadvantages of Positioning System Types

Ultrashort baseline (USBL) advantages:

- Low complexity.
- Easy to use.
- Good range accuracy.
- Ship-based system.

Ultrashort baseline (USBL) disadvantages:

- Detailed calibration of vessel-based transducer assembly required (usually not performed accurately).
- Requires vessel installation of a rigid pole for mounting of the receiver unit.

- Position accuracy depends on ship's gyro and vertical reference unit.
- Minimal redundancy.
- Large transducer/gate valve requiring accurate/repeatable orientation.

Short baseline (SBL) advantages:

- Good range accuracy.
- Redundancy.
- Ship-based system.
- Small transducers/gate valves.
- No orientation requirement on deployment poles. Easy 'over the side' cable mounted transducer deployment is possible.
- Good positioning accuracy is possible when operating from larger vessels or a dock (good transducer separation).

Short baseline (SBL) disadvantages:

- Poor positioning accuracy when operating from very small vessels (limited transducer separation).
- Requires deployment of three or more surface station transducers (more wiring than USBL).

Long baseline (LBL) advantages:

- Very good position accuracy independent of water depth.
- Redundancy.
- Wide area of coverage.
- Single, simple deployment pole.

Long baseline (LBL) disadvantages:

- Complex systems requiring more competent operators.
- Large arrays required.
- Longer deployment/recovery time.
- Calibration time required at each location.

4.2.8 Capabilities and Limitations of Acoustic Positioning

Acoustic positioning is capable of high-precision positioning in a number of frames of reference. Once the local sonic speed is computed, range is capable of being accurately measured. The larger the spacing between the transducers forming the baselines, the higher the possible bearing accuracy. The basic range/bearing resolution accuracy with LBL positioning is generally of a higher quality than the SBL and USBL techniques due to the inherent greater angular offset reception time spacing of the baselines. Range in all methods is quite accurate, assuming line of sight is maintained. Bearing accuracy of 1–3° for SBL and USBL is considered acceptable, while bearings of <1° are possible.

Acoustic positioning relies upon a line-of-sight sound signal from a transducer (generally mounted aboard the submersible) to a receiver, so that an accurate range and bearing can be resolved to give a position in some frame of reference.

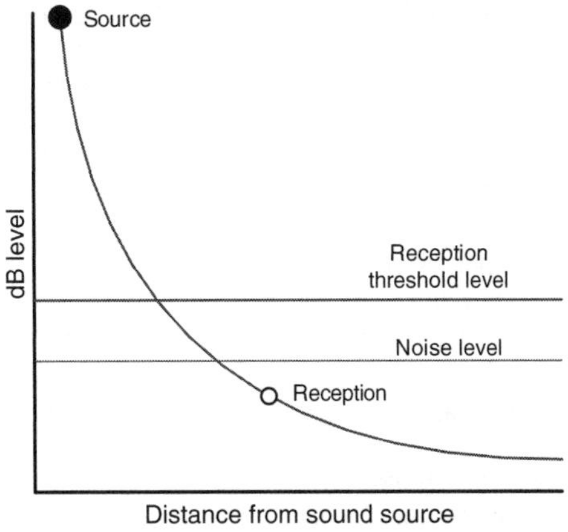

Figure 4.29 *An example of the reception of a signal below the noise and reception threshold level – signal is rejected.*

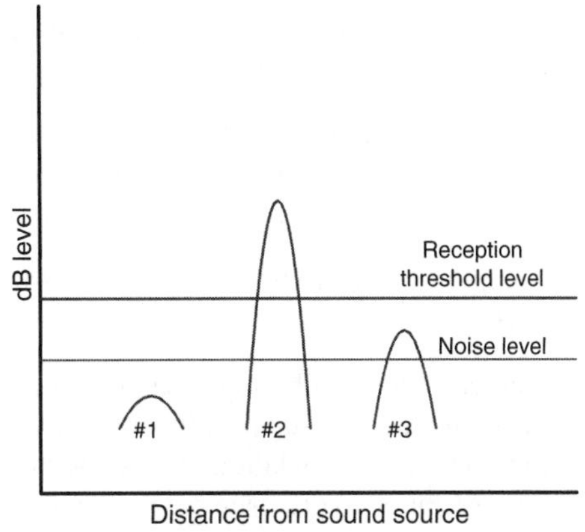

Figure 4.30 *Sample sound source relationships to the reception threshold and noise level.*

Just as picking out a voice is a crowd is difficult, so is picking out an acoustic positioning beacon in a noisy harbor environment. The challenge is balancing the reception sound threshold to pick up the line-of-sight transmission and not the false signal from noise or from multi-path sound reflections.

If the sound threshold level is placed too high, the possibility exists that the true line-of-sight signal from the beacon to the receiver is rejected (Figure 4.29). If the sound threshold reception level is set too low, false-positive readings from either noise or from multi-path reception will show erroneous positions for your vehicle.

In Figure 4.30, sound source 1 is below the noise level and below the reception threshold – the signal is not received. In signal source 2, the source level is above the reception threshold and above the noise level – the signal is received and is passed. In signal source 3, the source is above the noise level but below the reception threshold – signal is received but rejected.

Table 4.3 *Environment and typical acoustic system settings.*

Environment	*Noise/receiver sensitivity*	*Reverb/time spacing*
Lake	Low/high	Low/low
Pool	Low/high	High/high
Open ocean	High/low	Low/low
Harbor	High/low	High/high

The two relevant underwater acoustical environmental issues are:

- Background noise
- Reverberation/multi-path.

Background noise requires the transmitting beacon to possess a higher output in order to 'burn through' the noise and be recognized by the receiver. Reverberation/multi-path propagation causes problems by falsely recognizing signals other than the legitimate line-of-sight signal necessary to measure range to the source transducer. The two responses to these issues are to:

- *Increase* the dB output of the transmitter
- *Time space* the transmission of source signals more widely to allow reverberation of sound to dissipate before sending the next signal.

Some examples of environment types are listed below with their respective characteristics:

- *Pools* are characterized by low background noise yet with a high presence of reverberation. The acoustic positioning system should be set to high sensitivity (due to the absence of background noise) with larger spacing between signals to allow reverberation to dissipate before the next signal generation.
- *Harbors, constricted waters, and surf zones* are noisy acoustical environments with a great presence of reverberation. The acoustic positioning system would be set to low sensitivity (to reject false signals) with long signal spacing.
- *Open ocean* means moderate noise and low echoes. Set the positioning system to low sensitivity and short spacing.
- *Lakes* are quiet and contain few possibilities of reverberation. Set the positioning system to high sensitivity and short spacing.

A breakdown of environment versus settings is provided in Table 4.3.

4.2.9 Operational Considerations

4.2.9.1 *Operations in open ocean*

The open ocean can have high background noise due to broadband biological noise generation. Another factor effecting proper positional accuracy is the pitch and sway of the operation platform (i.e. the vessel). When deploying an SBL or USBL system from

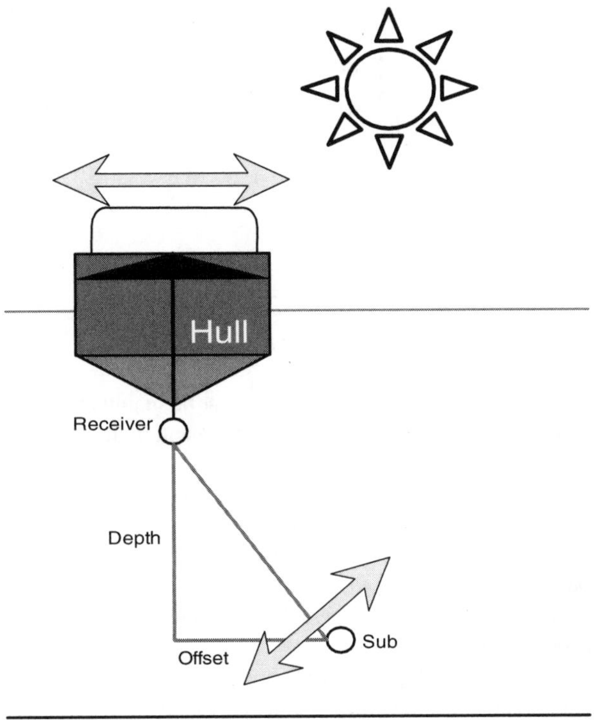

Figure 4.31 *Effect of a moving vessel on the submersible's estimated position.*

a rocking boat (with the frame of reference attached to the vessel), the baseline sensing station is constantly moving, disrupting a steady frame of reference (Figure 4.31).

If the transponder beacons are deployed anchored to the bottom (as in an LBL system), the frame of reference is steady, obviating the apparent bearing error from boat sway. The vessel motion errors are resolved and canceled by the VRU (vertical reference unit). The VRU determines the instantaneous platform orientation then corrects the out-of-plane errors to produce a clean position resolution.

4.2.9.2 *Operations in ports and harbors*

The port and harbor acoustic environment is characterized by noisy reverberant conditions with broadband machinery noises from vessels and various sources as well as strong reverberations within the water space. Tidal current flow also affects port and harbor water sound propagation. Acoustic positioning operations require close attention to the details of background noise as well as signal spacing in order to gain accurate position readings.

4.2.9.3 *Operations in close proximity to vessels and underwater structures*

Special consideration is necessary when operating in close proximity to vessels and underwater structures such as piers, submerged structures, and anchorages. As stated

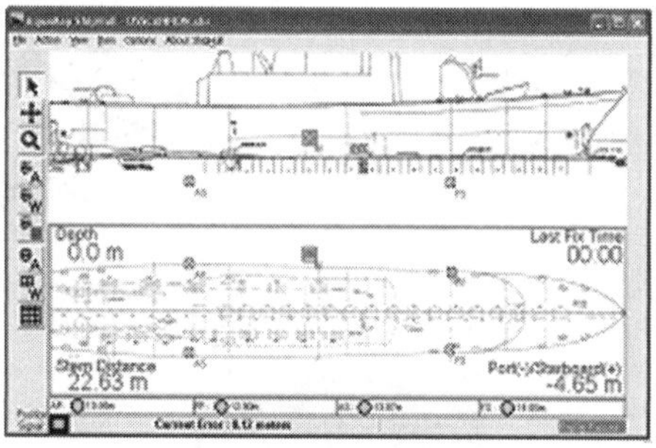

Figure 4.32 *Vessel-referenced positioning system.*

previously, line-of-sight reception must be maintained to gain accurate range measurements. It is not always possible to maintain line of sight while performing a pier or hull inspection.

SBL and USBL systems require that the submersible be operated near the deployment platform in order to see all of the surface units required to gain a full position resolution. If an LBL system is used with the transponders located on the bottom, a better angle is possible to 'see' around the vessel being inspected. With either method, operation in and around vessels and enclosed structures may not be conducive to acoustic positioning.

One acoustic ship hull inspection system uses a series of assumptions in order to gain accurate position measurements near ship hulls despite blockage and multi-path problems.

A series of acoustic transponders are hung over the side of a vessel to be inspected (Figure 4.32). The transponders are positioned below the level of the keel so that all are 'in sight' once the submersible is under the vessel. Each transponder has a separate frequency corresponding to its position on the vessel (i.e. forward port, aft port, forward starboard, aft starboard). Once the submersible travels toward the surface on one side of the vessel, it loses sight of the transponders on the opposite side of the vessel (an example would be traveling up toward the surface on the starboard side, losing sight of the port side transponders). Positional accuracy is then maintained with two transponders. The 'baseline ambiguity' is resolved by knowing that the submersible cannot be 'inside of' the vessel being inspected, thus putting the submersible in only one of the two possible positions. If only one transponder is received, it can be assumed that the submersible is in contact with the hull in the quadrant of the receiving transponder.

4.2.10 Position Referencing

4.2.10.1 *Geo-referencing of position*

The basic resolution of an acoustic positioning system is range and bearing. Input of the offset from the primary reference transducer to the center rotation point of the vessel will produce a reference point for the extended center line of the vessel (Figure 4.33). From this, convert relative bearing from the vessel's axis into magnetic bearing. By comparing the relative bearing (resolved by triangulation) from the vessel to the

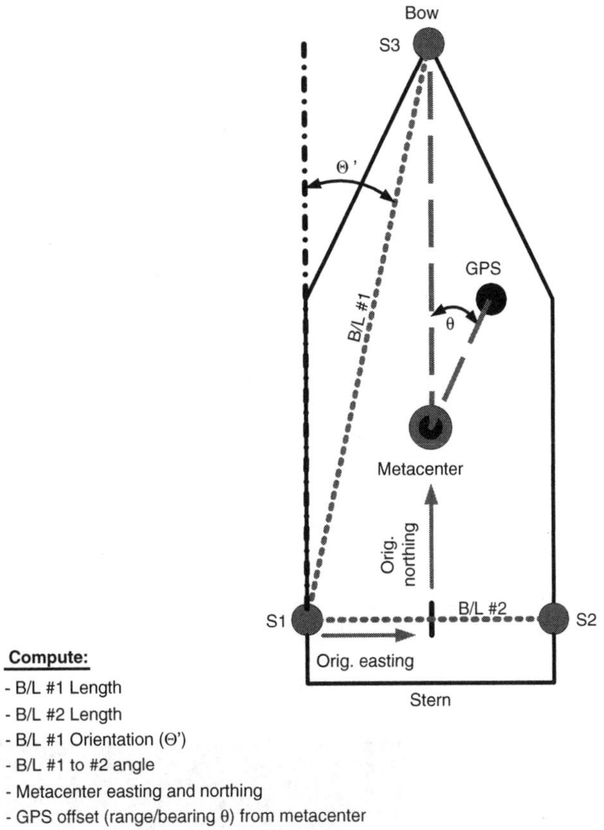

Figure 4.33 *Resolution of vehicle position to ship's coordinates.*

submersible, with the magnetic heading of the vessel, a resolution is then possible for magnetic bearing to the submersible. Then add the offset of the center pivot point of the vessel to the GPS antenna, and resolution of the range and magnetic bearing to place the submersible in three-dimensional space. Resolution to geographic coordinates is then possible (Figure 4.34).

Refer to the manufacturer's operations manual of the equipment to be operated to gain specific steps in setting up the equipment on your vessel. Figures 4.35 and 4.36 depict a typical screen readout of a geo-referenced positioning system.

4.2.10.2 *Vessel referencing of position*

In the geo-referenced system above, changing the submersible from relative reference (range/bearing from the vessel – more specifically to the transducer array) was complex due to the difficulty of resolving range/bearing from relative to geographical. To determine the position referenced to a ship (whose geographic coordinates are irrelevant), use a scaled drawing and place the transducer array on that scaled drawing to register position in reference to that vessel (Figure 4.37).

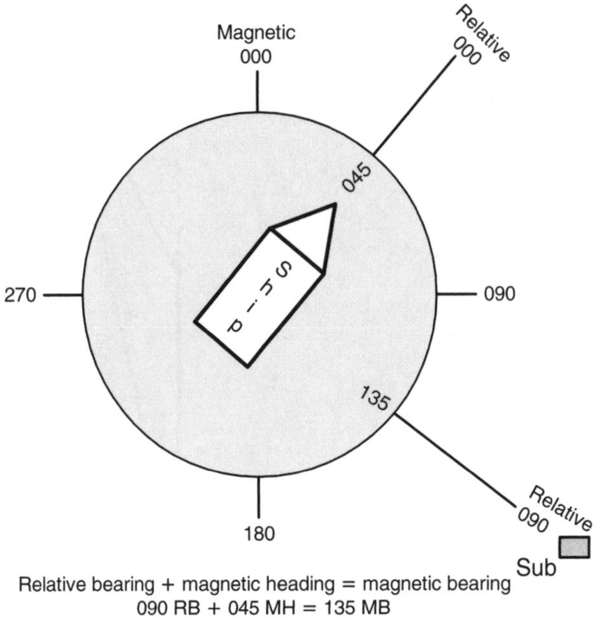

Figure 4.34 *Resolution of ship's vehicle position to geographic coordinates.*

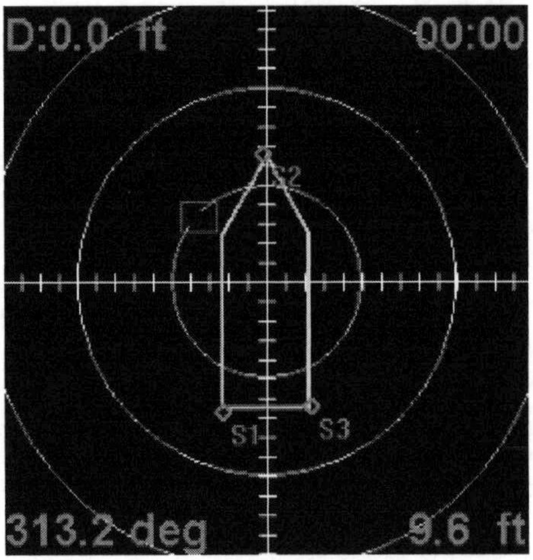

Figure 4.35 *Typical relative bearing display of acoustical positioning to the deployment vessel.*

4.2.10.3 *Relative referencing of position*

It is now possible to take the ship relative reference and make any reference system that is desired. Take a nautical chart and register the coordinates for length/width to create a scaled drawing depicting the estimated surface of the earth at that location.

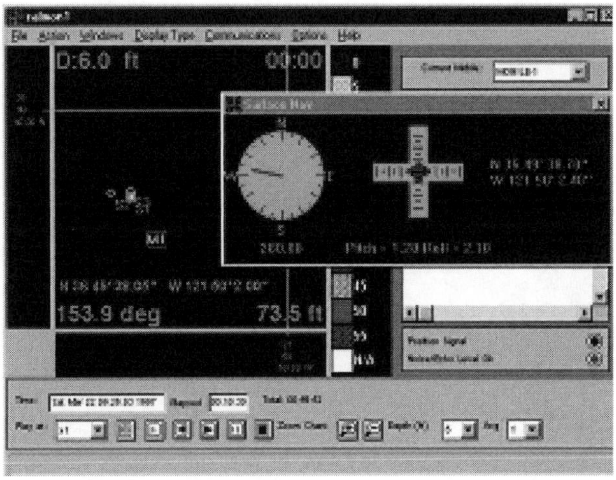

Figure 4.36 *Readout of geo-referenced coordinates (courtesy Desert Star Systems LLC).*

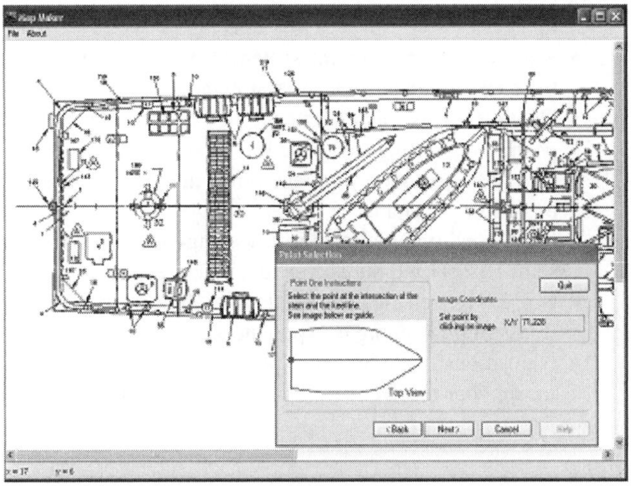

Figure 4.37 *Position referenced to ship.*

Take a drawing of a dock and scale as well. This creates a scaled dock drawing where the acoustic positioning system is now relatively referenced to the drawing of the dock. Any number of combinations is possible with an understanding of scaled drawings and relative reference with regards to a frame of reference. Figure 4.38 depicts a scaled drawing of a dam with positioning beacons registered on the drawing.

4.2.11 General Rules for Use of Acoustic Positioning Systems

The following are some guidelines to use for acoustic positioning systems:

- The best angular resolution is broadside to the longest baseline.
- Attention to detail in placement of baseline stations is paramount due to propagation of any baseline errors to the position resolution of the vehicle.

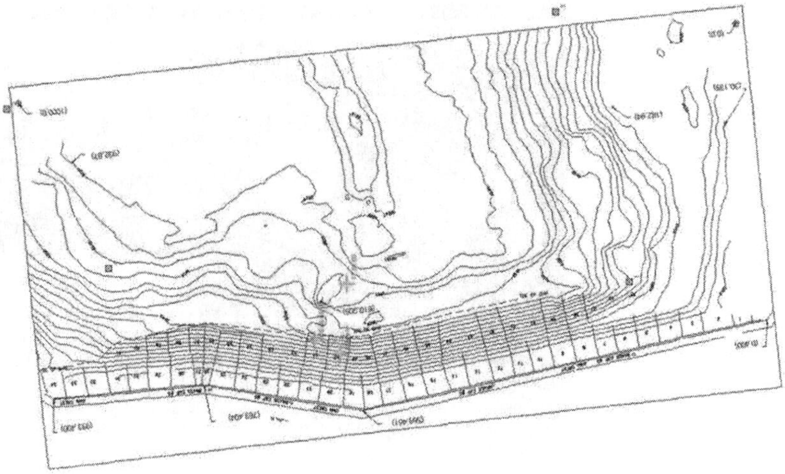

Figure 4.38 *Relative referenced dam map with acoustic positioning (courtesy Desert Star Systems LLC).*

- Most acoustic positioning systems have a noise test capability. If in doubt about the background noise level, test it and adjust accordingly.
- Operate all vessel equipment to be used during performance of acoustic positioning during any noise level test.
- Carefully consider your environment type when selecting settings on your positioning system.
- If the operating platform is a heaving vessel and the frame of reference is the same vessel (without access to a VRU), expect position dithering.
- Noise levels change constantly over the course of any operation. Consider this if loss of positioning is experienced.
- If positional dithering is experienced, multi-path is the most likely problem. Lower your receiver sensitivity to compensate for false (multi-path) readings.

Chapter 5
Sonar

In this section, the subject of sonar is presented with emphasis on 'imaging sonar' systems as deployed aboard many observation-class ROV systems. The field of sonar comprises a broad and very in-depth body of knowledge lightly dealt with in this chapter. For further reading on this subject, refer to the bibliography.

This chapter relies heavily upon contributions by experts in this field. The explanation of CHIRP technology comes courtesy of Maurice Fraser with Tritech International Limited. Special thanks go to Richard Marsh for arranging for its inclusion. The explanation of acoustic lens technology comes from a technical paper entitled 'Object Identification with Acoustic Lenses' courtesy of Edward Bletcher of Sound Metrics. And a special thanks goes to Willie Wilhelmsen, Jeff Patterson, and Mitch Henselwood of Imagenex Technology Corporation for their extensive contributions to this chapter over a 2-year period.

5.1 SONAR BASICS

5.1.1 Why Sound?

Sound propagates readily through water well beyond the range of light penetration (even at light's highest penetration wavelengths). Sound propagates best through liquid and solids, less well through gases, and not at all in a vacuum. As light reflections differentiate between objects by varying levels of reflection (light intensity) as well as changing wavelengths (light color), so does sonar characterize targets through reflected sound frequency and intensity. Through proper data interpretation, target information may be discerned to identify the object through active or passive sound energy.

Some typical applications for sonar technology include:

- Echo sounding for bathymetry
- Side-scan sonar for bathymetry and item location
- Underwater vehicle-mounted imaging sonar for target identification
- Geophysical research
- Underwater communications
- Underwater telemetry
- Military listening devises (passive sonar) for submarines and shipping identification
- Position fixing with acoustic positioning systems
- Fish finding

- Acoustic seabed classification
- Underwater vehicle tracking over bottom
- Measuring waves and currents.

5.1.2 Definition of Sonar

Sound transmission through water has been researched since the early 1800s. The technology rapidly matured, beginning in the 1930s. With the explosion of technology in the fields of beam-forming transducers and digital signal processing, today's sonar system encompasses a wide range of acoustical instruments lumped under the general heading 'Sonar.'

The term 'Sonar' is derived from '*so*und *na*vigation and *r*anging'. The purpose of this technology is to determine the range and reverberation characteristics of objects based upon underwater sound propagation.

5.1.3 Elements Required for Sonar Equipment

For sonar equipment to function, three key elements are necessary:

1. Source – A sound source is needed to produce the pulse energy for reflection (in active sonar systems) and/or reception (for passive sonar systems).
2. Medium – In the vacuum of space, sound does not travel. Some type of medium is required to transmit the sound wave energy between the source and receiver.
3. Receiver – Some type of receiver is needed to transmit the mechanical energy (sound waves) into electrical energy (electrical signal) in order to process the sound into signals for information processing.

The source vibrates (whether it is a sonar transducer, machinery noise from an engine aboard a vessel, or the mating call of whales in the ocean), causing a series of compressions and rarefactions, thus propagating the sound through the transmission medium.

5.1.4 Frequency and Signal Attenuation

Beginning in 1960 and culminating with the Heard Island feasibility tests in 1991, sound propagation tests were conducted off the coast of Australia with listening stations placed around the world's oceans. What was discovered was the extreme distance low-frequency sound in seawater would travel. As the frequency increased, however, sound attenuation increased dramatically. Table 5.1 presents a general working range of various sound frequencies in seawater.

Lower frequency sound propagates through higher density materials better than higher frequency sound waves. Very low frequency sound is used in applications such as seismic surveys of sub-bottom rock formations in search of hydrocarbons, since the sound waves will penetrate more dense strata before reflecting off of hard substrate. Low frequency is used to penetrate the mud level by sub-bottom profilers.

Table 5.1 *Sonar two-way working ranges.*

Frequency	*Wavelength*	*Distance*
100 Hz	15 m	1000 km or more
1 kHz	1.5 m	100 km or more
10 kHz	15 cm	10 km
25 kHz	6 cm	3 km
50 kHz	3 cm	1 km
100 kHz	1.5 cm	600 m
500 kHz	3 mm	150 m
1 MHz	1.5 mm	50 m

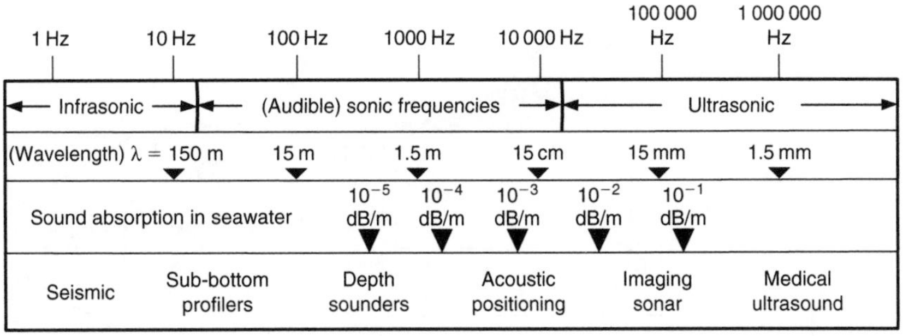

Figure 5.1 *Acoustic spectrum.*

In the very high frequency ranges, sound reflection gives only a surface paint of the target but gives higher target surface detail. An analysis of the sound frequency spectrum is provided in Figure 5.1.

5.1.5 Active Versus Passive Sonar

Range and bearing from a sound source can be derived either actively or passively. The transducer/receiver receives sound from some source then (through a series of computations) arrives at a range and bearing solution. *Passive* sonar essentially uses listening *only* to derive these computations. *Active* sonar (in a method similar to radar) uses a transmitter/receiver arrangement to send out an acoustic signal, then listens for a reflection (echo) of that signal back to the receiver over time to derive a range and signal strength plot.

Beam forming of sonar signals provides for various acoustic properties of the reflected signal (for active sonar), thus allowing analysis of the backscatter for target characterization. For imaging sonar, a so-called 'fan beam' is used to depict small details, thus building a clear pictorial image of the target being insonified (Figure 5.2). For depth sounding sonar, a broad conical beam gives an indication of the closest distance to the bottom over a broader area. The typical usage of a conical beam is

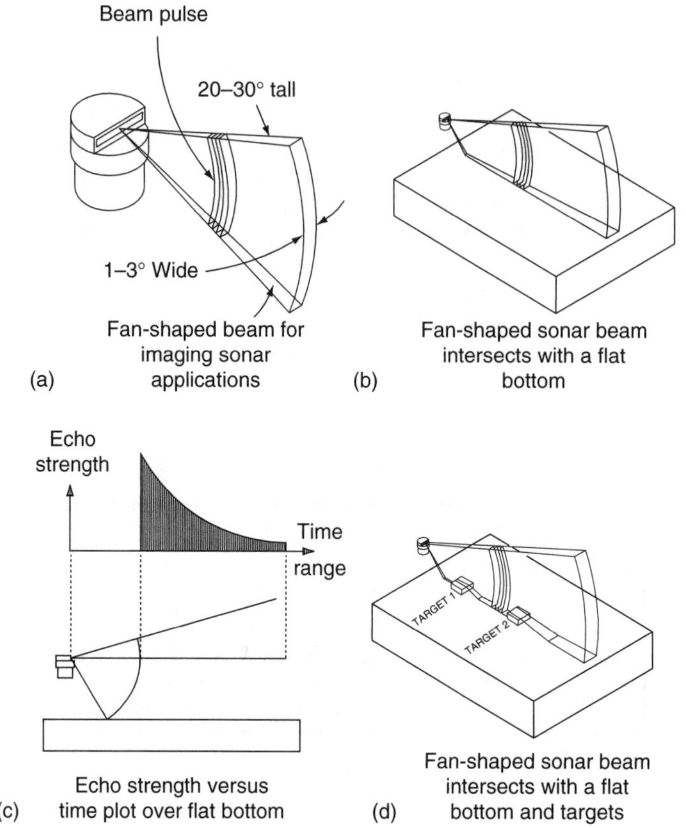

Figure 5.2 *Sonar signals are actively emitted from a transmitter and are plotted over time.*

for depth sounding. For ROV-mounted sonar applications, a fan beam is used for the imaging sonar while a conical beam is used for the 'altimeter' to determine the submersible's height above the bottom.

5.1.6 Transducers

This section will delve further into the transducer and its use in beam forming for gathering various backscatter characteristics of the target. By definition, a 'transducer' is any of a number of devices that convert some other form of energy to or from electrical energy. Examples of devices that fall under the category of 'transducer' are photoelectric cells, common stereo speakers, microphones, electric thermometers, any type of electronic pressure gauge, and an underwater piezoelectric transducer. This chapter will concentrate on the piezoelectric transducer and its usage with sonar systems. This analysis begins with the familiar land-based transducer, the home stereo speaker, then progresses into the unfamiliar underwater transducer.

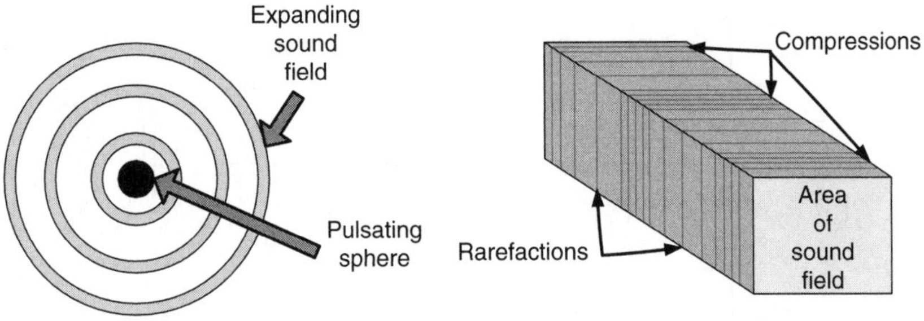

Figure 5.3 *Sound pulse traveling in wave fronts through a medium.*

In a stereo speaker, any number of technological twists is used to convert electrical energy to acoustic energy. The most basic means involves the varying of a voltage within a coil wrapped around an electromagnet moving a paper cone. The cone vibration produces acoustic energy in the form of sound waves propagating in an omnidirectional fashion as depicted in Figure 5.3. As shown in the figure, the sound moves through the medium (in this case air) through a series of compressions and rarefactions as the wave travels at sonic velocity away from the sound source. As the cone vibrates, it produces sound in a bidirectional pattern. In order to change the sound pattern from an omnidirectional wave pattern to a more focused sound projection, enclosures and sound projectors are used to focus the sound beam to a desired direction. If a home stereo speaker is disassembled, one would find three basic speakers (the woofer, midrange, and tweeter), each with frequency response based upon its size and nominal resonance frequency.

The most prevalent underwater transducer is the polycrystalline, 'piezoelectric' material PZT (lead zirconium titanate), which was discovered in the 1940s to be an effective transducer. As discussed in Chapter 4, piezoelectric underwater transducers emit acoustic pulses through vibration when an electric signal is sent through the silver deposits at the poles of the ceramic core. Piezoelectric underwater transducers are resonant; therefore, a transducer is sized to match a certain nominal narrowband frequency in which it is most sensitive. The sonar system is then designed around this nominal frequency.

On a typical active transducer, the transducer will vibrate at the frequency of the electrical signal being applied, not necessarily the resonant frequency. If that frequency is near the resonant frequency, the transducer is freer to respond, giving a larger amplitude signal than if the frequency is far from the resonant frequency. The length of time the transducer is activated determines the pulse length generated (Figures 5.4 and 5.5).

A short pulse length allows better discrimination between targets, but may not allow enough acoustic energy to reflect off targets as distance increases.

Directivity index for a transducer defines the ratio between an omnidirectional point source, i.e. a circular (2D) or spherical (3D) sound source with no directivity, to the source level intensity on the beam axis of a directional sound pulse. The reason this factor is important is that the beam form is defined by its directivity index as well as its vector (Figure 5.6).

As the directivity index of a sound beam becomes more focused, some side-effects become pronounced, thus requiring consideration. The main beam of a directional

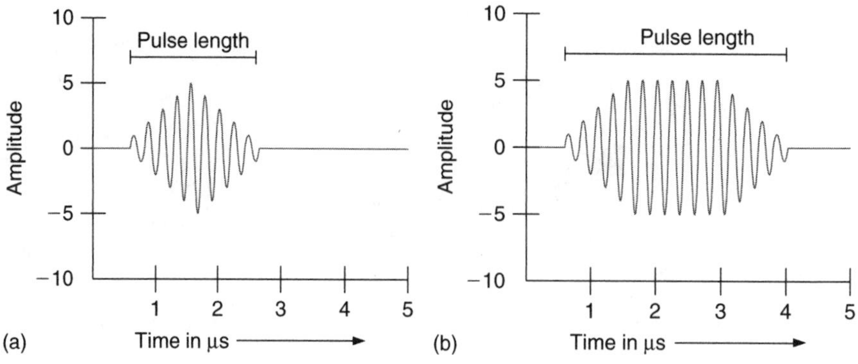

Figure 5.4 *Pulse length as a function of time.*

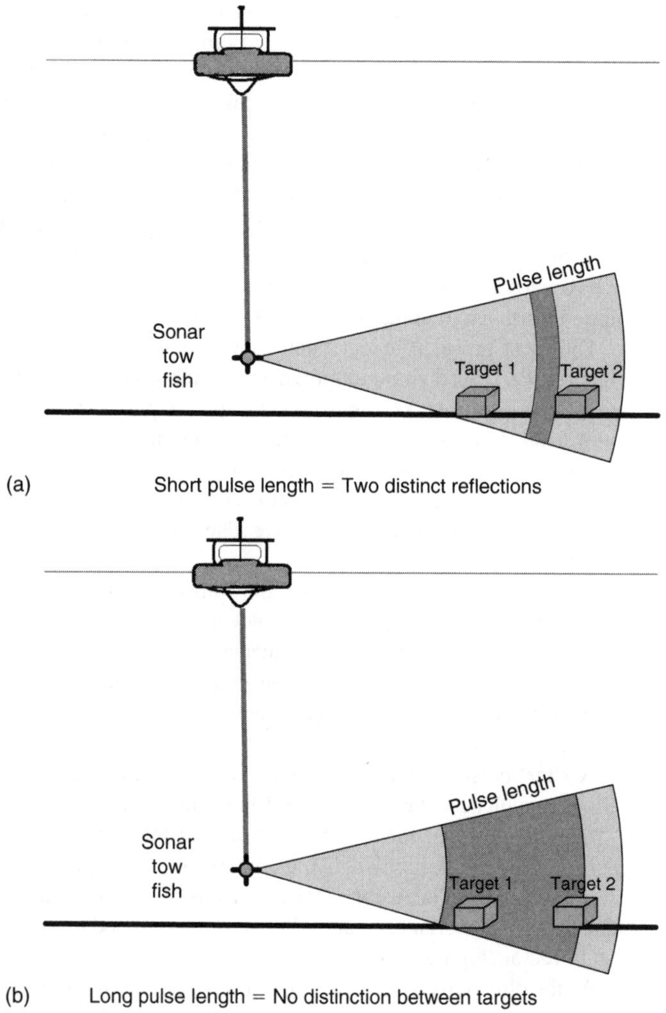

Figure 5.5 *Pulse length for discrimination between targets.*

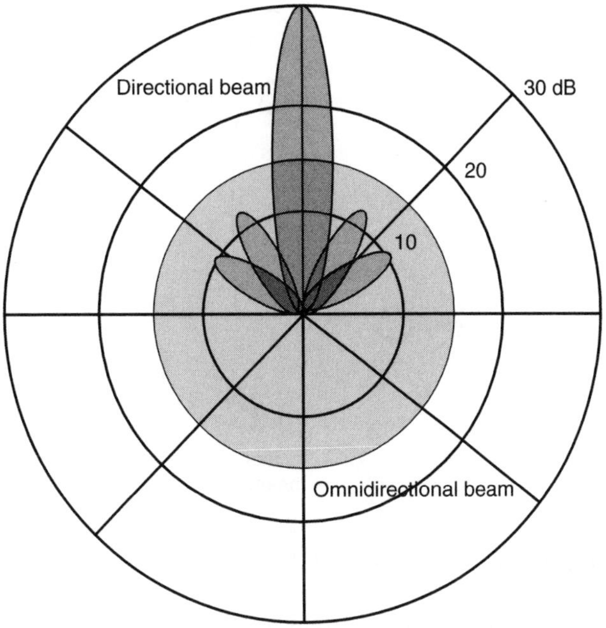

Figure 5.6 *Directivity index for a circular as well as a directional sound beam.*

sound source is called the 'main lobe' and is the primary pulse directed towards the target. As depicted in Figures 5.7 and 5.8, a by-product of this main lobe, side lobes, develop that are also subject to backscatter off secondary targets. An analogy to this concept is the simple push of the hand on the surface of a swimming pool directing a wave pulse in a specific direction. As the hand is pushed in one direction, it can be noticed that the main pulse pushes in the direction of the wave but side waves emanate on either side of the main wave.

5.1.7 Active Sonar

The majority of civilian sonar systems fall under the classification of active sonar. Active sonar systems make use of beam forming, frequency shifting, and backscatter analysis to characterize targets insonified. There are a vast number and category of active sonar products on the market to fit various applications (with the technology advancing daily). Any number of these technologies is available for mounting on ROV equipment for delivery to the work site.

5.1.8 Terminology

The following provides an explanation of the basic terms used to describe sonar techniques:

- *Angle of incidence*: The angle at which a sound reflects upon a surface (in degrees). Example: If a flashlight is shined on a mirror, the light will reflect

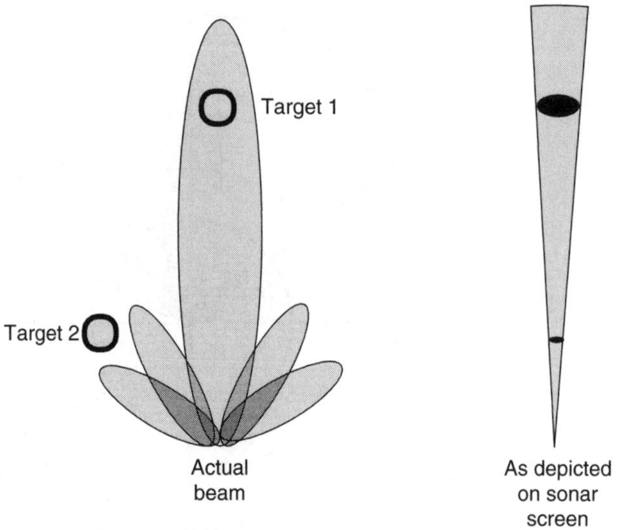

Figure 5.7 *Target in the side lobe of this beam will produce an echo at the receiver.*

directly back to the flashlight if there is a zero angle of incidence, reflected 90° with a 45° angle of incidence, etc. The same principle applies to a sound reflector.

- *Beam forming*: The principle of forming the sound propagation wave to provide the desired data from the wave front.
- *Color*: The different colors used to represent the varying echo return strengths.
- *Echo*: The reflected sound wave.
- *Echo return*: The time required for the echo to return to the source of the sound.
- *Pulse width*: The width of a sound propagation wave (in feet, meters, or microseconds) generated from some source. This factor has an influence on the quality of the image generated by the sonar receiver.
- *Target*: The object requiring characterization.
- *Target strength (TS)*: A measure of the reflectivity of the target to an active sonar signal.
- *Reflectivity*: The ability of a material to reflect sound energy.
- *Transmission (sound) loss (TL)*: Total of all sound losses incurred between the sound source and the ultimate receiver. Transmission losses come in two main types: spreading loss and attenuation loss.
- *Spreading loss*: Sound energy loss due to geometrical spreading of the wave over an increasingly large area as the sound propagates. Spreading losses are considered on either a two-dimensional cylindrical plane (horizontal radiation only, or thermal layer, or large ranges compared to depth) or a three-dimensional spherical profile (omnidirectional point source).
- *Attenuation loss*: Generally considered to be the lumped together sum of losses produced through absorption, spreading, scattering, reflection, and refraction.
- *Absorption loss*: The process of converting acoustic energy into heat. Absorption loss increases with higher frequency.
- *Scattering and reverberation*: Sound energy bouncing or reflecting off items or surfaces.

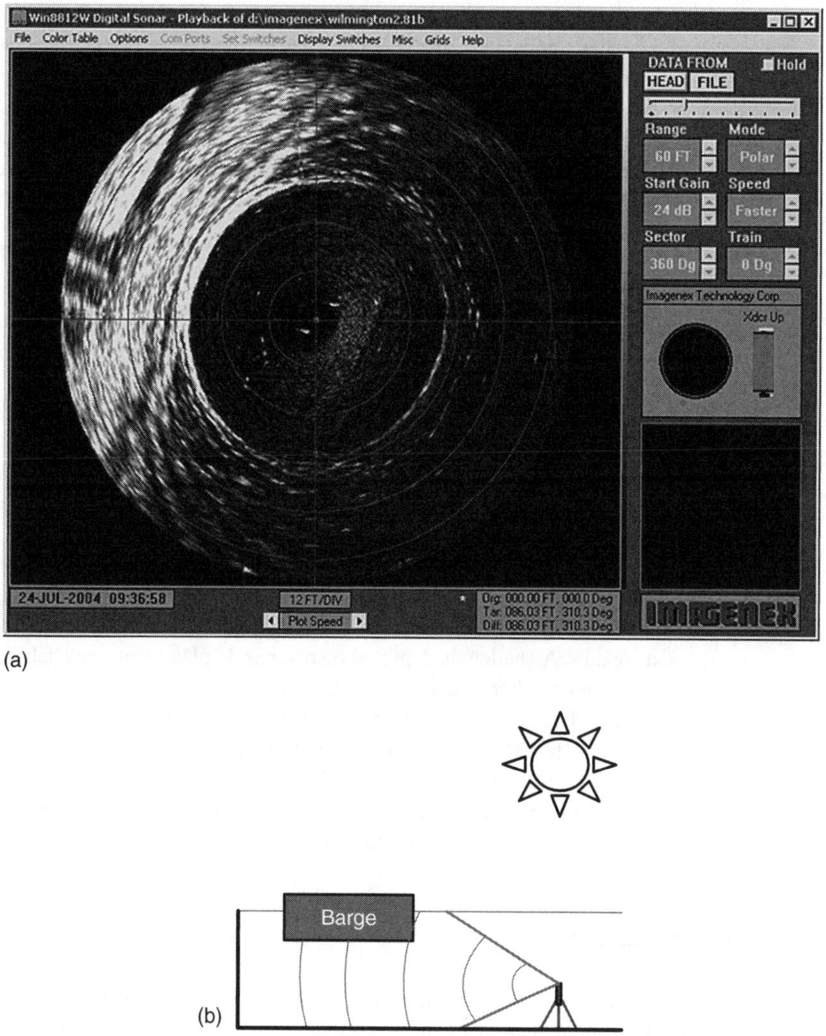

Figure 5.8 *Mechanically scanning sonar viewing barge with scalloped lines from side lobe backscatter.*

- *Noise level (NL)*: Total noise from all sources potentially interfering with reception of the sound source reception. Therefore, Noise level (NL) = Self noise (SN) + Ambient noise (AN).
- *Self noise (SN)*: Noise generated from the sonar reception platform potentially interfering with reception of the sound source. Examples of this are machinery noise (pumps, reduction gears, power plant, etc.) and flow noise (high hull speed, hull fouling, such as barnacles or other animal life attached to hull), and propeller cavitations.
- *Ambient noise (AN)*: Background noise in the medium potentially causing interference with signal reception. Ambient noise can be either hydrodynamic (caused

by the movement of water, such as tides, current, storms, wind, rain, etc.), seismic (i.e. movement of the earth – earthquakes), biological (i.e. produced by marine life), or by ongoing ocean traffic (i.e. noise caused by shipping).

- *Source level (SL)*: Sound energy level of sound source upon transmission.
- *Directivity index (DI)*: The ratio of the logarithmic relationship between the intensity of the acoustic beam versus the intensity of an omnidirectional source.
- *Signal-to-noise ratio (SNR)*: The ratio to the received echo from the target to the noise produced by everything else.
- *Reverberation level (RL)*: The slowly decaying portion of the backscattered sound from the sound source.
- *Detection threshold (DT)*: The minimum level of received signal intensity required for an experienced operator or automated receiver system to detect a target signal 50 percent of the time.
- *Figure of merit (FOM)*: The maximum allowable one-way transmission loss in passive sonar, and the maximum two-way transmission loss in active for a detection probability of 50 percent. These are termed active ($AFOM$) and passive ($PFOM$). This concept is especially pertinent to the anti-submarine warfare community.
- *Shadows*: The so-called 'shadowgraph effect' is an area of no sonar reflectivity due to blockage. The best analogy for this would be to shine a light onto an object a few feet away. The object will block the light directly behind it and cast a shadow. A shadow is depicted on a sonar display by an area behind a blocking mechanism where there is no sonar reflection.
- *Sonar*: The principle used to measure the distance between a source and a reflector (target) based on the echo return time.
- *Sound losses (absorption, spreading, scattering, attenuation, reflection, and refraction)*: Sound energy loss through various factors influencing the reception and display of reflected sound waves.

5.1.9 Sonar Equations

Sonar equations look at signal losses compared to signal sources to determine the likelihood of detection of the sound source. For the passive sonar, only one-way sound transmission and noise levels are considered in deriving the minimum detection threshold, whereas on active sonar systems the sound must travel first from the sound source (the transducer of the active sonar system) through the water column to the target then (after the backscatter) back to the receiver (Figure 5.9).

The passive sonar equation:

$$SL - TL - NL + DI \geq DT.$$

The active sonar equations:

Ambient noise limited:

$$SL - 2TL + TS - NL + DI \geq DT.$$

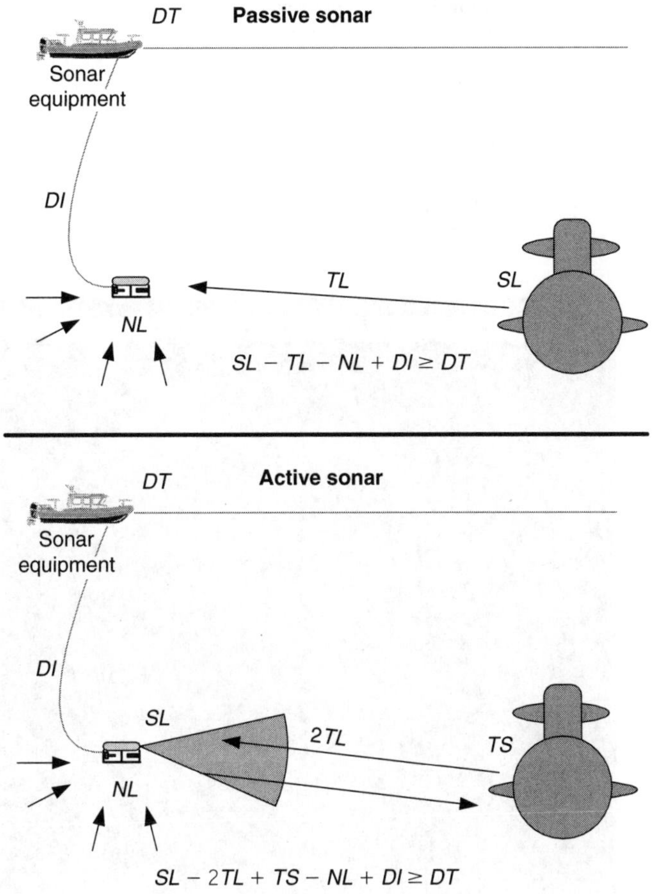

Figure 5.9 *Passive and active sonar equations.*

Reverberation noise limited (reverb > ambient noise):

$$RL > NL + DI$$

$$SL - 2TL + TS - RL \geq DT$$

The figure of merit equations:

$$PFOM = SL - NL + DI - DT$$

$$AFOM = SL + TS - NL + DI - DT.$$

5.1.10 Reflectivity and Gain Setting

Different materials reflect sound with different efficiency (Table 5.2). Mud will reflect sound very poorly while water will not reflect sound at all. The closer the substance's

Table 5.2 *Sample reflectivity indexes.*

Substance	*Reflectivity*
Mud	Low
Sand	Medium
Rock	High
Air/air-filled	Very high

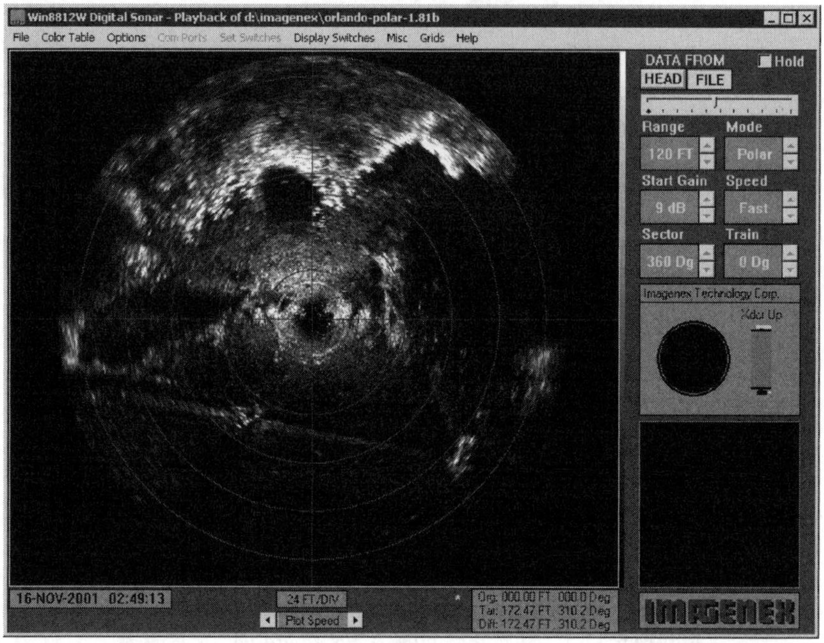

Figure 5.10 *Combination of sand and mud bottom showing differing reflectivity based upon bottom composition (gain setting 9 dB).*

consistency is to water, the lower the reflectivity index. Therefore, a very good feel for the target's makeup can be gained simply by the target's level of reflectivity.

The gain setting on the sonar system will allow the operator to pick up detail within the reflection. If the gain setting is set high while surveying a sandy bottom, the screen will display no contrast between targets, since everything will show a high reflectivity value. Likewise, if the gain is set too low with a mud bottom, no detail will display, since practically all of the reflections from the bottom will be below the display setting and will be rejected.

In Figure 5.10, differing compositions on this combination of sand and mud bottom allows for discrimination between bottom makeup as well as targets standing proud of the bottom with differing makeup. The bottom of this figure displays a practice course as well as two targets easily discernible from the mud bottom. This figure depicts differing bottom composition through varying echo strength with a fixed gain setting.

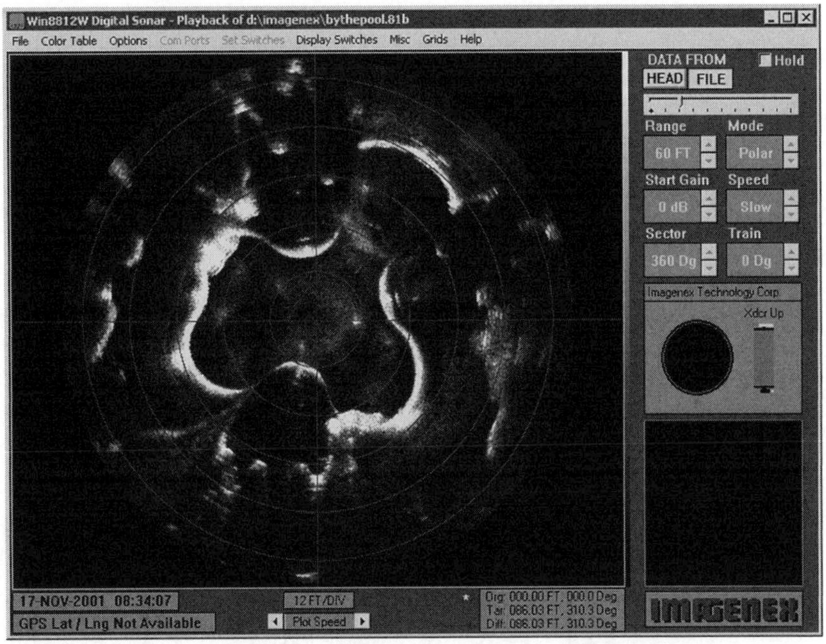

(a)

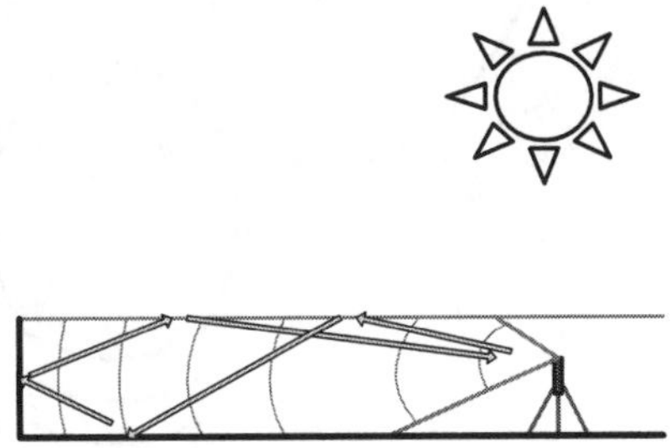

(b)

Figure 5.11 *Sonar picture of a recreational swimming pool – reflections outside of pool are reverberations (0 dB gain setting).*

In Figure 5.11, an example of a swimming pool is used to illustrate the effect of varying angles of incidence on sonar reflectivity. The tripod-mounted sonar was placed at the center of a varying shaped pool (see Figure 5.11(c) next page for analysis). The walls of the pool have a very high reflectivity at the zero incidence point with much lower backscatter on the higher angles of incidence.

As a further example of incidence versus reflectivity, consider a sonar source placed in an underwater room (Figure 5.12). The sonar will insonify the room with the highest target strength coming from the zero angle of incidence along with another 'bright spot' located in the corners. The corners will not be depicted as square due to the sound multi-path as the beam approaches the corner point. Instead of a square corner, the sonar display will depict a rounded corner due to the sound reflection as the beam sweeps the corner.

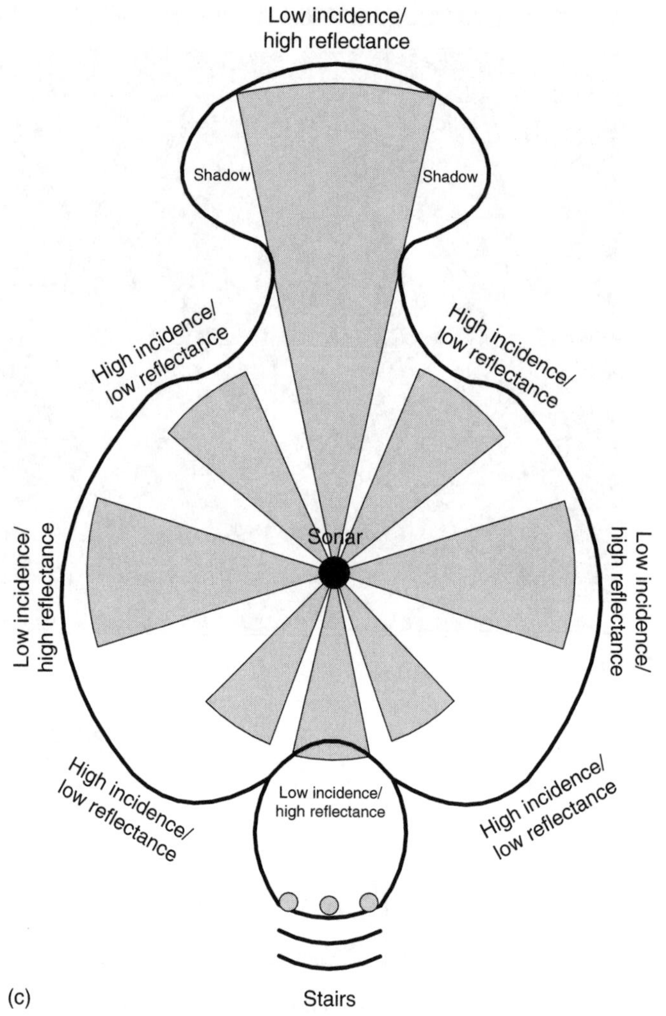

Figure 5.11 *(Continued)*

5.1.11 Sound Backscatter

Backscatter is the reflected sound from any object being insonified – the analysis of which is the subject of active sonar systems. Backscatter is analyzed for any number of parameters to solve particular operational needs. Some examples of backscatter analysis follow:

- Multiple bounce depth sounder backscatter is analyzed with acoustic seabed classification systems to determine the texture and makeup of the sea bottom (sand, mud, rock, oyster bed, kelp, etc.) for environmental monitoring as well as vessel navigation.

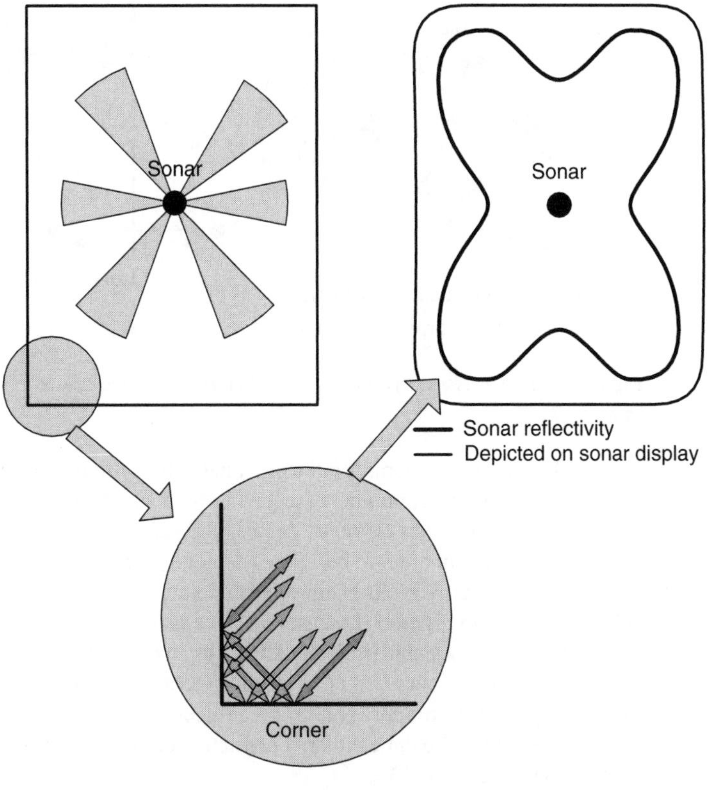

Figure 5.12 *Sonar picture of an underwater Rectangular room.*

- Doppler shift backscatter is used for vehicle speed over ground as well as current/wave profiling.
- Frequency shift backscatter is analyzed in CHIRP sonar systems (see Figures 5.22–5.24) to discriminate between objects in close proximity.
- Simple high-frequency backscatter can characterize a target as to aspect, texture, surface features, and orientation.

5.1.12 Single-Versus Multi-Beam

Active imaging or profiling sonar systems generally fall into three basic categories:

1. Multi-beam.
2. Mechanically or side-scanning sonar (single beam).
3. Single-beam directional sonar.

Single-beam sonars are simply one pulse with one reception on a single receiving element. The single-beam mechanically scanning system comprises the lion's share of ROV-mounted sonar systems currently on the market today due to their simplicity as well as their reduced cost.

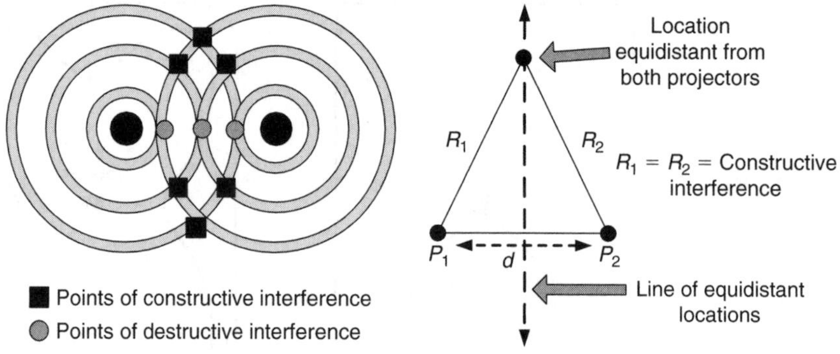

Figure 5.13 *Constructive and destructive interference of two point sources with spacing d.*

Multi-beam sonar systems, on the other hand, transmit one wide pulse and receive the backscatter on a large number of receiving elements. Accurate sensing of the delays in sound arrival between elements enables the sonar to distinguish the direction of the received sound, in order to build up a detailed profile of the area being insonified.

In its simplest form, multi-beam sonars operate on the principle of additive sound pulses. Recall from Figure 5.3 that sound spreads in an isotropic fashion (the same in all directions) from a sound projector known as an isotropic source (a sound projector projecting a sound source equally in all directions). Placing two isotropic sound sources in proximity proportional to the projector's wavelength, the additive pulses (termed interference) can be either constructive or destructive based upon the phase of the pulses (Figure 5.13).

From this constructive interference, a highly directional pulse can be constructed along the line of equidistance with spacing $\lambda/2$. This forms the axis of a pulse in a three-dimensional directional beam (Figure 5.14).

If several projectors are aligned in a row, the beam can become more focused and highly directional.

By varying the time delay between the projector transmissions, the axis of the beam can be steered to project multiple beams. With use of the same concept upon reception, the time variance can also be analyzed to only accept sound sources from a narrowly directive reception cone, thus forming the multiple beam transmission and reception. The reception array is perpendicular to the projector array so that reception is only allowed at the crossing points, allowing for clean multi-beam transmission and reception.

Table 5.3 analyzes some of the benefits of multi-beam technology. Multi-beam sonar systems are an amazing recent technological development brought on by advances in digital signal processing of sonar returns as well as advances in highly directional transducer receiving elements. With the technology trickling down to the observation-class ROV system through lower costs and increased functionality, look for more multi-beam sonar systems aboard these vehicles.

5.1.13 Frequency Versus Image Quality

Determination of the optimum sonar frequency for the particular application at hand is a trade-off of desired image quality, depth of penetration into the medium as well

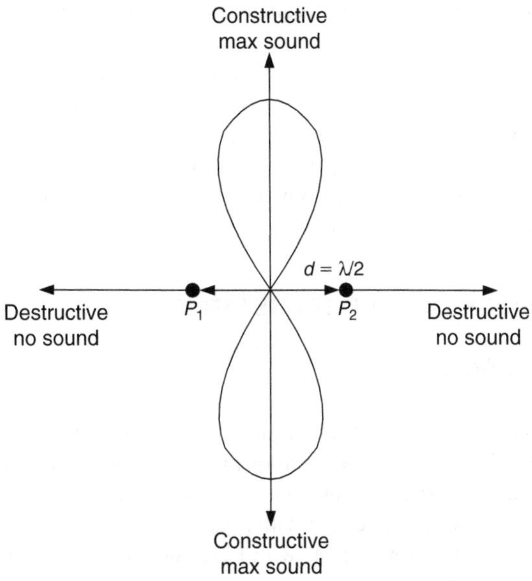

Figure 5.14 *Constructive dipole projection of sound along a beam axis.*

Table 5.3 *Some advantages and disadvantages of ROV-mounted single- and multi-beam sonar systems.*

Sonar system	*Advantages*	*Disadvantages*
Single-beam	• Able to view 360° around vehicle	• Slow image generation
Multi-beam	• Up to 10 frames per second image update rate at short range	• ROV-mounted projector allows narrow reception beam width • Higher computing power needed drive cost upward

as range of desired coverage area. As discussed above, the higher the frequency of the source, the higher the attenuation of the signal, thus lowering the effective range of the beam. The lower the frequency, the longer the propagation distance and the penetration depth into the medium being insonified. The higher the frequency, the less penetration into the medium, allowing for a more detailed surface paint of the target for target shape and texture classification.

For small-item surface discrimination, the wavelength of the pulse must be no larger than the target – and much shorter for any detailed imaging of the target. For better small-object discrimination, the higher the frequency the better. What suffers at the very high frequencies (700 kHz and higher) is the useful range of the sonar system. For searching a wide area for small-object location, a range of only 20 meters will require an extended search time to cover any appreciable area. On a more human note, medical ultrasound functions in the 2–3 MHz range (and higher) for very-small-object discrimination. At these frequencies, extremely high detail can be achieved over very short distances.

Low-frequency sonar is used for such applications as seismic surveys of geological strata (very low frequency – in the 60–100 Hz range) as well as sub-bottom profiling below the mud line (low frequency – in the 10–100 kHz range depending upon desired mud penetration depth). These systems can gain a broad classification of the overall area, but small details of individual areas are not possible due to the wavelength being larger than any small target.

5.2 SONAR TYPES AND INTERPRETATION

5.2.1 Imaging Sonar

A fan-shaped sonar beam scans surfaces at shallow angles, usually through an angle in the horizontal plane, then displays color images or pictures. The complete echo strength information for each point is displayed primarily for visual interpretation. As depicted in Figure 5.15 with imaging sonar a fan-shaped sonar beam scans a given area, by either rotating (as with ROV-mounted systems) or moving in a straight line (as with side-scan sonar systems).

A pulse of sound traveling through the water generates a backscatter intensity (or amplitude) which varies with time, and is digitized to produce a time series of points. The points are assigned a color or grayscale intensity. The different colored points, representing the time (or slant range) of each echo return, plot a line on a video display screen. The image (Figure 5.16), consisting of the different colored points or pixels, depicts the various echo return strengths.

The following characteristics are necessary to produce a visual or video image of the sonar image:

- The angle through which the beam is moved is small
- The fan-shaped beam has a narrow angle
- The transmitted pulse is short
- The echo return information is accurately treated.

These visual images provide the viewer with enough data to draw conclusions about the environment being scanned. The operator should be able to recognize sizes, shapes, and surface-reflecting characteristics of the chosen target. The primary purpose of the imaging sonar is to act as a viewing tool.

5.2.2 Profiling Sonar

Narrow pencil-shaped sonar beams scan surfaces at a steep angle (usually on a vertical plane). The echo is then displayed as individual points or lines accurately depicting cross-sections of a surface. Echo strength for each point, higher than a set threshold, digitizes a data set for interfacing with external devices.

The data set is small enough to be manipulated by computer (primarily a measurement tool). In profiling, a narrow pencil-shaped sonar beam scans across the surface of a given area, generating a single profile line on the display monitor (Figure 5.17).

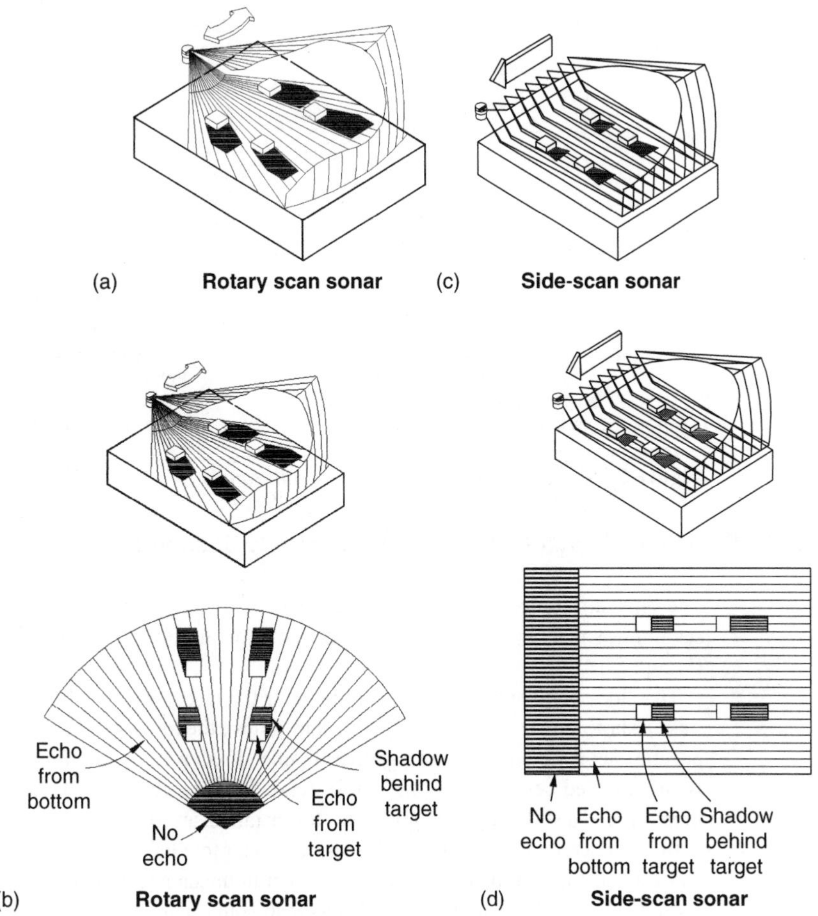

Figure 5.15 *An imaging sonar builds a picture via rotation of the head (i.e. ROV) or with motion through the water (i.e. side scan).*

This line, consisting of a few thousand points, accurately describes the cross-section of the targeted area.

A key to the profiling process is the selection of the echo returns for plotting. The sonar selects the echo returns, typically one or two returns for each 'shot', based on a given criterion for the echo return strength and the minimum profiling range. The information gathered from the selection criteria forms a data set containing the range and bearing figures. An external device, such as a personal computer or data logger, accesses the data set through a serial (or other communications protocol) interface with the sonar.

The profile data is useful for making pen plots of bottom profiles, trench profiles, and internal and external pipeline profiles. The primary purpose of the profiling sonar is as a quantitative measuring tool such as a depth sounder or for bottom characterization.

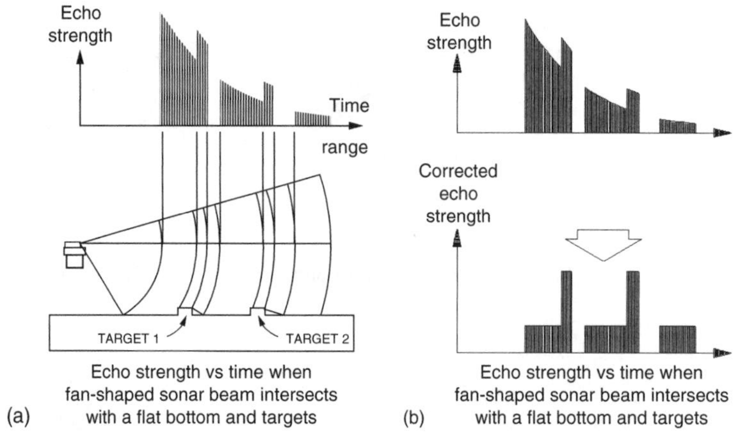

Figure 5.16 *Echo strength versus time and range.*

5.2.3 Side Scan Versus Mechanically/Electrically Scanning

The only difference between the side-scan sonar and the mechanically/electrically scanning sonar systems is the means of locomotion of the transducer head (Figure 5.19).

The side-scan sonar generates an image of the target area by the simple means of 'take a shot', move the tow fish, 'take a shot', move the tow fish, etc. A highly directive pulse of acoustic energy is transmitted/bounced/received as the transducer platform is either towed behind a moving platform (surface or submerged vessel) or mounted to the side of a self-propelled platform (for instance, an AUV). Problems encountered with image generation by a towed side-scan sonar tow fish involve the constant surge and stall of the tow fish as the tow platform bounces and pitches over a heavy sea state. AUVs (with vehicle-mounted side-scan sonars) are perceived to be potentially superior as an image generation platform due to the decoupling of the sonar platform from the surface sea state, allowing for steady and smooth image generation without image smear.

As depicted in Figure 5.18, the mechanically or electronically scanning sonar is mounted to some other relatively stable platform (ROV, tripod, tool, dock, etc). For the mechanically scanning sonar system, the transducer head is rotated via a stepper motor timed to move as the transmit/receive cycle completes for that shot line. An image is generated as the sonar head builds the sonar lines in either a polar fashion or in a sector scan (Figure 5.20).

The electrically scanning sonar can either be a spaced steerable array, a multi-beam sonar, or a focused array:

- A steerable array involves separate point sources timed to produce a dipole beam. The time spacing of the pulses allows the beam axis to be steered due to additive pulse fronts.
- The multi-beam sonar system is normally fixed with a nominal beam width of no more than 180°. The single sonar pulse is generated with multiple high directivity index receivers discriminating the backscatter return over its entire beam width to produce an image of the area under investigation.

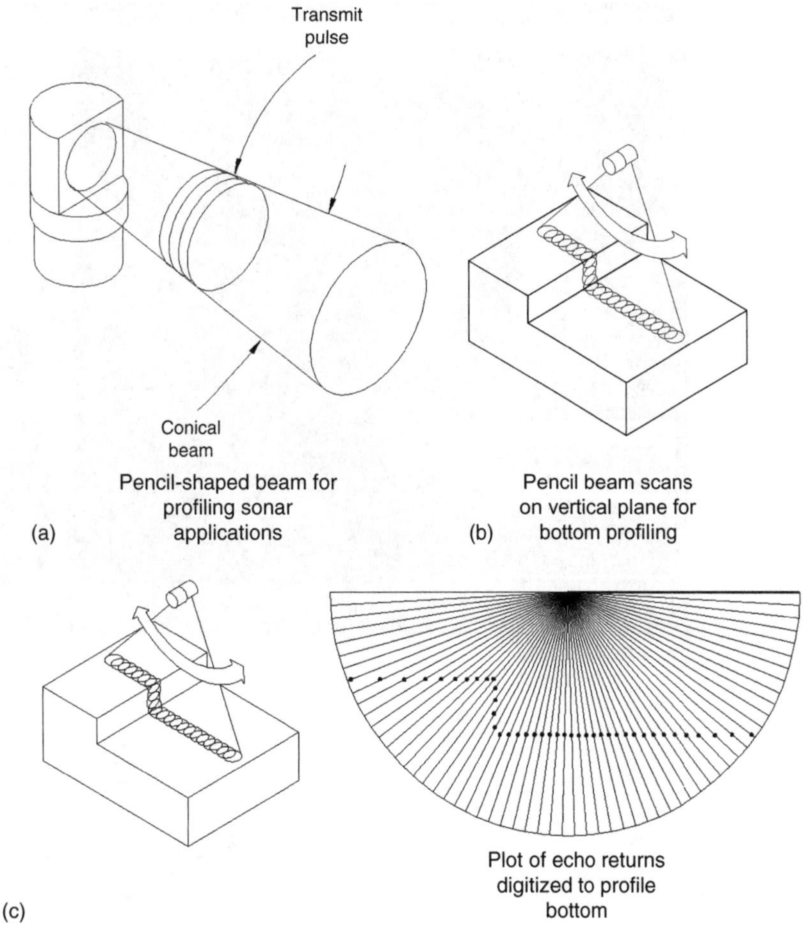

Figure 5.17 *A profiling sonar is used to scan a wider area to determine average distances (such as for a depth sounder).*

- The focused array and acoustic lens sonar technologies are both incredible technologies that focus the acoustic beams onto a localized point at very high frequencies, generating near picture quality images with a high frame rate. More on this technology.

5.2.4 Single/Dual/Multi-Frequency Versus Tunable Frequency

The single/dual/multi-frequency, CHIRP (*C*ompressed *H*igh *I*ntensity *R*adar *P*ulse), and tunable sonars all have their advantages and disadvantages. Definitions and discussion for each follows:

1. Single frequency – The single-frequency sonar system transmits and receives on one frequency and is the simplest sonar design due to its transducer selection on

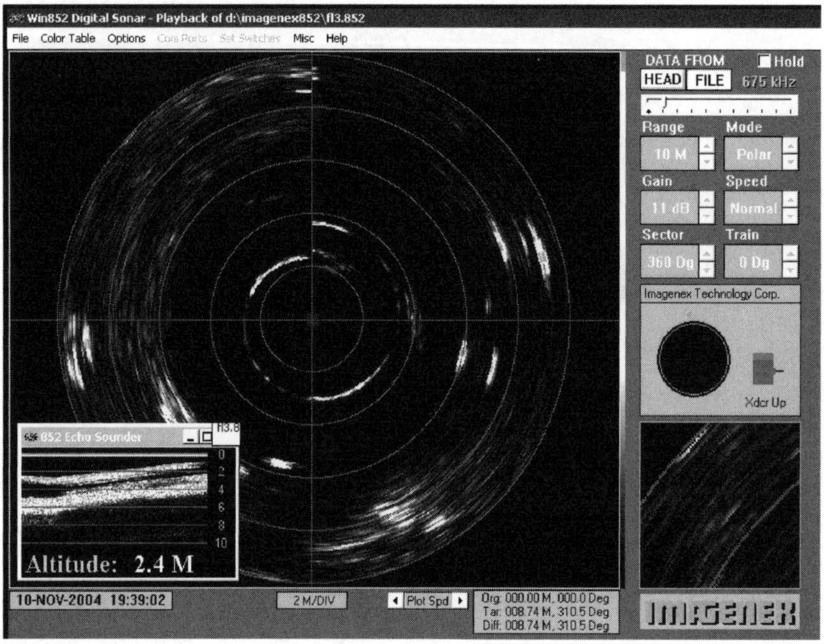

(a)

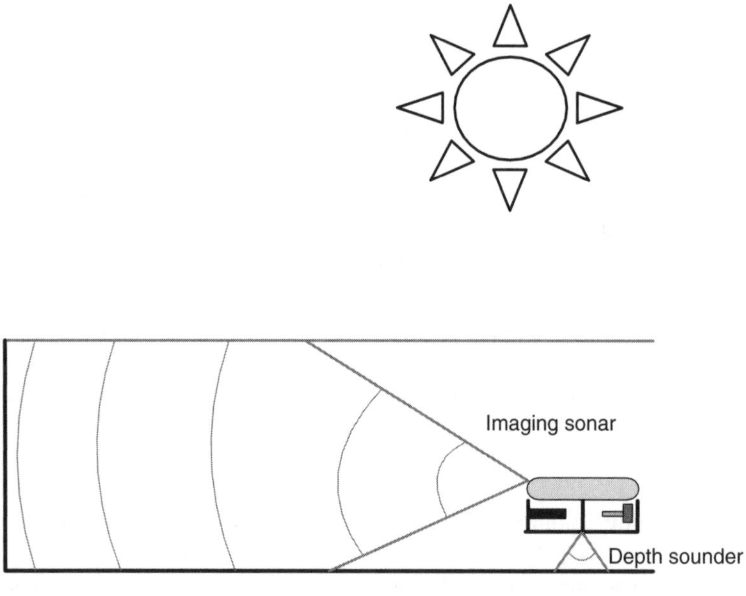

(b)

Figure 5.18 *Here is an example of a dual imaging sonar and altimeter combination (image on screen and altitude on left).*

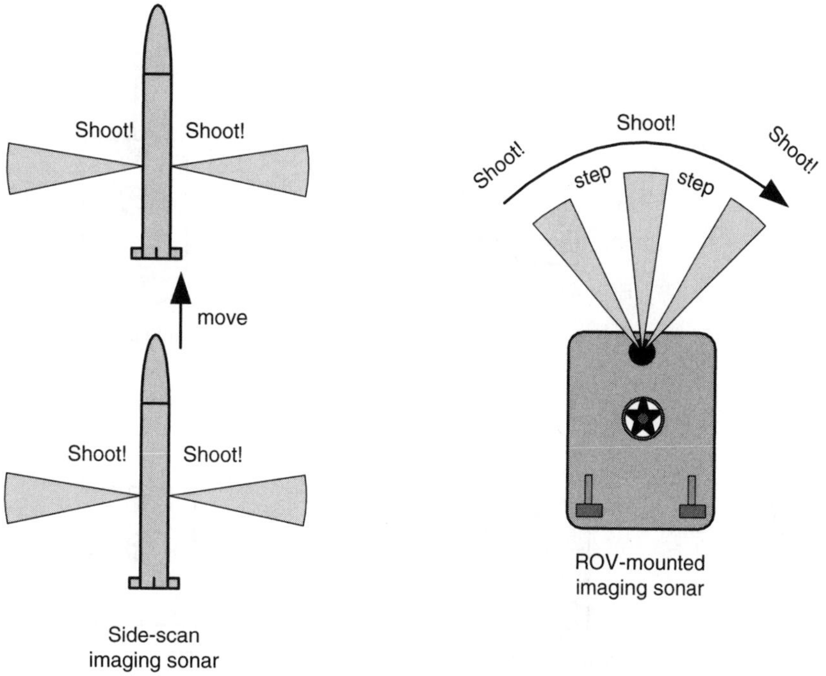

Figure 5.19 *Side-scan sonar versus ROV-mounted mechanically scanning sonar.*

its nominal resonance frequency. Unfortunately, only a limited amount of target data can be ascertained from use of a single-frequency backscatter analysis.

2. Dual frequency – This system arrangement allows gleaning of differing data parameters from the same target area based upon simultaneous or alternating transmissions of dual-frequency acoustic beams, generating each frequency's backscatter characteristics of the sample area. An excellent example of the advantages of dual-frequency sonar systems came recently during an oil spill in the Gulf of Mexico. A barge filled with heavy (heavier than water) heating oil struck a submerged unmarked platform destroyed by one of the storms of 2005. The oil sank to the bottom, requiring various governmental agencies to track the plume. To discriminate the submerged oil plume from the loosely consolidated mud bottom, a dual-frequency (100 kHz/500 kHz) towed Klein 5000 sonar was used. The low-frequency band penetrated through the relatively light viscosity oil while the higher frequency allowed for differentiation of the relatively rough mud bottom from the smooth oil as it migrated on the bottom (Figure 5.21). Once the search area was covered and the sonar data analyzed, an observation-class ROV visually characterized the bottom based upon sonar returns. Excellent tracking results were obtained of the oil plume allowing for a higher incidence of recovery.
3. CHIRP sonar – This frequency-shifting sonar technology (along with its numerous technological advantages) is described more fully in the next section.
4. Tunable frequency – This type of transmitter/receiver combination uses frequency tuning to glean differing characteristics of the target area based upon

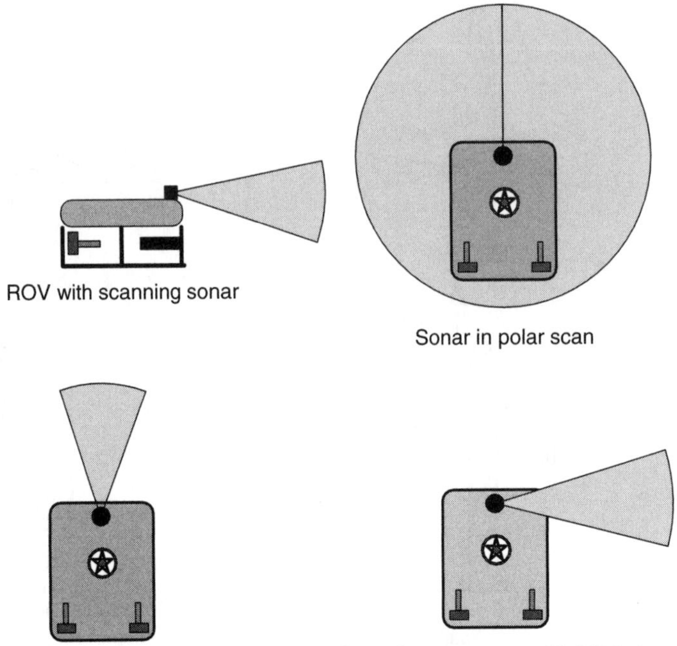

Figure 5.20 *Scan modes of a mechanically scanning sonar system.*

the backscatter characteristics of the frequency transmitted. The strength of this technique is the ability to insonify the target then analyze backscatter at the various frequencies to gain better target characterization. The weakness is the rapid degradation in efficiency of the transducer as the frequency departs the nominal transducer design range.

5.2.5 CHIRP Technology and Acoustic Lens Systems

5.2.5.1 *CHIRP sonar*

CHIRP techniques have been used for a number of years above the water in many commercial and military radar systems. The techniques used to create an electromagnetic CHIRP pulse have now been modified and adapted to commercial acoustic imaging sonar systems.

To understand the benefits of using CHIRP acoustic techniques, one needs to analyze the limitations using conventional monotonic techniques. An acoustic pulse consists of an on/off switch modulating the amplitude of a single carrier frequency (Figure 5.22).

The ability of the acoustic system to resolve targets is determined by the pulse length; This, however, has its drawbacks. To get enough acoustic energy into the water for good target identification and over a wide variety of ranges, the transmission pulse

(a)

(b)

Figure 5.21 *Dual-frequency sonar used during oil spill plume tracking. Figure 5.21(a) courtesy of Marty Klein (L-3 Klein Associates)*

length has to be relatively long. The equation for determining the range resolution of a conventional monotonic acoustic system is given by:

$$\text{Range resolution} = \text{velocity of sound}/(\text{band width} \times 2).$$

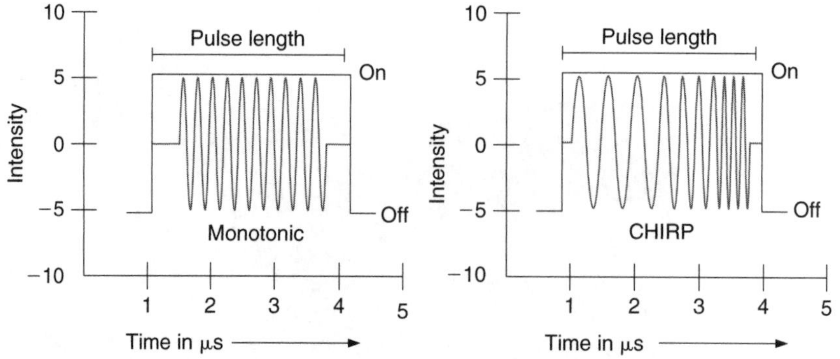

Figure 5.22 *Comparison of monotonic versus CHIRP sonar techniques.*

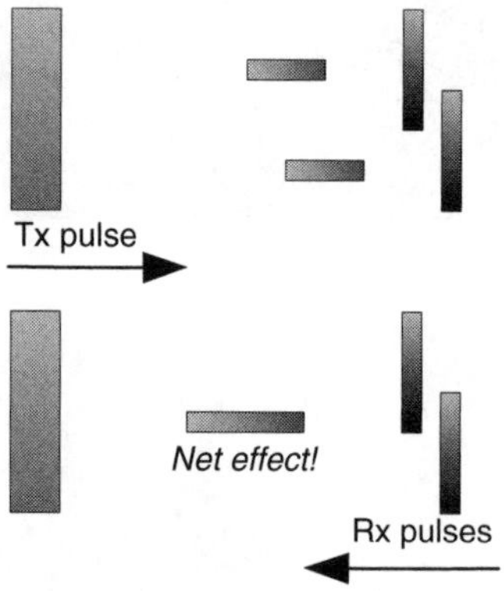

Figure 5.23 *Example of inability of monotonic sonar to distinguish between close proximity targets.*

In a conventional monotonic system at moderate range, a typical pulse length is 50 μs and velocity of sound in water (VOS) 1500 m/s (typical). Therefore, the range resolution = 37.5 mm.

This result effectively determines the range resolution (or ability to resolve separate targets) of the monotonic acoustic imaging system (Figure 5.23).

Using the example above, if two targets are less than 37.5 mm apart then they cannot be distinguished from each other. The net effect is that the system will display a single large target, rather than multiple smaller targets.

CHIRP signal processing overcomes these limitations. Instead of using a burst of a single carrier frequency, the frequency within the burst is swept over a broad range throughout the duration of transmission pulse. This creates a 'signature' acoustic pulse; The sonar knows what was transmitted and when. Using 'pattern-matching' techniques, it can now look for its own unique signature being echoed back from targets.

In a CHIRP system, the critical factor determining range resolution is now the bandwidth of the CHIRP pulse. The CHIRP range resolution is given by:

$$\text{Range resolution} = 2 \times \text{Velocity of sound bandwidth}.$$

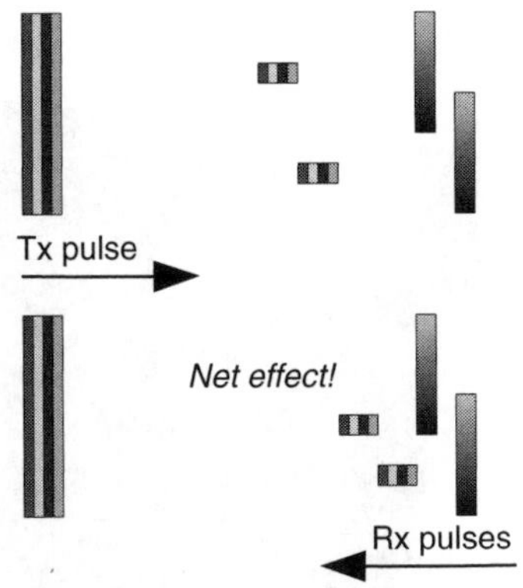

Figure 5.24 *CHIRP sonar successfully distinguishes between close proximity targets.*

The bandwidth of a typical commercial CHIRP system is 100 kHz.

With velocity of sound in water (VOS) 1500 m/s (typical), our new range resolution = 7.5 mm – a theoretical improvement by a factor of 5!

This time, when two acoustic echoes overlap, the signature CHIRP pulses do not merge into a single return. The frequency at each point of the pulse is different and the sonar is able to resolve the two targets independently (Figure 5.24).

The response from the 'pattern-matching' algorithms in the sonar results in the length of the acoustic pulse no longer affecting the amplitude of the echo on the sonar display. Therefore, longer transmissions (and operating ranges) can be achieved without a loss in range resolution.

Additionally, CHIRP offers improvements in background noise rejection, as the sonar is only looking for a swept frequency echo, and removes random noise or out-of-band noise.

In summary, CHIRP techniques provide the following advantages:

- Greatly improved range resolution compared to fixed-frequency sonars
- Larger transmission pulse lengths for increased operating ranges
- Improved discrimination between closely spaced targets
- Improved noise rejection and signal-to-noise ratios
- Reduced power consumption, from high-speed digital circuitry.

5.2.5.2 *Acoustic lens sonar*

For an explanation of acoustic lens sonars, the concept will first be introduced followed by an example of a commercially available product using this technology.

From Loeser (1992) comes a general explanation of acoustic lens technology:

> *'Acoustic lenses simplify the beam-forming process. The liquid lens is a spherical shell, filled with a refractive medium, that focuses sound energy in the same manner an optical lens focuses light energy. Sound waves incident on the lens are refracted to form a high intensity focal region. The refraction is caused by a difference in acoustic*

Figure 5.25 *The DIDSON acoustic lens sonar.*

wave speed in the lens media and surrounding water. The focusing ability is set by its diameter as measured in wavelengths for the frequency of interest.

A single hydrophone located in the focal region of the lens forms a highly directive beam pattern without the necessity of auxiliary beam-forming electronics.'

The Sound Metrics name for their acoustic lens products are MIRIS and DIDSON, referenced here as an example of the acoustic lens technology (Figure 5.25).

MIRIS and DIDSON use acoustic lenses to form very narrow beams during transmission of pulses and reception of their echoes. Conventional sonars use delay lines or digital beam-forming techniques on reception and generally transmit one wide beam on transmission that covers the entire field of view. Acoustic lenses have the advantage of using no power for beam forming, resulting in a sonar that requires only 30 watts to operate. A second advantage is the ease to transmit and receive from the same beam. The selective dispersal of sound and two-way beam patterns make the images cleaner due to reduced acoustic crosstalk and sharper due to higher resolution.

Lenses

Figure 5.26 shows a photograph of a DIDSON sonar with the lens housing removed. The acoustic lenses and transducer array are shown above the electronics housing. The small cylinder under the side of the lens housing contains the focus motor and mechanism that moves the second lens forward and aft. This movement focuses the sonar on objects at ranges from 1 to 40 m. The front lens is actually a triplet made with two plastic (polymethylpentene) and one liquid (3M FC-70) components. The plastic lenses as well as the transducer array are separated by ambient water when the lens system is submerged. Optical lens design programs determined the lens curvatures. The designs were analyzed by custom software to evaluate the beam-patterns over the field of view of interest.

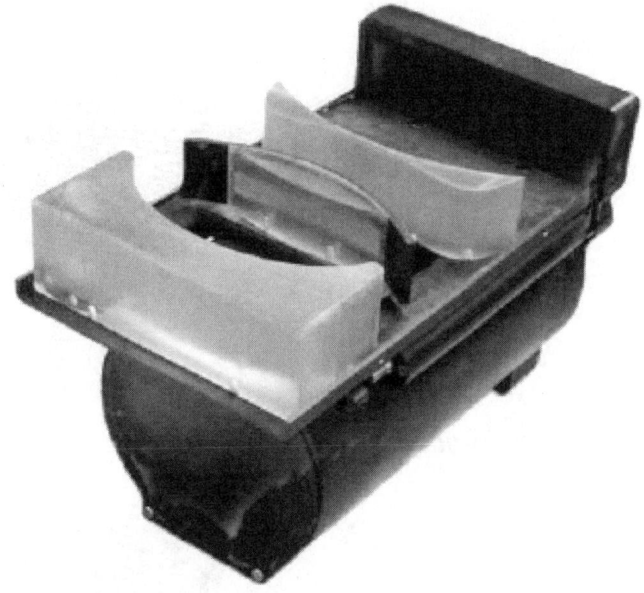

Figure 5.26 *DIDSON internal workings.*

Transducer array

The DIDSON and MIRIS transducer arrays are linear arrays. DIDSON has 96 elements with a pitch of 1.6 mm and a height of 46 mm. The elements are made with PZT 3:1 composite material constructed by the dice-and-fill method. The 3:1 composite provides a wide bandwidth, allowing DIDSON to operate at 1.8 or 1.0 MHz, the upper and lower ends of the transducer passband. The composite also allows the transducers to be curved in the height direction to aid in the formation of the elevation beam pattern. All 96 elements are used when operating at 1.8 MHz. Only the 'even' 48 elements are used when operating at 1.0 MHz.

Beam Formation

Figure 5.27 shows a ray diagram of the MIRIS system. A plane wave entering the left side through the front triplet L1 and single lenses L2 and L3 is focused to a line perpendicular to the page at the transducer T. If the normal to the plane wave is perpendicular to the front lens surface at the center, the acoustic line is formed at 0° in the diagram. If the normal is 9° off from perpendicular, the line is formed at 9° in the diagram. When a focused line of sound coincides with a long, thin transducer element, the acoustic energy is transformed into electrical energy and processed. The DIDSON beam former loses approximately 10 dB in sensitivity each way with beams 15° off-axis. Even with this reduction, DIDSON images fill the 29° field of view as shown in Figure 5.28. The average beam width in the vertical direction for both MIRIS and DIDSON is 14° (one-way). The lenses form the horizontal beam width and the curved transducer element forms the vertical beam width.

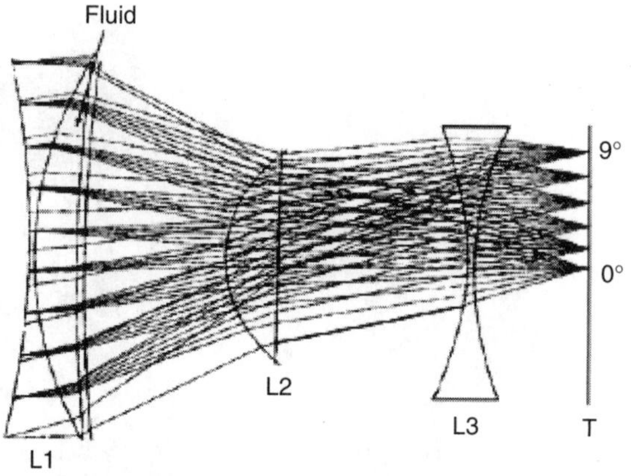

Figure 5.27 *MIRIS internal workings.*

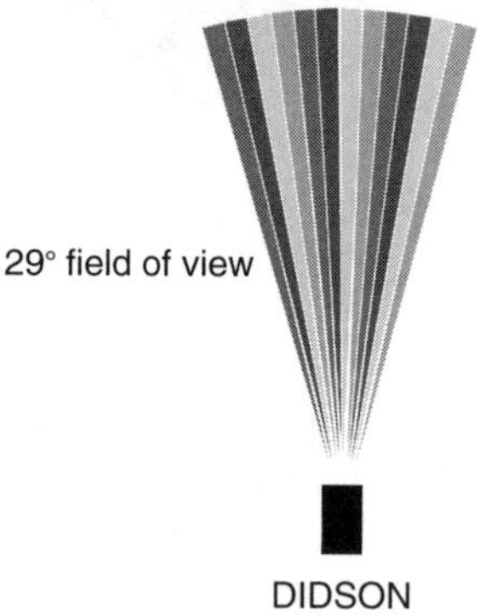

Figure 5.28 *DIDSON beam forming.*

Image Formation

The sonar transmits a short pulse and then receives its echo as it sweeps along the stripe. The echo amplitude varies in time as the reflectance varies with range along the ensonified surface. Echoes from 96 adjacent lines, which together map the reflectance of the ensonified sector-shaped area, are used to form a DIDSON image.

The difference between optical video and images from these sonars is more than the usage of light or sound. The sonar must be oriented to project beams with a small grazing angle to the surface of interest. The resulting image appears to be viewed from a direction perpendicular to the surface and the shadows indicate a source off to the side. Optical video could image a surface with the camera view perpendicular to the surface. If the sonar beams were perpendicular to the surface, the resulting image would show a single line perpendicular to the center beam axis. The line would be located at the range the surface is from the sonar. The sonar images are formed with 'line-focused' beams that provide good images in many conditions, but not in all conditions. If objects were at the same range in the same beam but at different

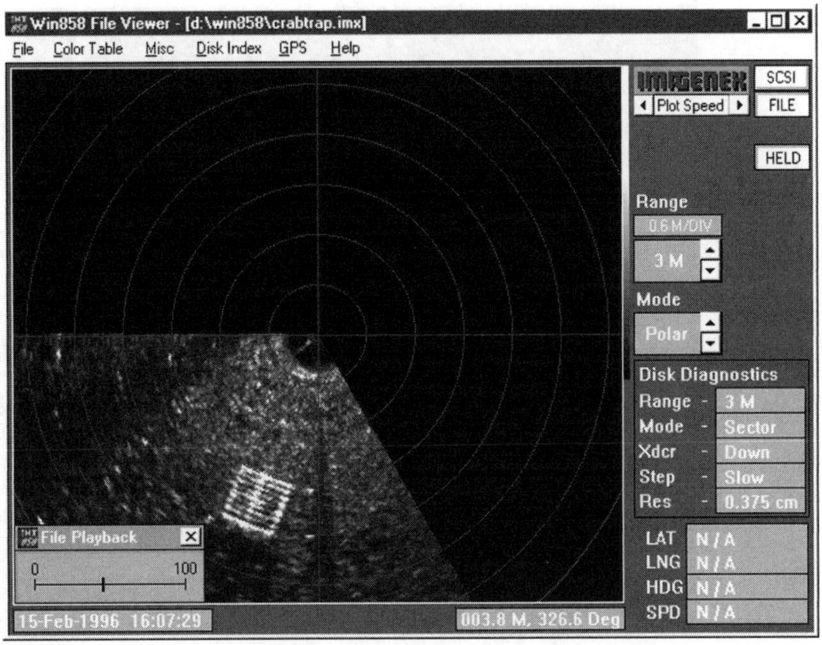

Figure 5.29 *Sonar scan of a crab trap with imaging sonar.*

elevations, this type of imaging could not sort them out. An example would be trying to image an object in a pile of debris on the ocean floor. If the object were imbedded in the pile, the acoustic images from MIRIS and DIDSON would be confusing. Video using 'point-focused' optics could meaningfully image the object imbedded in the pile as long as it was not totally covered.

Fortunately, the great majority of imaging tasks does not have multiple objects in the same beam, at the same range, but at different elevations. In most cases, MIRIS and DIDSON provide unambiguous, near-photographic quality images.

5.3 SONAR TECHNIQUES

5.3.1 Using an Imaging Sonar on an ROV

The imaging sonar is a useful addition to a positioning system on an ROV. Without an imaging sonar, an ROV operator must rely on flying the submersible underwater to bring new targets into view. With an imaging sonar, instead of traveling, it is more useful to spend some time with the vehicle sitting on the bottom while the sonar scans the surrounding area. Scanning a large area takes only a short time. The vehicle pilot can quickly assess the nature of the surrounding area, thereby eliminating objects that are not of interest. The ability to 'see' a long distance underwater allows the pilot to use natural (or man-made) features and targets as position references (Figures 5.29 and 5.30).

If the ROV pilot is searching for a particular object, recognition can take place directly from the sonar image. In other cases, a number of potential targets may

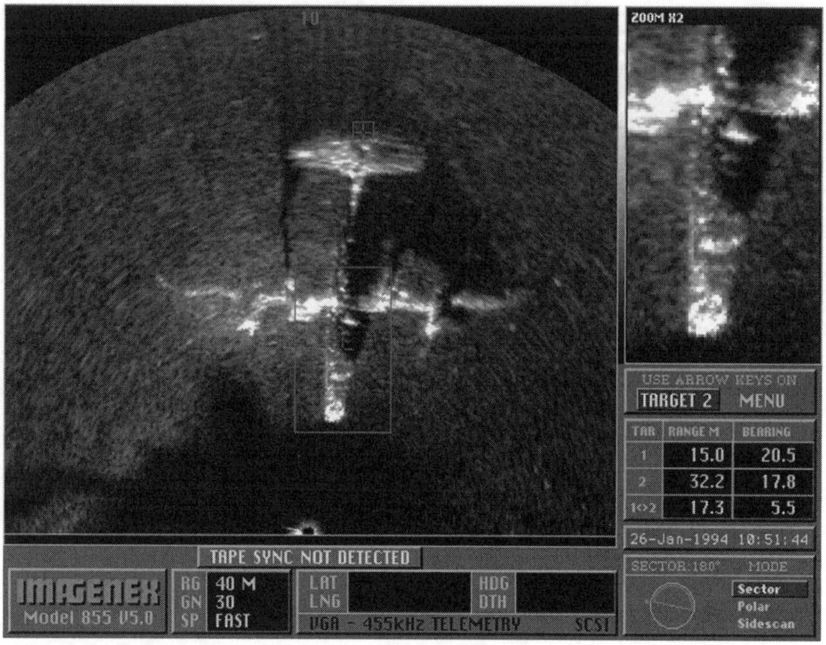

Figure 5.30 *Convair PB4Y-2 Privateer in Lake Washington, WA.*

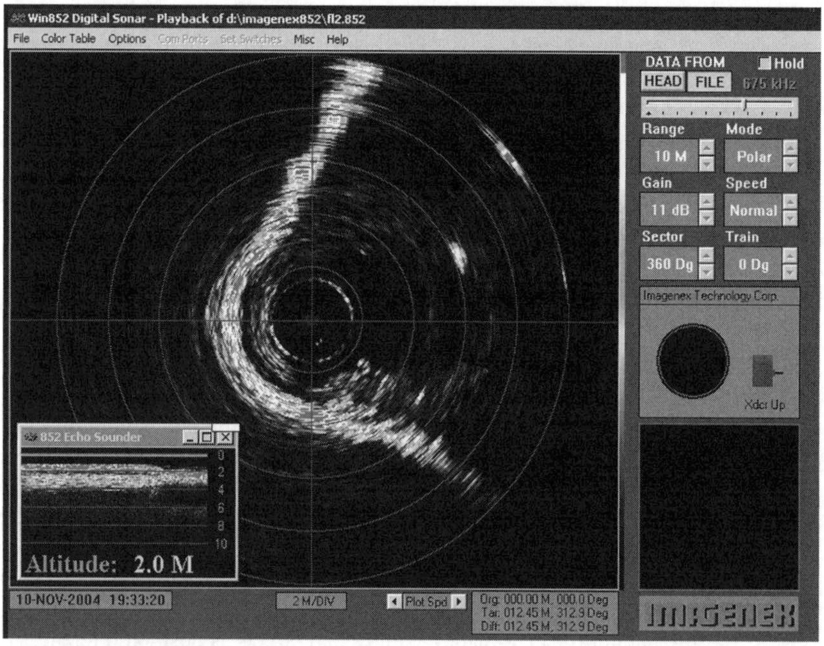

Figure 5.31 *This image displays the sonar image of a straight wall distorted due to movement of the submersible before image was allowed to generate.*

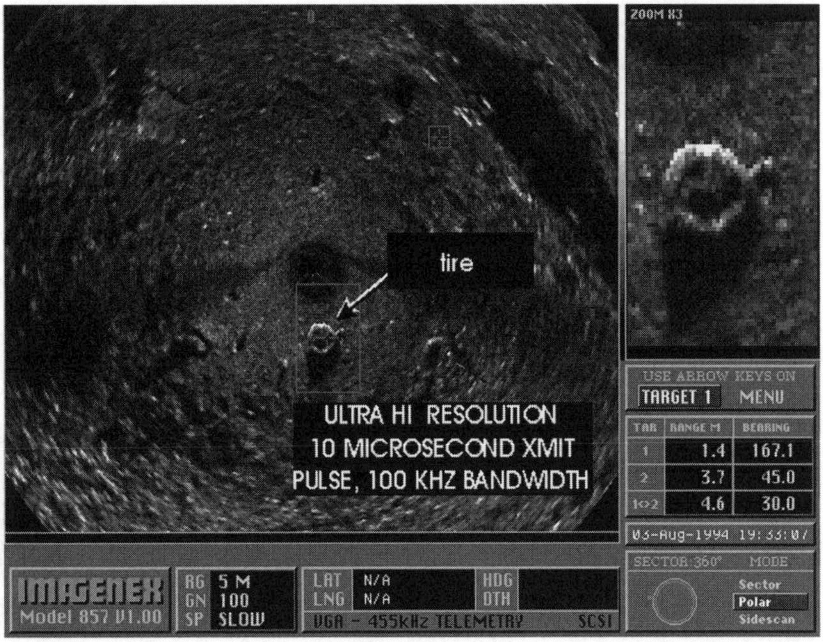

Figure 5.32 *Ultra-high-resolution sonar image.*

be seen. A pilot can sharpen sonar interpretation skills by viewing these targets with the vehicle's video camera in order to correctly identify/characterize them.

A word of caution regarding mechanically scanning sonar systems mounted on small ROVs. An image is generated as the sonar transducer is rotated around its axis. If the sonar platform, i.e. the submersible, is moved before the image is allowed to generate, 'image smear' will occur (Figure 5.31). This phenomenon distorts the 2D display, which will not depict the correct placement of the items on the screen in x/y perspective.

5.3.2 Technique for Locating Targets with ROV-Mounted Scanning Sonar

The accepted technique for locating then identifying items of interest insonified by sonar is a four-step process:

- Place the vehicle in a very stable position on or near the bottom to allow generation of a 360° (or wide angle) image. Depending upon the range, sector angle, and scan speed selected, the image may take up to 30 seconds to generate (Figure 5.32).
- Identify the relative bearing of the item of interest.
- Turn the vehicle to place the item at a zero bearing (in line with the bow of the vehicle), then narrow the sector scan of the sonar system to approximately 45°. By narrowing the sector, the sonar scans only the area of the target to maintain contact.
- Maintain contact with the item on sonar as the vehicle is driven forward toward the target until the item is in view.

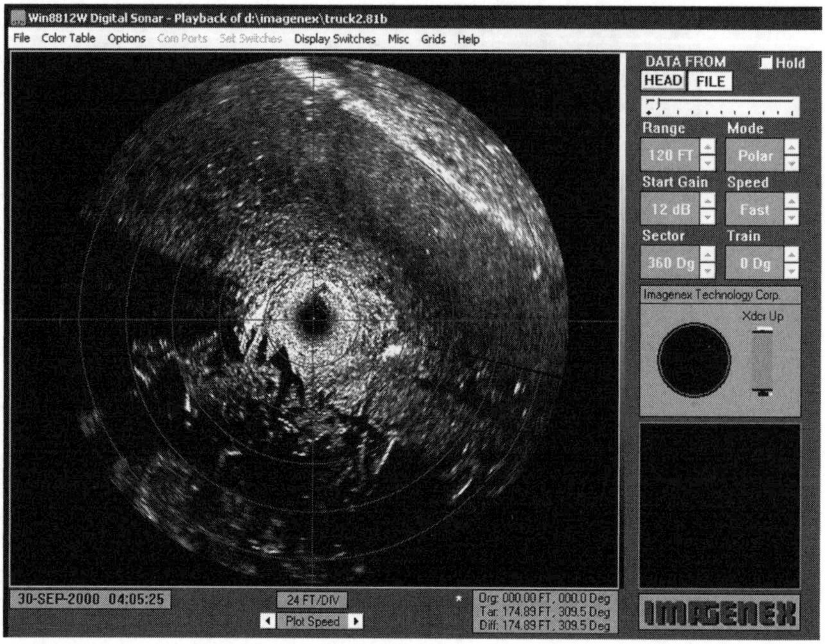

Figure 5.33 *On the left side of the sonar display is a Ford F-150 pickup truck in 40 feet of water completely reflecting sonar signal (only shadow is recognized).*

5.3.3 Interpretation of Sonar Images

In many cases, the sonar image of a target will closely resemble an optical image of the same object. In other cases, the sonar image may be difficult to interpret and unlike the expected optical image.

The scanning process used to create a sonar image is different from the process used by the human eye or a camera to produce optical images. A sonar image will always have less resolution than an optical image, due to the nature of the ultrasonic signals used to generate it. Generally, rough objects reflect sound well in many directions and are therefore good sonar targets. Smooth angular surfaces may give a very strong reflection in one particular direction, but almost none at all in other directions (Figure 5.33). They can also act as a perfect mirror (so-called specular reflectors), reflecting the sonar pulse off in unexpected directions, never to return. This happens to people visually when they see an object reflected in a window. The human eye deals with such reflections daily, but it is unexpected to see the same thing occur with a sonar image.

As with normal vision, it is useful to scan targets from different positions to help identify them. A target unrecognizable from one direction may be easy to identify from another. It is important to note that the ranges shown to the targets on the sonar image are 'slant' ranges. Usually the relative elevations of the targets are not known, only the range from the transducer. This means that two targets, which are displayed in the same location on the screen, may be at different elevations. For example, you might see a target on the bottom and a target floating on the surface in the same place (Figure 5.34).

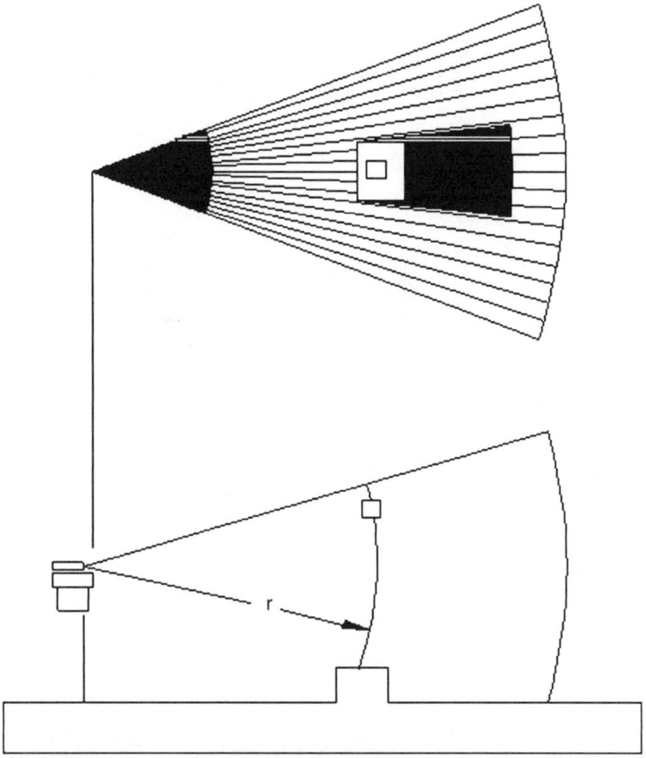

(a)

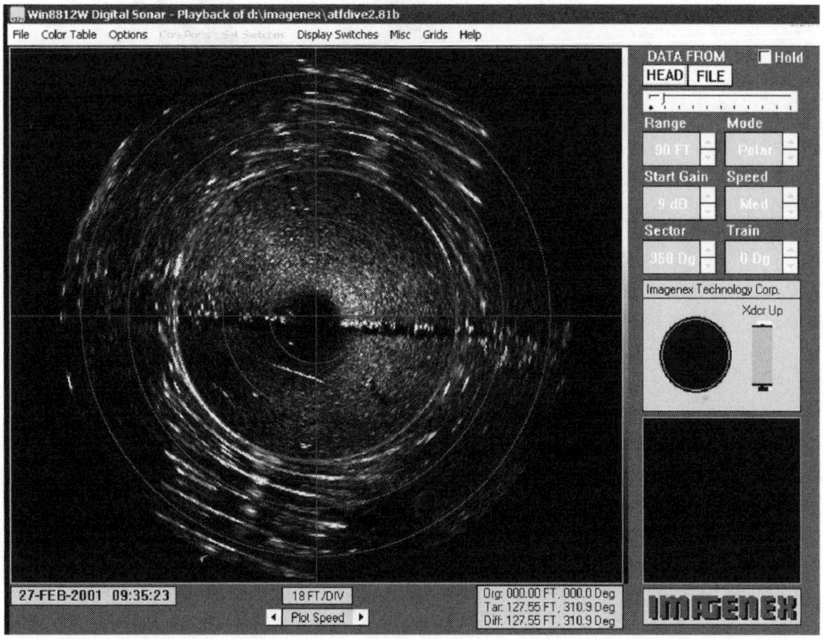

(b)

Figure 5.34 *Graphic illustration of two targets (one on surface and one on bottom) displaying co-location.*

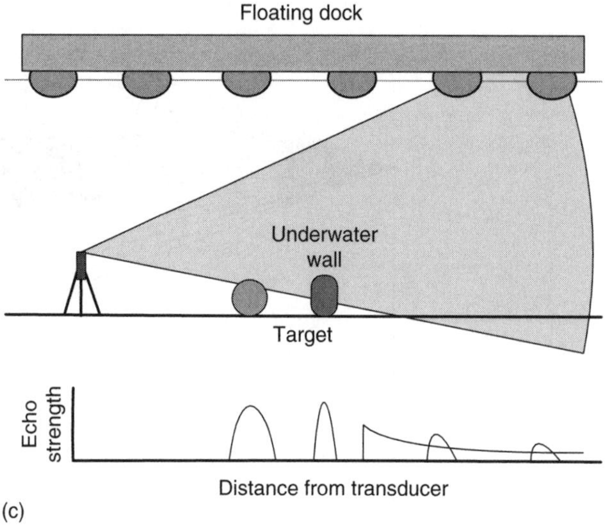

Figure 5.34 *(Continued)*

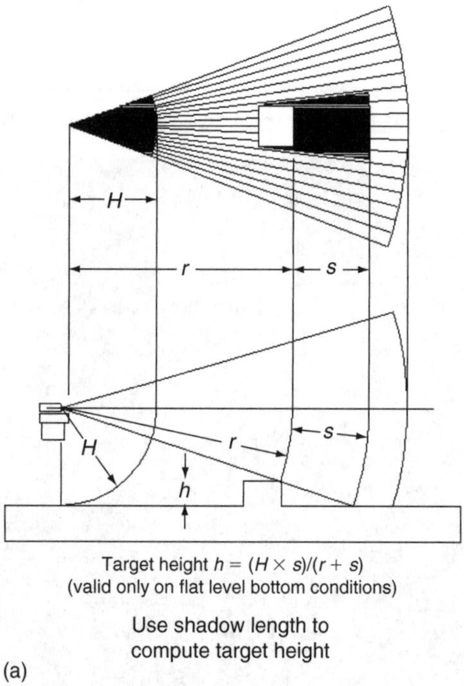

Figure 5.35 *Target height = (length of shadow × height of transducer) divided by range to the end of shadow.*

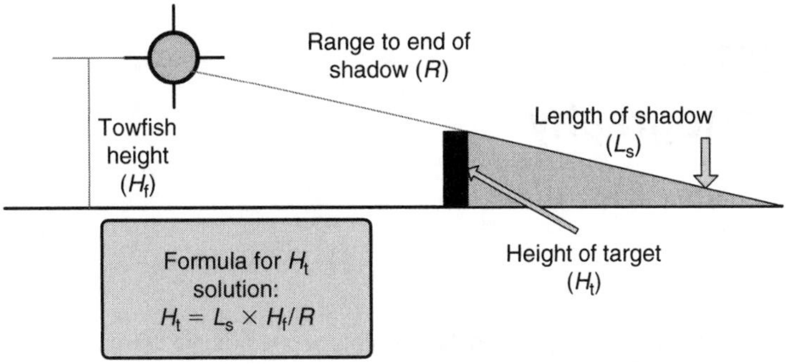

(b)

Figure 5.35 *(Continued)*

By analyzing the shadows, an estimation of the height of objects above the bottom can be ascertained. An example of this calculation is shown in Figure 5.35.

The diagrams in this chapter are examples of the sonar scanning process. Studying the diagrams will help to better understand the images seen. A basic knowledge of this process will help users interpret what otherwise might be confusing images.

Chapter 6
Oceanography

In order to comprehend the concepts of operating in the ocean world, an understanding of the details of this environment is needed. The content of this chapter explores the makeup of fresh water and sea water, then goes into the interaction of this substance with the world of robotics. We will explore the basic concepts of water density, ocean circulation, currents, tides, and how each of these affect the operation of ROV equipment. Armed with a general knowledge of oceanography, work site predictions may be made on such variables as turbidity (affecting camera optics), temperature/salinity (affecting acoustic equipment and vehicle buoyancy), tide and current flows (affecting drag computations on the submersible/tether combination), and dissolved gases (affecting biological population). This section condenses information from complete college curriculums; Therefore, for further details, please consult the references in the bibliography. Special thanks goes to Steve Fondriest of Fondriest Environmental, Inc. for his contribution to the fundamentals of environmental monitoring and data collection instrumentation.

6.1 DISTRIBUTION OF WATER ON EARTH

Earth is the only planet known to have water resident in all three states (solid, liquid, and gas). It is also the only planet to have known liquid water currently at its surface. Distribution of the earth's water supply is provided in Table 6.1.

As shown in Table 6.1, most (97 percent) of the world's water supply is in the oceans. Water can dissolve more substances (and in greater quantities) than any other liquid. It is essential to sustain life, a moderator of our planet's temperature, a major contributor to global weather patterns, and is, of course, essential for operation of an ROV.

The oceans cover 70.8 percent of the earth's surface, far overreaching earth's land mass. Of the ocean coverage, the Atlantic covers 16.2 percent, the Pacific 32.4 percent, the Indian Ocean 14.4 percent, and the margin and adjacent areas the balance of 7.8 percent. It is also interesting to note that the Pacific Ocean alone covers 3.2 percent more surface area on earth than all of the land masses combined.

6.2 PROPERTIES OF WATER

Water is known as the 'universal solvent'. While pure water is the basis for life on earth, as more impurities are added to that fluid the physical and chemical properties change drastically. The chemical makeup of the water mixture in which the ROV operates will directly dictate operational procedures and parameters if a successful operation is to be achieved.

Table 6.1 *Earth's water supply.*

Water source	*Water volume, in cubic miles*	*Water volume, in cubic kilometers*	*Percent of freshwater*	*Percent of total water*
Oceans, seas, and bays	321 000 000	1 338 000 000	–	96.5
Ice caps, glaciers, and permanent snow	5 773 000	24 064 000	68.7	1.74
Groundwater				
Fresh	2 526 000	10 530 000	30.1	0.76
Saline	3 088 000	12 870 000	–	0.94
Soil moisture	3959	16 500	0.05	0.001
Ground ice and permafrost	71 970	300 000	0.86	0.022
Lakes				
Fresh	21 830	91 000	0.26	0.007
Saline	20 490	85 400	–	0.006
Atmosphere	3095	12 900	0.04	0.001
Swamp water	2752	11 470	0.03	0.0008
Rivers	509	2120	0.006	0.0002
Biological water	269	1120	0.003	0.0001

Source: US Geological Survey.

Two everyday examples of water's physical properties and their effect on our lives are (1) ice floats in water and (2) we salt our roads in wintertime to 'melt' snow on the road. Clearly, it is important to understand the operating environment and its effect on ROV operations. To accomplish this, the properties and chemical aspects of water and how they're measured will be addressed to determine their overall effect on the ROV.

The early method of obtaining environmental information was by gathering water samples for later analysis in a laboratory. Today, the basic parameters of water are measured with a common instrument named the 'CTD sonde'. Some of the newest sensors can analyze a host of parameters logged on a single compact sensing unit.

Fresh water is an insulator, with the degree of electrical conductivity increasing as more salts are added to the solution. By measuring the water's degree of electrical conductivity, a highly accurate measure of salinity can be derived. Temperature is measured via electronic methods and depth is measured with a simple water pressure transducer. The CTD probe measures 'conductivity/temperature/depth', which are the basic parameters in the sonic velocity equation. Newer environmental probes are available for measuring any number of water quality parameters such as pH, dissolved oxygen and CO_2, turbidity, and other parameters.

The measurable parameters of water are needed for various reasons. A discussion of the most common measurement variables the commercial or scientific ROV pilot will encounter, the information those parameters provide, and the tools/techniques to measure them follow.

6.2.1 Chlorophyll

In various forms, chlorophyll is bound within the living cells of algae and other phytoplankton found in surface water. Chlorophyll is a key biochemical component in the molecular apparatus that is responsible for photosynthesis, the critical process in which the energy from sunlight is used to produce life-sustaining oxygen. In the photosynthetic reaction, carbon dioxide is reduced by water, and chlorophyll assists this transfer.

6.2.2 Circulation

The circulation of the world's water is controlled by a combination of gravity, friction, and inertia. Winds push water, ice, and water vapor around due to friction. Water vapor rises. Fresh and hot water rise. Salt and cold water sink. Ice floats. Water flows downhill. The high-inertia water at the Equator zooms eastward as it travels toward the slower-moving areas near the Poles (coriolis effect – an excellent example of this is the Gulf Stream off the East Coast of the USA). The waters of the world intensify on the Western boundary of oceans due to the earth's rotational mechanics (the so-called 'Western intensification' effect). Add into this mix the gravitational pull of the moon, other planets and the sun, and one has a very complex circulation model for the water flowing around our planet.

In order to break this complex model into its component parts for analysis, oceanographers generally separate these circulation factions into two broad categories, 'currents' and 'tides'. Currents are broadly defined as any horizontal movement of water along the earth's surface. Tides, on the other hand, are water movement in the vertical plane due to periodic rising and falling of the ocean surface (and connecting bodies of water) resulting from unequal gravitational pull from the moon and sun on different parts of the earth. Tides will cause currents, but tides are generally defined as the diurnal and semi-diurnal movement of water from the sun/moon pull.

A basic understanding of these processes will arm the ROV pilot with the ability to predict conditions at the work site, thus assisting in accomplishing the work task.

Per Bowditch (2002), 'currents may be referred to according to their forcing mechanism as either wind-driven or thermohaline. Alternatively, they may be classified according to their depth (surface, intermediate, deep, or bottom). The surface circulation of the world's oceans is mostly wind driven. Thermohaline currents are driven by differences in heat and salt. The currents driven by thermohaline forces are typically subsurface.' If performing a deep dive with an ROV, count on having a surface current driven by wind action and a subsurface current driven by thermohaline forces – plan for it and it will not ruin the day.

An example of the basic differences between tides and currents follow:

- In the Bay of Fundy's Minas Basin (Nova Scotia, Canada), the highest tides on planet earth occur near Wolfville. The water level at high tide can be as much as 45 feet (16 meters) higher than at low tide! Small Atlantic tides drive the Bay of Fundy/Gulf of Maine system near resonance to produce the huge tides. High tides happen every 12 hours and 25 minutes (or nearly an hour later each day) because of the changing position of the moon in its orbit around the earth. Twice a day at this location, large ships are alternatively grounded and floating. This is an extreme example of tides in action.

- At the Straits of Gibraltar, there is a vertical density current through the Straits. The evaporation of water over the Mediterranean drives the salinity of the water in that sea slightly higher than that of the Atlantic Ocean. The relatively denser high salinity waters in the Mediterranean flow out of the bottom of the Straits while the relatively lower (less dense) salinity waters from the Atlantic flow in on the surface. This is known as a density current. Trying to conduct an ROV operation there will probably result in a very bad day.
- Currents flow from areas of higher elevation to lower elevation. By figuring the elevation change of water over the area, while computing the water distribution in the area, one can find the volume of water that flows in currents past a given point (volume flow) in the stream, river, or body of water. However, the wise operator will find it much easier to just look it up in the local current/tide tables. There are people who are paid to make these computations on a daily basis, which is great as an intellectual exercise, but it is not recommended to 'recreate the wheel'.

6.2.2.1 *Currents*

The primary generating forces for ocean currents are wind and differences in water density caused by variations in heat and salinity. These factors are further affected by the depth of the water, underwater topography, shape of the basin in which the current is running, extent and location of land, and the earth's rotational deflection. The effect of the tides on currents is addressed in the next section.

Each body of water has its peculiar general horizontal circulation and flow patterns based upon a number of factors. Given water flowing in a stream or river, water accelerates at choke points and slows in wider basins per the equations of Bernoulli. Due to the momentum of the water at a river bend, the higher volume of water (and probably the channel) will be on the outside of the turn. Vertical flow patterns are even more predictable with upwelling and downwelling patterns generally attached to the continental margins.

Just as there are landslides on land, so are there mudslides under the ocean. Mud and sediment detach from a subsea ledge and flow downhill in the oceans, bringing along with it a friction water flow known as a turbidity current. Locked in the turbidity current are suspended sediments. This increase in turbidity can degrade camera optics if operating in an area of turbidity currents – take account of this during project planning.

Currents remain generally constant over the course of days or weeks, affected mostly by the changes in temperature and salinity profiles caused by the changing seasons.

Of particular interest to ROV operators is the wind-driven currents culled into the so-called 'Ekman spiral' (Figure 6.1). The model was developed by physicist V. Walfrid Ekman from data collected by arctic exploration legend Fridtjof Nansen during the voyage of the *Fram*. From this model, wind drives idealized homogeneous surface currents in a motion 45° from the wind line to the right in the Northern hemisphere and to the left in the Southern hemisphere. Due to the friction of the surface water's movement, the subsurface water moves in an ever-decreasing velocity

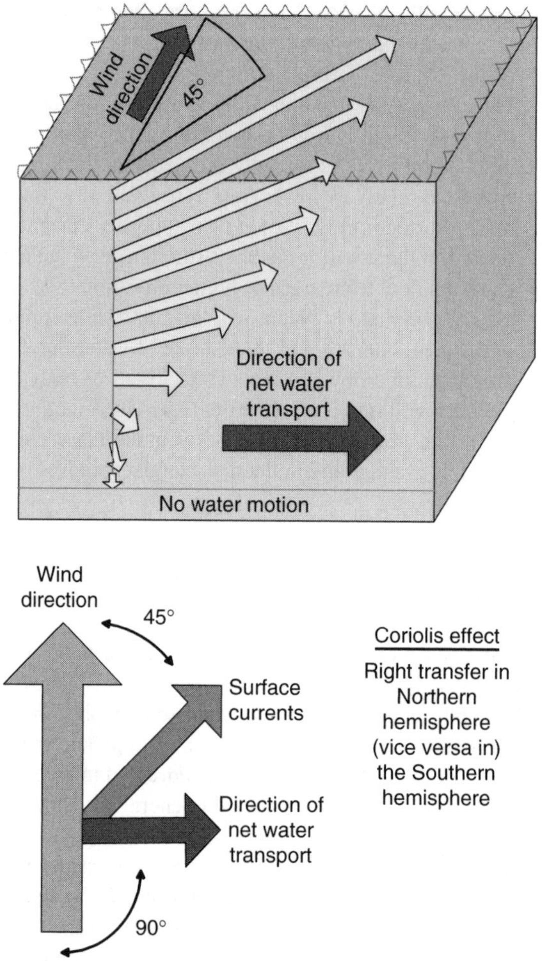

Figure 6.1 *Ekman spiral.*

(and ever increasing vector) until the momentum imparted by the surface lamina is lost (termed the 'depth of frictional influence'). Although the 'depth of frictional influence' is variable depending upon the latitude and wind velocity, the Ekman frictional transfer generally ceases at approximately 100 m depth. The net water transfer is at a right angle to the wind.

6.2.2.2 *Tides*

Tides are generally referred to as the vertical rise and fall of the water due to the gravitational effects of the moon, sun, and planets upon the local body of water (Figure 6.2). Tidal currents are horizontal flows caused by the tides. Tides rise and fall. Tidal currents flood and ebb. The ROV pilot is concerned with the amount and time of the tide, as it affects the drag velocity and vector computations on water flow

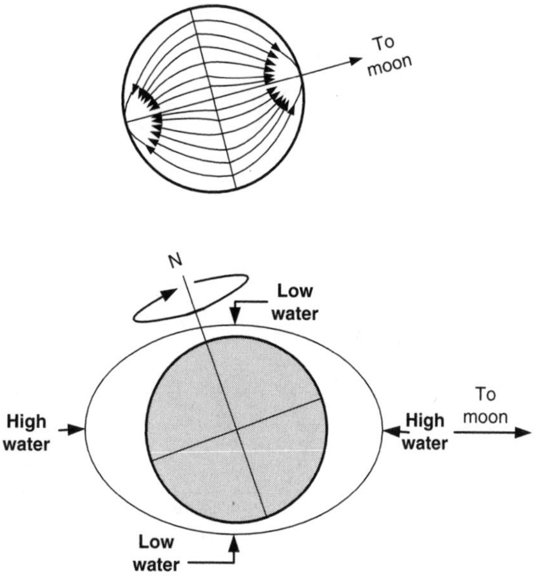

Figure 6.2 *Tidal movement in conjunction with planets.*

across the work site. Tidal currents are superimposed upon non-tidal currents such as normal river flows, floods, and freshets. Put all of these factors together to find what the actual current will be for the job site (Figure 6.3).

On a vertical profile, the tide may interact with the general flow pattern from a river or estuary – the warm fresh water may flow from a river on top of the cold salt water (a freshet, as mentioned above) as the salt water creeps for a distance up the river. According to Van Dorn (1993), fresh water has been reported over 300 miles at sea off the Amazon. If a brackish water estuary is the operating area, problems to be faced will include variations in water density, water flow vector/speed, and acoustic/turbidity properties.

6.2.3 Compressibility

For the purposes of observation-class ROV operations, sea water is essentially incompressible. There is a slight compressibility factor, however, that does directly effect the propagation of sound through water. This compressibility factor will affect the sonic velocity computations at varying depths (see 'Sonic velocity and sound channels' later in this chapter).

6.2.4 Conductivity

Conductivity is the measure of electrical current flow potential through a solution. In addition, because conductivity and ion concentration are highly correlated, conductivity measurements are used to calculate ion concentration in solutions.

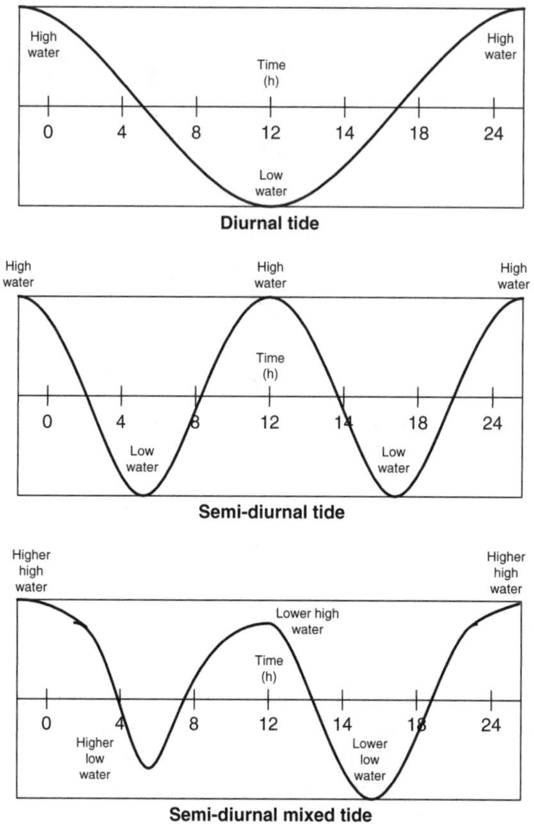

Figure 6.3 *Tide periods over a 24-hour span.*

Commercial and military operators observe conductivity for gauging water density (for vehicle ballasting and such) and for determining sonic velocity profiles (for acoustic positioning and sonar use). Water quality researchers take conductivity readings to determine the purity of water, to watch for sudden changes in natural or waste water, and to determine how the water sample will react in other chemical analyses. In each of these applications, water quality researchers count on conductivity sensors and computer software to sense environmental waters, log and analyze data, and present this data.

6.2.5 Density

Density is mass per unit volume measured in SI units in kilograms per cubic meter (or, on a smaller scale, grams per cubic centimeter). The density of sea water depends upon salinity, temperature, and pressure. At a constant temperature and pressure, density varies with water salinity. This measure is of particular importance to the ROV pilot for determination of neutral buoyancy for the vehicle.

The density range for sea water is from 1.02200 to 1.030000 g/cm^3 (Thurman, 1994). In an idealized stable system, the higher density water sinks to the bottom while the lower density water floats to the surface. Water under the extreme pressure of depth will naturally be denser than surface water, with the change in pressure (through motion between depths) being realized as heat. Just as the balance of pressure/volume/temperature is prevalent in the atmosphere, so is the temperature/salinity/pressure model in the oceans.

A rapid change in density over a short distance is termed a 'pycnocline' and can trap any number of energy sources from crossing this barrier, including sound (sonar and acoustic positioning systems), current, and neutrally buoyant objects in the water column (underwater vehicles). Changing operational area from lower latitude to higher latitude produces a mean temperature change in the surface layer. As stated previously, the deep ocean is uniformly cold due to the higher density cold water sinking to the bottom of the world's oceans. The temperature change from the warm surface at the tropics to the lower cold water can be extreme, causing a rapid temperature swing within a few meters of the surface. This surface layer remains near the surface, causing a small tight 'surface duct' in the lower latitudes. In the higher latitudes, however, the difference between ambient surface temperature and the temperature of the cold depth is less pronounced. The thermal mixing layer, as a result, is much larger (over a broader range of depth between the surface and the isothermal lower depths) and less pronounced (Figure 6.4). In Figure 6.4(a), density profiles by latitude and depth are examined to display the varying effects of deep-water temperature/density profiles versus ambient surface temperatures. Figures 6.4(b) and (c) look at the same profiles only focusing on temperature and salinity. Figure 6.4(d) demonstrates a general profile for density at low to midlatitudes (the mixed layer is water of constant temperature due to the effects of wave mechanics/mixing).

A good example of the effect of density on ROV operations comes from a scientific mission conducted in 2003 in conjunction with *National Geographic* magazine. The mission was to the cenotes (sink holes) of the Northern Yucatan in Mexico. Cenotes are a series of pressure holes in a circular arrangement, centered around Chicxulub (the theoretical meteor impact point), purportedly left over from the K-T event from 65 million years ago that killed the dinosaurs.

The top water in the cenote is fresh water from rain runoff, with the bottom of the cenote becoming salt water due to communication (via an underground cave network) with the open ocean. This column of still water is a near perfect unmixed column of fresh water on top with salt water below. A micro-ROV was being used to examine the bottom of the cenote as well as to sample the salt-water/freshwater (halocline) layer. The submersible was ballasted to the fresh water on the top layer. When the vehicle came to the salt water layer, the submersible's vertical thruster had insufficient downward thrust to penetrate into the salt water below and kept 'bouncing' off the halocline. The submersible had to be re-ballasted for salt water in order to get into that layer and take the measurements, but the vehicle was useless on the way down due to it being too heavily ballasted to operate in fresh water.

6.2.6 Depth

Depth sensors, discussed below, measure the distance from the surface of a body of water to a given point beneath the surface, either the bottom or any intermediate level.

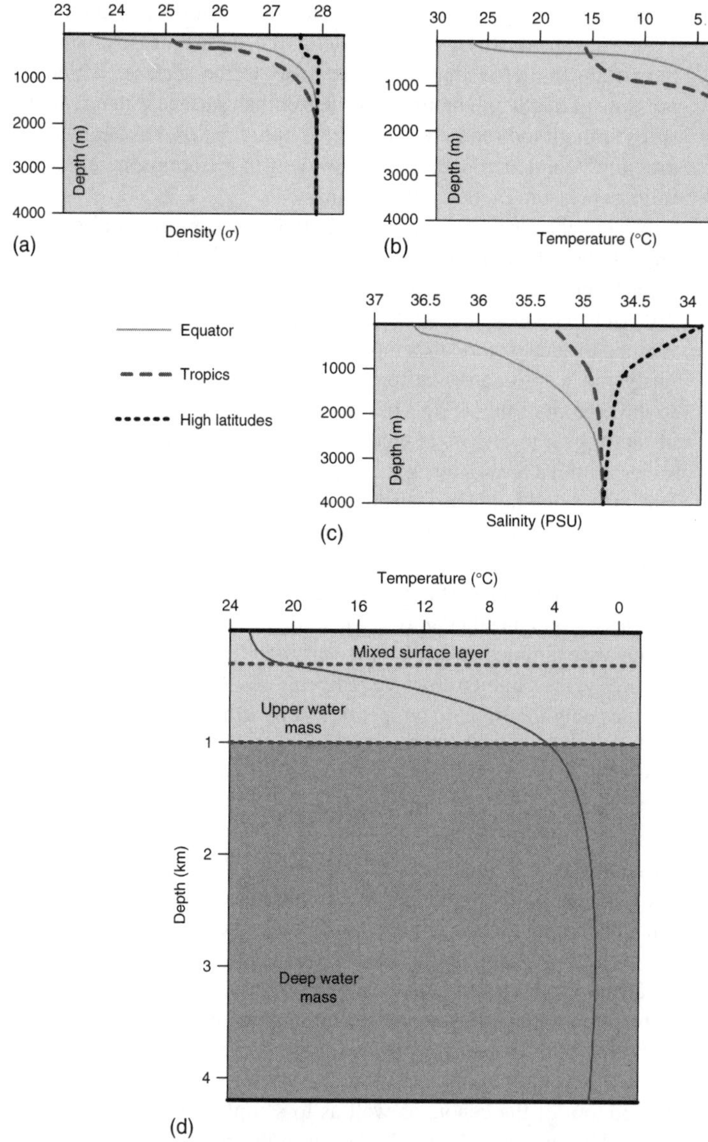

Figure 6.4 *Density profiles with varying latitudes along with a general density curve.*

Depth readings are used by researchers and engineers in coastal and ocean profiling, dredging, erosion and flood monitoring, and construction applications.

Bathymetry is the measurement of depth in bodies of water. Further, it is the underwater version of topography in geography. Bottom contour mapping details the shape of the sea floor, showing the features, characteristics, and general outlay. Tools for bathymetry and sea bottom characterization follow.

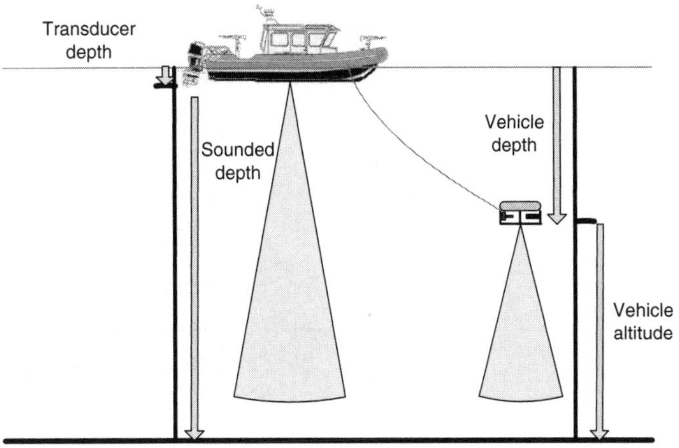

Figure 6.5 *Vessel-mounted and sub-mounted sounders.*

6.2.6.1 *Echo sounder*

An echo sounder measures the round trip time it takes for a pulse of sound to travel from the source at the measuring platform (surface vessel or on the bottom of the submersible) to the sea bottom and return. When mounted on a vessel, this device is generally termed a fathometer and when mounted on a submersible it is termed an altimeter.

According to Bowditch (2002), 'the major difference between various types of echo sounders is in the frequency they use. Transducers can be classified according to their beam width, frequency, and power rating. The sound radiates from the transducer in a cone, with about 50 percent actually reaching the sea bottom. Beam width is determined by the frequency of the pulse and the size of the transducer. In general, lower frequencies produce a wider beam, and at a given frequency, a smaller transducer will produce a wider beam. Lower frequencies penetrate deeper into the water, but have less resolution in depth. Higher frequencies have a greater resolution in depth, but less range, so the choice is a trade-off. Higher frequencies also require a smaller transducer. A typical low-frequency transducer operates at 12 kHz and a high-frequency one at 200 kHz.' Many smaller ROV systems have altimeters on the same frequency as their imaging system for easier software integration (same software can be used for processing both signals) and reduced cost purposes (Figure 6.5).

Computation of depth as determined by an echo sounder is determined via the following formula:

$$D = (V \times T/2) + K + D_r,$$

where D is depth below the surface (or from the measuring platform), V is the mean velocity of sound in the water column, T is time for the round trip pulse, K is the system index constant, and D_r is the depth of the transducer below the surface.

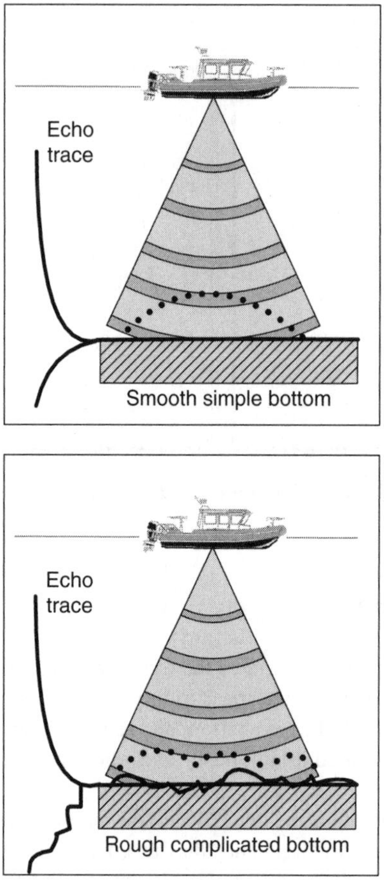

Figure 6.6 *Acoustic seabed classification.*

6.2.6.2 *Optic-acoustic seabed classification*

Traditional sea floor classification methods have, until recently, relied upon the use of mechanical sampling, aerial photography, or multi-band sensors (such as Landsat™) for major bottom-type discrimination (e.g. mud, sand, rock, sea grass, and corals). Newer acoustic techniques for collecting hyperspectral imagery are now available through processing of acoustic backscatter.

Acoustic seabed classification analyzes the amplitude and shape of acoustic backscatter echoed from the sea bottom for determination of bottom texture and makeup (Figure 6.6). Sea floor roughness causes impedance mismatch between water and the sediment. Further, reverberation within the substrate can be analyzed in determining the overall composition of the bottom being insonified. Acoustic data acquisition systems and a set of algorithms that analyze the data allow for determining the seabed acoustic class.

Table 6.2 *Distribution of gases in the atmosphere and dissolved in sea water.*

	Percentage of gas phase by volume		
Gas	*Atmosphere*	*Surface oceans*	*Total oceans*
Nitrogen (N_2)	78	48	11
Oxygen (O_2)	21	36	6
Carbon dioxide (CO_2)	0.04	15	83

Source: Segar (1998).

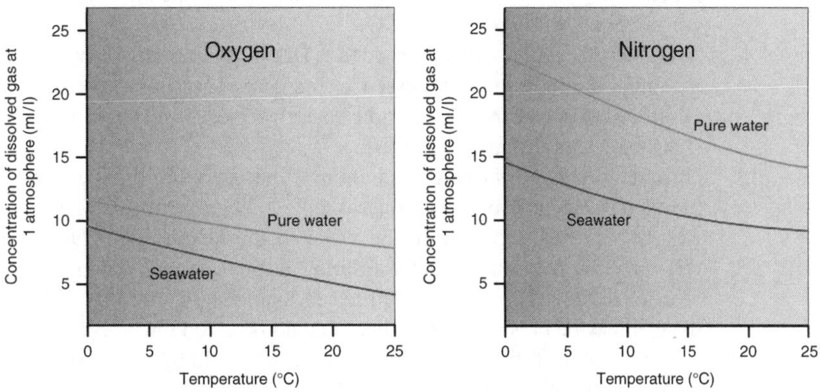

Figure 6.7 *Solubility of two sea-level pressure gases based upon temperature.*

6.2.7 Dissolved Gases

Just as a soda dissolves CO_2 under the pressure of the soda bottle, so does the entire ocean dissolve varying degrees of gases used to sustain the life and function of the aquatic environment. The soda bottle remains at gas/fluid equilibrium under the higher-than-atmospheric pressure condition until the bottle is opened and the pressure within the canister is lowered. At that point, the gas bubbles out of solution until the gas/air mixture comes back into balance. If, however, that same bottle were opened in the high-pressure condition of a saturation diving bell deep within the ocean, that soda would (instead of bubbling) absorb CO_2 into solution until again saturated with that gas.

The degree of dissolved gases within a given area of ocean is dependent upon the balance of all gases within the area. The exchange of gases between the atmosphere and ocean can only occur at the air/ocean interface, i.e. the surface. Gases are dissolved within the ocean and cross the air/water interface based upon the balance of gases between the two substances (Table 6.2). A certain gas is said to be at local saturation if its distribution within the area is balanced given the local environmental conditions. The degree upon which a gas is able to dissolve within the substance is termed its solubility, and in water is dependent upon the temperature, salinity, and pressure of the surrounding fluid (Figure 6.7). Once a substance is at its maximum gas content (given the local environmental conditions) the substance is saturated with that gas. The degree to which a substance can either accept or reject transfer of gas into substance is deemed its saturation solubility.

The net direction of transfer of gases between the atmosphere and ocean is dependent upon the saturation solubility of the water. If the sea is oversaturated with a certain gas, it will off-gas back into the atmosphere (and vice versa). Once equilibrium is again reached, net gas transfer ceases until some environmental variable changes.

Oxygen and CO_2 are dissolved in varying degrees within the world's oceans. Most marine life requires some degree of either of these dissolved gases in order to survive. In shallow water, where photosynthesis takes place, plant life consumes CO_2 and produces O_2. In the deep waters of the oceans, decomposition and animal respiration consumes O_2 while producing CO_2.

Dissolved oxygen, often referred to as DO, is the amount of oxygen that is dissolved in water. It is normally expressed in mg/l (mg/l = ppm) or percent air saturation. Dissolved oxygen can also enter the water by direct absorption from the atmosphere (transfer across the air/water interface).

Aquatic organisms (plants and animals) need dissolved oxygen to live. As water moves past the gills of a fish, microscopic bubbles of oxygen gas, called dissolved oxygen, are transferred from the water to the bloodstream. Dissolved oxygen is consumed by plants and bacteria during respiration and decomposition. A certain level of dissolved oxygen is required to maintain aquatic life. Although dissolved oxygen may be present in the water, it could be at too low a level to maintain life.

The amount of dissolved oxygen that can be held by water depends on the water temperature, salinity, and pressure.

- Gas solubility increases when temperature decreases (colder water holds more O_2).
- Gas solubility increases when salinity decreases (fresh water holds more O_2 than sea water).
- Gas solubility decreases as pressure decreases (less O_2 is absorbed in water at higher altitudes).

There is no mechanism for replenishing the O_2 supply in deep water while the higher pressure of the deep depths allows for greater solubility of gases. As a result, the deep oceans of the world contain huge amounts of dissolved CO_2. The question that remains is to what extent the industrial pollutants containing CO_2 will be controlled before the deep oceans of the world become saturated with this gas. Water devoid of dissolved oxygen (DO) will exhibit lifeless characterization.

Three examples of contrasting DO levels follow:

1. During an internal wreck survey of the USS *Arizona* in Pearl Harbor performed in conjunction with the US National Park Service, it was noted that the upper decks of the wreck were in a fairly high state of metal oxidation. However, as the investigation moved into the lower spaces (less and stable oxygenation due to little or no water circulation), the degree of preservation took a significant turn. On the upper decks, there were vast amounts of marine growth that had decayed the artifacts and encrusted the metal. However, on the lower living

quarters, fully preserved uniforms were found where they were left on that morning over 60 years previous, still neatly pressed in closets and on hangars.

2. During an internal wreck survey of a B-29 in Lake Mead, Nevada (again done with the US National Park Service), the wreck area at the bottom of the reservoir was anaerobic (lacking significant levels of dissolved oxygen). The level of preservation of the wreck was amazing, with instrument readings still clearly visible, aluminum skin and structural members still in a fully preserved state, and the shiny metal data plate on the engine still readable.
3. During the Cenote project in Northern Yucatan (mentioned earlier in this chapter), the fresh water above the halocline was aerobic and alive with all matter of fish, plant, and insect life flourishing. However, once the vehicle descended into the anaerobic salt waters below the halocline, the rocks were bleached, the leaves dropped into the pit were fully preserved and nothing lived.

6.2.8 Fresh Water

A vast majority of the world's freshwater supply is locked within the ice caps and glaciers of the high Arctic and Antarctic regions. Fresh water is vital to man's survival, thus most human endeavors surround areas of fresh water. Due to the shallow water nature of the freshwater collection points, man has placed various items of machinery, structures, and tooling in and around these locations. The observation-class ROV pilot will, in all likelihood, have plenty of opportunity to operate within the freshwater environment.

The properties of water directly affect the operation of ROV equipment in the form of temperature (affecting components and electronics), chemistry (affecting seals, incurring oxidation, and degrading machinery operation), and specific gravity (buoyancy and performance). These parameters will determine the buoyancy of vehicles, the efficiency of thrusters, the amount and type of biological specimens encountered, as well as the freezing and boiling points of the operational environment. The water density will further affect sound propagation characteristics, directly impacting the operation of sonar and acoustic positioning equipment.

Fresh water has a specific gravity of 1.000 at its maximum density. As water temperature rises, molecular agitation increases the water's volume, thus lowering the density of the fluid per unit volume. In the range between 3.98°C and the freezing point of water, the molecular lattice structure (in the form of ice crystals) again increases the overall volume, thus lowering its density per unit volume (remember, ice floats). The point of maximum density for fresh water occurs at 3.98°C (the point just before formation of ice crystals). At the freezing point of water, the lattice structure rapidly completes, thus significantly expanding the volume per unit mass and lowering the density at that temperature point. A graphic describing the temperature/density relationship of pure water is shown in Figure 6.8.

Fresh water has a maximum density at approximately 4°C (see Figure 6.11) yet ocean water has no maximum density above the freezing point. As a result, lakes and rivers behave differently at the freezing point than ocean water. As the weather cools with the approach of winter, the surface water of a freshwater lake is cooled and its density is increased. Surface water sinks and displaces bottom water upward to be cooled in turn. This convection process is called 'overturning'. This overturning

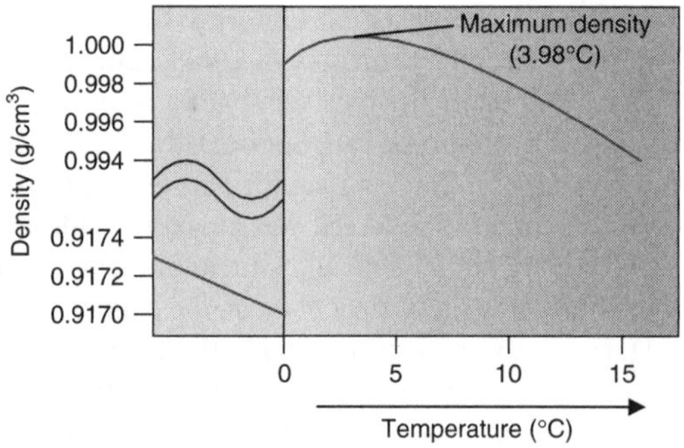

Figure 6.8 *Water density versus temperature.*

process continues until the maximum density is achieved, thus stopping the convection churning process at 4°C. As the lake continues to cool, the crystal structure in the water forms, thus allowing the cooler water at the surface to decrease in density, driving still further the cooler (less dense) surface water upward, allowing ice to form at the surface.

6.2.9 Ionic Concentration

There are four environmentally important ions: nitrate (NO_3^-), chloride (Cl^-), calcium (Ca^{2+}), and ammonium (NH_4^+). Ion-selective electrodes used for monitoring these parameters are described below.

- *Nitrate* (NO_3^-). Nitrate ion concentration is an important parameter in nearly all water quality studies. Nitrates can be introduced by acidic rainfall, fertilizer runoff from fields, and plant or animal decay or waste.
- *Chloride* (Cl^-). This ion gives a quick measurement of salinity of water samples. It can even measure chloride levels in ocean salt water or salt in food samples.
- *Calcium* (Ca^{2+}). This electrode gives a good indication of hardness of water (as Ca^{2+}). It is also used as an endpoint indicator in EDTA-Ca/Mg hard water titrations.
- *Ammonium* (NH_4^+). This electrode measures levels of ammonium ions introduced from fertilizers. It can also indicate aqueous ammonia levels if sample solutions are acidified to convert NH_3 to NH_4^+.

6.2.10 Light and Other Electromagnetic Transmissions Through Water

Light and other electromagnetic transmission through water is affected by the following three factors:

1. Absorption
2. Refraction
3. Scattering.

All of these factors, which can be measured by the ROV using light sensors, are normally lumped under the general category of attenuation.

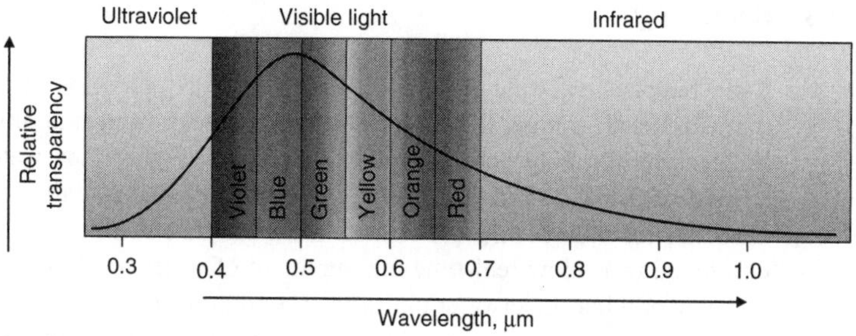

Figure 6.9 *Light transparency through water (by wavelength).*

6.2.10.1 *Absorption*

Electromagnetic energy transmission capability through water varies with wavelength. The best penetration is gained in the visible light spectrum (Figure 6.9). Other wavelengths of electromagnetic energy (radar, very low frequency RF, etc.) are able to marginally penetrate the water column (in practically all cases only a few wavelengths), but even with very high intensity transmissions only very limited transmission rates/depths are possible under current technologies. Submerged submarines are able to get RF communications in deep water with very low frequency RF, but at that frequency it may take literally minutes to get through only two alphanumeric characters.

In the ultraviolet range as well as in the infrared wavelengths, electromagnetic energy is highly attenuated by sea water. Within the visible wavelengths, the blue/green spectrum has the greatest energy transparency, with other wavelengths having differing levels of energy transmission. Disregarding scattering (which will be considered below), within 1 meter of the surface, fully 60 percent of the visible light energy is absorbed, leaving only 40 percent of original surface levels available for lighting and photosynthesis. By the 10-meter depth range, only 20 percent of the total energy remains from that of the surface. By 100 meters, fully 99 percent of the light energy is absorbed, leaving only 1 percent visible light penetration – practically all in the blue/green regime (Duxbury and Alison, 1997). Beginning with the first meter of depth, artificial lighting becomes increasingly necessary to bring out the true color of objects of interest below the surface.

Why not put infrared cameras on ROVs? The answer is simple – the visible light spectrum penetration in water favors the use of optical systems (Figure 6.9). IR cameras can certainly be mounted on ROV vehicles, but the effective range of the sensor suffers significantly due to absorption. The sensor may be effective at determining reflective characteristics, but the sensor would be required to be placed at an extremely close range, negating practically all benefits from non-optical IR reflectance.

6.2.10.2 *Refraction*

Light travels at a much slower speed through water, effectively bending (refracting) the light energy as it passes through the medium. This phenomenon is apparent not only with the surface interaction of the sea water, but also with the air/water interface of the ROV's camera system.

6.2.10.3 *Scattering*

Light bounces off water molecules and suspended particles in the water (scattering), further degrading the light transmission capability (in addition to absorption) by blocking the light path. The scattering agents (other than water molecules) are termed suspended solids (e.g. silt, single-cell organisms, salt molecules, etc.) and are measured in mg/l on an absolute scale. Modern electronic instruments have been developed that allow real-time measurement of water turbidity from the ROV submersible or other underwater platform. The traditional physical measure of turbidity, however, is a simple measure of the focal length of a reflective object as it is lost from sight. Termed Secchi depth, a simple reflective Secchi disk (coated with differing colors and textures) is lowered into the water until it just disappears from view.

All of the above issues and parameters will aid in determining the submersible's capability to perform the assigned task within a reasonable timeframe.

6.2.11 Pressure

The SI unit for pressure is the kilopascal (expressed as kPa). One pascal is equal to one newton per square meter. However, oceanographers normally use ocean pressure with reference to sea level atmospheric pressure. The imperial unit is one atmosphere. The SI unit is the bar. The decibar is a useful measure of water pressure and is equal to 1/10 bar. Sea water generally increases by one atmosphere of pressure for every 33 feet of depth (approximately equaling 10 meters). Therefore, one decibar is approximately equal to one meter of depth in sea water (Figure 6.10).

As an ROV pilot, sea-water pressure directly affects all aspects of submersible operation. The design of the submersible's air-filled components must withstand the pressures of depth, the floatation must stand up to the pressure without significant deformation (thus losing buoyancy and sinking the vehicle), and tethers must be sturdy enough to withstand the depth while maintaining their neutral buoyancy.

6.2.12 Salt Water and Salinity

The world's water supply consists of everything from pure water to water plus any number of dissolved substances due to water's soluble nature. Water quality researchers measure salinity to assess the purity of drinking water, monitor salt water intrusion into freshwater marshes and groundwater aquifers, and to research how the salinity will affect the ecosystem.

The two largest dissolved components of typical sea water are chlorine (56 percent of total) and sodium (31 percent of total), with the total of all lumped under the designation of 'salts'. Components of typical ocean water dissolved salts are comprised of major constituents, minor constituents and trace constituents. An analysis of one kilogram of sea water (detailing only the major constituents of dissolved salts) is provided in Table 6.3.

The total quantity of dissolved salts in sea water is expressed as salinity, which can be calculated from conductivity and temperature readings. Salinity was historically expressed quantitatively as grams of dissolved salts per kilogram of water (expressed

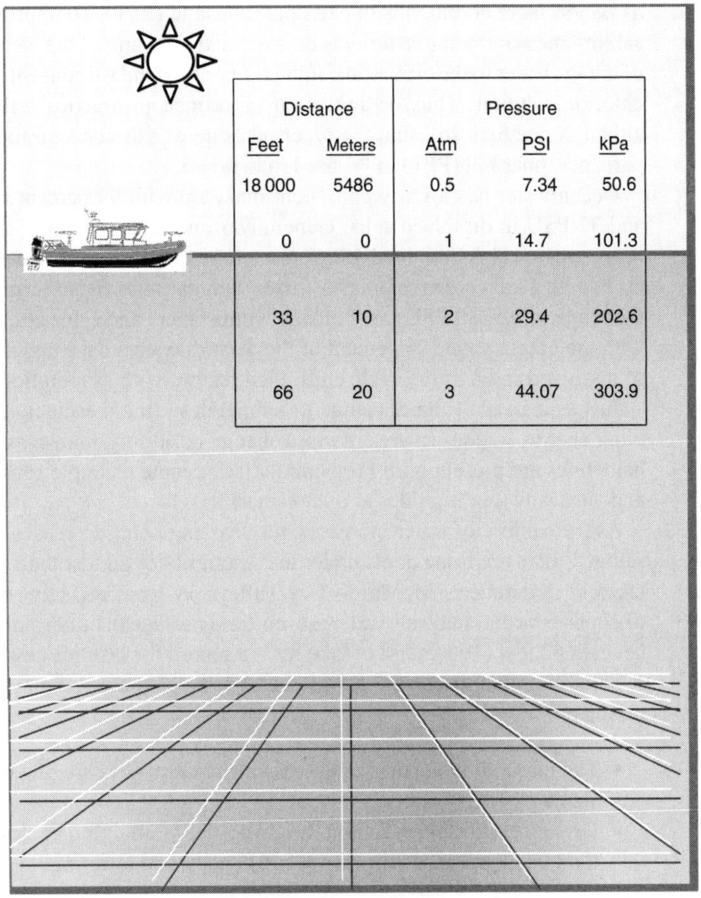

Distance
Pressure
Feet
Meters
Atm
PSI
kPa
18 000 5486 0.5 7.34 50.6
0 0 1 14.7 101.3
33 10 2 29.4 202.6
66 20 3 44.07 303.9

Figure 6.10 *Pressure in atmospheres from various levels.*

Table 6.3 *Dissolved salts in water.*

Component	*Weight in grams*
Pure water	965.31
Major constituents	
Chlorine	19.10
Sodium	10.62
Magnesium	1.28
Sulfur	2.66
Calcium	0.40
Potassium	0.38
Minor constituents	0.24
Trace constituents	0.01
Total (in grams)	1000.00

as ‰) or, more commonly, in parts per thousand (PPT). To improve the precision of salinity measurements, salinity is now defined as a ratio of the electrical conductivity of the sea water to the electrical conductivity of a standard concentration of potassium chloride solution. Thus, salinity is now defined in practical salinity units (PSU), although one may still find the older measure of salt concentration in a solution as parts per thousand (PPT) or ‰ used in the field.

Ocean water has a fairly consistent makeup, with 99 percent having between 33 and 37 PSU in dissolved salts. Generally, rain enters the water cycle as pure water then gains various dissolved minerals as it travels toward the ocean. Water enters the cycle with a salt content of 0 PSU, mixes with various salts to form brackish water (in the range of 0.5–30 PSU) as it blends within rivers and estuaries, then homogenizes with the ocean water (75 percent of the ocean's waters have between 33 and 34 PSU of dissolved salts) as the cycle ends, then renews with evaporation.

Just as a layer of rapid change in temperature (the thermocline) traps sound and other energy, so does an area of rapid change in salinity, known as a halocline. These haloclines are present both horizontally (see cenote example earlier in this chapter) and vertically (e.g. rip tides at river estuaries).

As the salinity of water increases, the freezing point decreases. As an anecdote to salinity, there are brine pools under the Antarctic ice amidst the glaciers in the many lakes of Antarctica's McMurdo Dry Valleys. A team recently found a liquid lake of super-concentrated salt water, seven times saltier than normal sea water, locked beneath 62 feet (19 meters) of lake ice – a record for lake ice cover on earth.

Salts dissolved in water change the density of the resultant sea water for these reasons:

- The ions and molecules of the dissolved substances are of a higher density than water.
- Dissolved substances inhibit the clustering of water molecules (particularly near the freezing point), thus increasing density and lowering the freezing point.

Unlike fresh water, ocean water continues to increase in density up to its freezing point of approximately −2°C. At 0 PSU (i.e. fresh water), maximum density is approximately 4°C with a freezing point of 0°C. At 24.7 PSU and above, ocean water has a freezing point of its maximum density; Therefore, there is no maximum density temperature above the freezing point. The maximum density point scales in a linear fashion between 0 and 24.7 PSU (see Figure 6.11). Thus, ocean water continues to increase in density as it cools and sinks in the open ocean. The deep waters of the world's oceans are uniformly cold as a result.

Comparisons of salinity and temperature effects upon water density yield the following:

- At a constant temperature, variation of the salinity from 0 to 40 PSU changes the density by about 0.035 specific gravity (or about 3.5 percent of the density for 0 PSU water at 4°C).
- At a constant salinity (i.e. 0 PSU), raising the temperature from 4°C (maximum density) to 30°C (highest temperature generally found in surface water) yields a decrease in density of 0.0043 (1.000 to 0.9957) for a change of 0.4 percent.

Clearly, salinity has a much higher effect upon water density than does temperature.

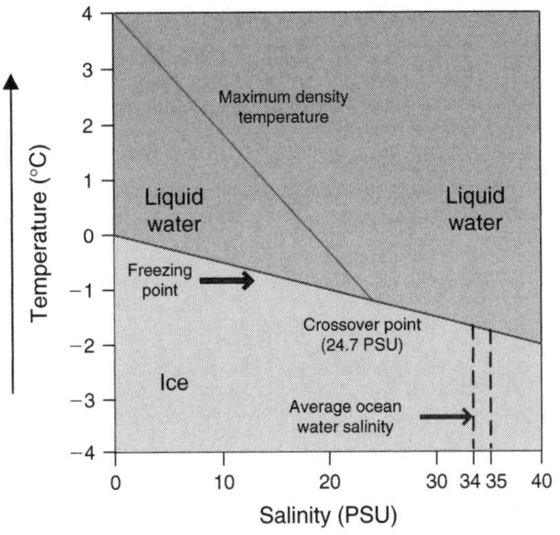

Figure 6.11 *Salt-water density, salinity versus temperature.*

As a practical example, suppose a 100-kilogram ROV is ballasted for 4°C fresh water at exactly neutral buoyancy. If that same submersible were to be transferred to salt water, an additional ballasting weight of approximately 3.5 kilograms would be required to maintain that vehicle at neutral buoyancy.

Ice at 0°C has a density of 0.917 g/cm^3, which is about 8 percent less than that of water at the same temperature. Obviously, water expands when it freezes, bursting pipes and breaking apart water-encrusted rocks, thus producing revenues for marine and land plumbing contractors.

6.2.13 Solar Radiation

Solar radiation is the electromagnetic radiation emitted by the sun and is measured in some underwater scientific applications.

6.2.14 Sonic Velocity and Sound Channels

Sound propagation (vector and intensity) in water is a function of its velocity. And velocity is a function of water density and compressibility. As such, sound velocity is dependent upon temperature, salinity, and pressure, and is normally derived expressing these three variables (Figure 6.12). The speed of sound in water changes by 3–5 meters per second per °C, by approximately 1.3 meters per second per PSU salinity change, and by about 1.7 meters per second per 100 meters change in depth (compression). The speed of sound in sea water increases with increasing pressure, temperature, and salinity (and vice versa).

The generally accepted underwater sonic velocity model was derived by W.D. Wilson in 1960. A simplified version of Wilson's (1960) formula on the speed of sound in water follows:

$$c = 1449 + 4.6T - 0.055T^2 + 0.0003T^3 + 1.39(S - 35) + 0.017D,$$

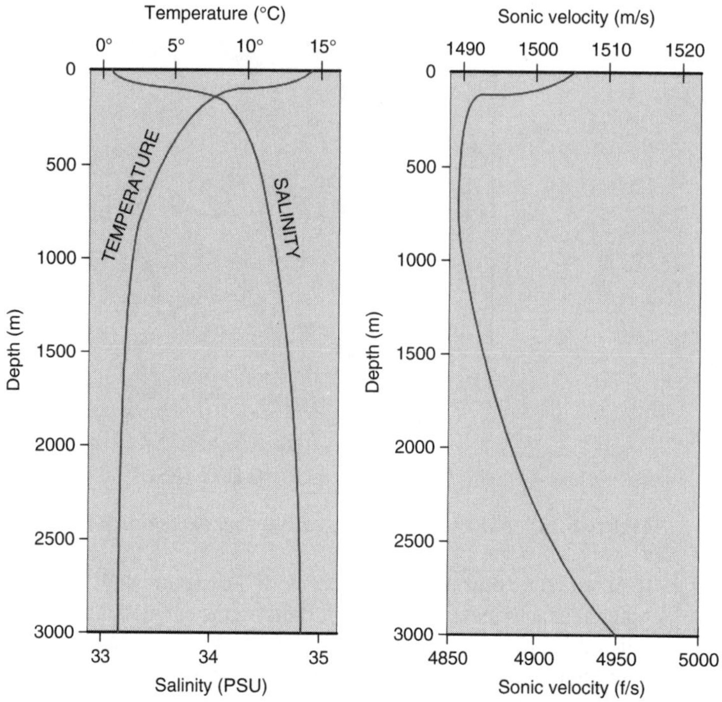

Figure 6.12 *Sonic velocity profiles with varying temp/salinity/depth.*

where c is the speed of sound in meters per second, T is the temperature in °C, S is the salinity in PSU, and D is the depth in meters.

Temperature/salinity/density profiles are important measurements for sensor operations in many underwater environments, and they have a dramatic effect on the transmission of sound in the ocean. A change in overall water density over short range due to any of these three variables (or in combination) is termed a pycnocline. Overall variations of pressure and temperature are depicted graphically in Figure 6.13.

This layering within the ocean, due to relatively impervious density barriers, causes the formation of sound channels within bodies of water. These 'channels' trap sound, thus channeling it over possibly long ranges. Sound will also refract based upon its travel across varying density layers, bending toward the more dense water and affecting both range and bearing computations for acoustics (Figure 6.14). Over short ranges (tens or hundreds of meters) this may not be a substantial number and can possibly be disregarded, but for the longer distances of some larger ROV systems, this becomes increasingly a factor to be considered.

6.2.15 Turbidity

Turbidity (which causes light scattering – see the 'Scattering' section above), the measure of the content of suspended solids in water, is also referred to as the 'cloudiness' of the water. Turbidity is measured by shining a beam of light into

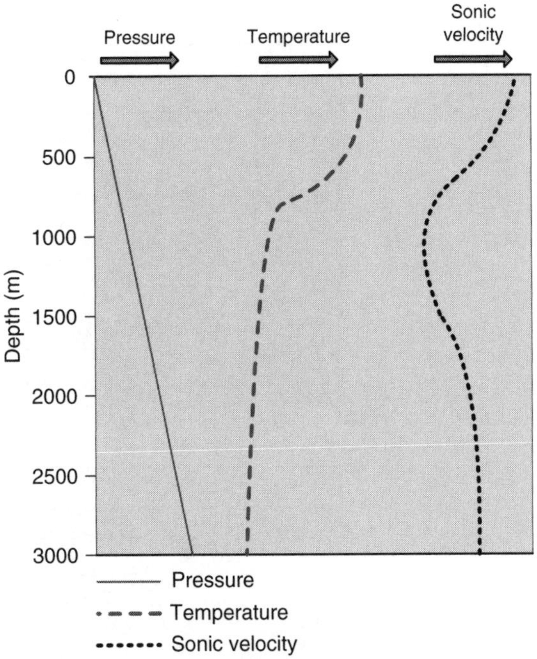

Figure 6.13 *Variations of pressure and temperature with depth producing sound velocity changes.*

the solution. The light scattered off the particles suspended in the solution is then measured, and the turbidity reading is given in nephelometric turbidity units (NTU). Water quality researchers take turbidity readings to monitor dredging and construction projects, examine microscopic aquatic plant life, and to monitor surface, storm, and wastewater.

6.2.16 Viscosity

Viscosity is a liquid's measure of internal resistance to flow, or resistance of objects to movement within the fluid. Viscosity varies with changes in temperature/salinity, as does density. Sea water is more viscous than fresh water, which will slightly effect the computations of vehicle drag.

6.2.17 Water Flow

Water flow is the rate at which a volume of water moves or flows across a certain cross-sectional area during a specified time, and is typically measured in cubic feet/meters per second (cfs/cms). The flow rate changes based upon the amount of water and the size of the river or stream being monitored. Environmental researchers monitor water flow in order to estimate pollutant spread, to monitor groundwater flow, to measure river discharge, to manage water resources, and to evaluate the effects of flooding.

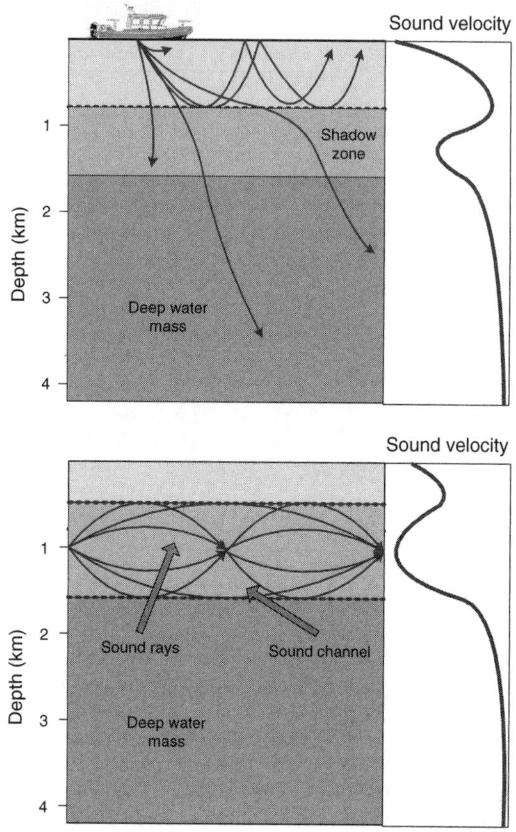

Figure 6.14 *Sound velocity profiles and channeling at various depths and layers.*

6.2.18 Water Quality

Water quality researchers count on sensors and computer software to sense environmental waters, log and analyze data. Factors to be considered in water quality measurement are discussed below.

6.2.18.1 *Alkalinity and pH*

The acidity or alkalinity of water is expressed as pH (potential of hydrogen). This is a measure of the concentration of hydrogen (H^+) ions. Water's pH is expressed as the logarithm of the reciprocal of the hydrogen ion concentration, which increases as the hydrogen ion concentration decreases (and vice versa). When measured on a logarithmic scale of 0 to 14, a pH of 0 is the highest acidity, a pH of 14 is the highest alkalinity, and a pH of 7 is neutral (Figure 6.15). Pure water is pH neutral, with sea water normally at a pH of 8 (mildly alkaline).

pH measurements help determine the safety of water. The sample must be between a certain pH to be considered drinkable, and a rise or fall in pH may indicate a

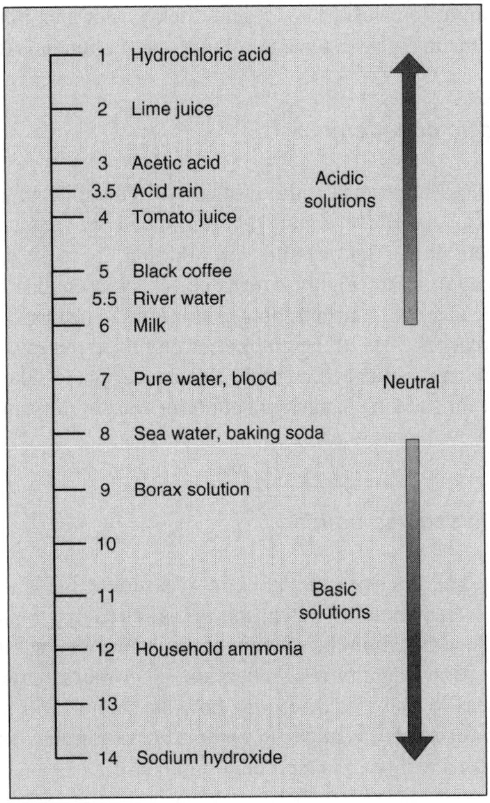

Figure 6.15 *Graphical presentation of common items with their accompanying pH.*

chemical pollutant. Changes in pH affect all life in the oceans; Therefore, it is most important to aquatic biology to maintain a near neutral pH. As an example, shellfish cannot develop calcium carbonate hard shells in an acidic environment.

6.2.18.2 *ORP (oxidation reduction potential)*

ORP is the measure of the difference in electrical potential between a relatively chemically inert electrode and an electrode placed in a solution. Water quality researchers use ORP to measure the activity and strength of oxidizers (those chemicals that accept electrons) and reducers (those that lose electrons) in order to monitor the reactivity of drinking water and groundwater.

6.2.18.3 *Rhodamine*

Rhodamine, a highly fluorescent dye, has the unique quality to absorb green light and emit red light. Very few substances have this capability, so interference from other compounds is unlikely, making it a highly specific tracer. Water quality researchers use rhodamine to investigate surface water, wastewater, pollutant time of travel, groundwater tracing, dispersion and mixing, circulation in lakes, and storm water retention.

6.2.18.4 *Specific conductance*

Specific conductance is the measure of the ability of a solution to conduct an electrical current. However, unlike the conductivity value, specific conductance readings compensate for temperature. In addition, because specific conductance and ion concentration are highly correlated, specific conductance measurements are used to calculate ion concentration in solutions. Specific conductance readings give the researcher an idea of the amount of dissolved material in the sample. Water quality researchers take specific conductance readings to determine the purity of water, to watch for sudden changes in natural or wastewater, and to determine how the water sample will react in other chemical analyses.

6.2.18.5 *Total dissolved solids*

TDS (total dissolved solids) is the measure of the mass of solid material dissolved in a given volume of water, and is measured in grams per liter. The TDS value is calculated based on the specific conductance reading and a user-defined conversion factor. Water quality researchers use TDS measurements to evaluate the purity or hardness of water, to determine how the sample will react in chemical analyses, to watch for sudden changes in natural or wastewater, and to determine how aquatic organisms will react to their environment.

6.2.19 Water Temperature

Water temperature is a measure of the kinetic energy of water and is expressed in degrees Fahrenheit (F) or Celsius (C). Water temperature varies according to season, depth and, in some cases, time of day. Because most aquatic organisms are cold-blooded, they require a certain temperature range to survive. Some organisms prefer colder temperatures and others prefer warmer temperatures. Temperature also affects the water's ability to dissolve gases, including oxygen. The lower the temperature, the higher the solubility. Thermal pollution, the artificial warming of a body of water because of industrial waste or runoff from streets and parking lots, is becoming a common threat to the environment. This artificially heated water decreases the amount of dissolved oxygen and can be harmful to cold-water organisms.

In limnological research, water temperature measurements as a function of depth are often required. Many reservoirs are controlled by selective withdrawal dams, and temperature monitoring at various depths provides operators with information to control gate positions. Power utility and industrial effluents may have significant ecological impact with elevated temperature discharges. Industrial plants often require water temperature data for process, use, and heat transfer calculations.

Pure water freezes at 32°F (0°C) and boils at 212°F (100°C). ROV operations do not normally function in boiling water environments; Therefore, the focus here will be

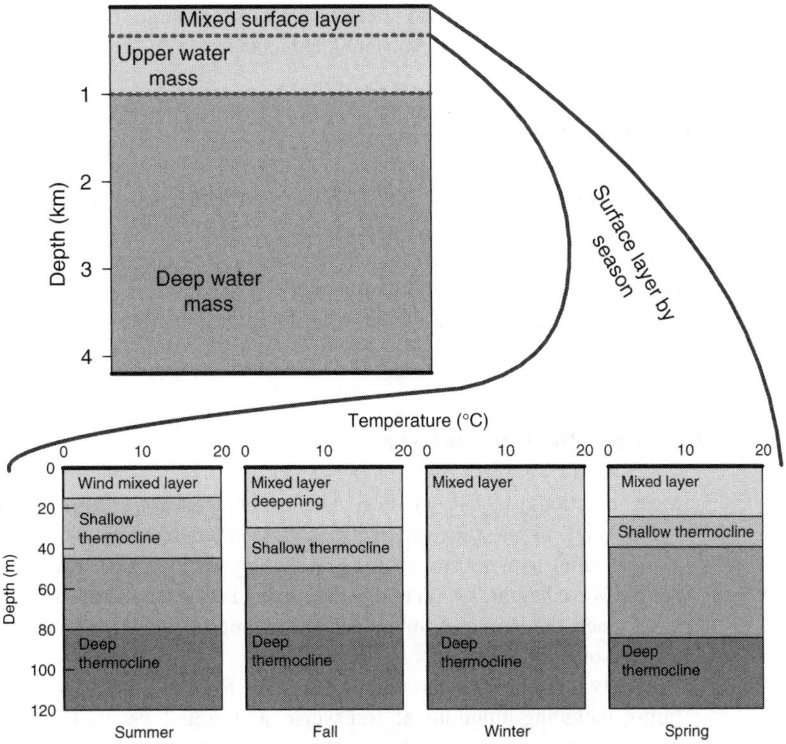

Figure 6.16 *Surface layer mixing by season.*

upon the temperature ranges in which most ROV systems operate (0–30°C). The examination of salinity will be in the range from fresh water to the upper limit of sea water.

Temperature in the oceans varies widely both horizontally and vertically. On the high temperature side, the Persian Gulf region during summertime will achieve a maximum of approximately 32°C. The lowest possible value is at the freezing point of −2°C experienced in polar region(s).

The vertical temperature distribution nearly everywhere (except the polar regions) displays a profile of decreasing water temperature with increasing depth. Assuming constant salinity, colder water will be denser and will sink below the warmer water at the surface.

There is usually a mixed layer of isothermal (constant temperature) water from the surface to some near-surface depth due to wind mixing and convective overturning (thermally driven vertical density mixing) that changes with the seasons (Figure 6.16). The layer is thin at the equator and thick at the poles. The layer where there is a rapid change in temperature over a short distance is termed a thermocline and has some interesting characteristics. Due to the rapid temperature gradient, this thermocline forms a barrier that can trap sound energy, light energy, and any number of suspended particles. The degree of perviousness of the barrier is determined by the relative strength or degree of change over distance. For the ROV pilot, a thermocline in the area of operation can hinder the function of acoustic positioning, sonar, and any sounding equipment aboard attempting to burn through the layer. It is especially of concern to anti-submarine warfare technicians.

6.2.20 Water Velocity

Water velocity is the measure of the speed at which water travels, or the distance it travels over a given time, and is measured in meters per second. Hydrologists and other researchers measure water velocity for monitoring current in rivers, channels, and streams, to measure the effect of vessel traffic in harbors and ports, and to calculate water flow. To account for drift, water velocity readings are key in knowing where and when to deploy buoys and other environmental devices to ensure their correct location.

6.2.21 Waves and the Beaufort Scale

Most operations manuals will designate launch, recovery, and operational parameters, as they relate to sea state, which is measured in the Beaufort scale (Table 6.4). Waves are measured with several metrics, including wave height, wave length, and wave period. Wave height, length and period depend upon a number of factors, such as the wind speed, the length of time it has blown, and its fetch (the straight distance it has traveled over the surface).

Energy is transmitted through matter in the form of waves. Waves come in several forms, including longitudinal, transverse, and orbital waves. The best example of a longitudinal wave is a sound wave propagating through a medium in a simple back and forth (compression and rarefaction) motion. A transverse wave moves at right angles to the direction of travel, such as does a guitar string. An orbital wave, however, moves in an orbital path as between fluids of differing densities.

A visual example of all three waves can be gained with a simple spring anchored between two points demonstrated as follows:

- Longitudinal wave – Hit one end with your hand and watch the longitudinal wave travel between end points.
- Transverse wave – Take an end point and quickly move it at a 90° axis to the spring and watch the transverse wave travel between end points.
- Orbital wave – Now rotate the end points in a circular fashion to propel the circular wave toward its end point. Orbital waves have characteristics of both longitudinal and transverse waves.

Described here are the basic components of ocean waves in an idealistic form to allow the analysis of each component individually. As described in Thurman (1994):

As an idealized progressive wave passes a permanent marker, such as a pier piling, a succession of high parts of the wave, crests, will be encountered, separated by low parts, troughs. If the water level on the piling were marked when the troughs passes, and the same for the crests, the vertical distance between the marks would be the wave height (H). *The horizontal distance between corresponding points on successive waveforms, such as from crest to crest, is the*

Table 6.4 *Beaufort scale.*

Beaufort No.	*Surface winds in m/s (mph)*	*Seaman's description*	*Effect at sea*
0	< 1 (< 2)	Calm	Sea like mirror
1	0.3–1.5 (1–3)	Light air	Ripples with appearances of scales, no foam crests
2	1.6–3.3 (4–7)	Light breeze	Small wavelets, crests of glassy appearance, no breaking
3	3.4–5.4 (8–12)	Gentle breeze	Large wavelets, crests beginning to break, scattered whitecaps
4	5.5–7.9 (13–18)	Moderate breeze	Small waves, becoming longer, numerous whitecaps
5	8.0–10.7 (19–24)	Fresh breeze	Moderate waves, taking longer to form, many whitecaps, some spray
6	10.8–13.8 (25–31)	Strong breeze	Large waves begin to form, whitecaps everywhere, more spray
7	13.9–17.1 (32–38)	Near gale	Sea heaps up and white foam from breaking waves begins to be blown in streaks
8	17.2–20.7 (39–46)	Gale	Moderately high waves of greater length, edges of crests begin to break into spindrift, foam is blown in well-marked streaks
9	20.8–24.4 (47–54)	Strong gale	High waves, dense streaks of foam and sea begins to roll, spray may affect visibility
10	24.5–28.4 (55–63)	Storm	Very high waves with overhanging crests, foam is blown in dense white streaks, causing the sea to appear white, the rolling of the sea becomes heavy, visibility reduced
11	28.5–32.6 (64–72)	Violent storm	Exceptionally high waves (small- and medium-sized ships might be for a time lost to view behind the waves), the sea is covered with white patches of foam, everywhere the edges of the wave crests are blown into froth, visibility further reduced
12	32.7–36.9 (73–82)	Hurricane	The air is filled with foam and spray, sea completely white with driving spray, visibility greatly reduced

Source: Thurman (1994).

wavelength (L). *The ratio of* H/L *is wave steepness. The time that elapses during the passing of one wavelength is the period* (T). *Because the period is the time required for the passing of one wavelength, if either the wavelength or period of a wave is known, the other can be calculated, since* L *(m)* = *1.56 (m/s)*T^2.

Speed (S) = L/T.

For example : *Speed* (S) = L/T = $156\,m/10\,s = 15.6\,m/s$.

Another characteristic related to wavelength and speed is frequency. Frequency (f) *is the number of wavelengths that pass a fixed point per unit of time and is equal to 1/*T. *If six wavelengths pass a point in one minute, using the same wave system in the previous example, then:*

Speed (S) = Lf = $156\,m \times 6/min = 936\,m/min \times (1\,min/60\,s) = 15.6\,m/s$.

Because the speed and wavelength of ocean waves are such that less than one wavelength passes a point per second, the preferred unit of time for scientific measurements, period (rather than frequency), is the more practical measurement to use when calculating speed.

The complexities of air/wave as well as air/land and fixed structure interaction, while quite important to the operation of ROV equipment in a realistic environment, are beyond the scope of this text. Please consult the bibliography for a more in-depth study of the subject.

A more practical system of gauging the overall sea state/wind combination was developed in 1806 by Admiral Sir Francis Beaufort of the British Navy. The scale runs from 0 to 12, calm to hurricane, with typical wave descriptions for each level of wind speed. The Beaufort scale reduces the wind/wave combination into category ranges based upon the energy of the combined forces acting upon the sea surface.

6.3 COASTAL ZONE CLASSIFICATIONS AND BOTTOM TYPES

General coastal characteristics tend to be similar for thousands of kilometers. Most coasts can be classified as either erosional or depositional depending upon whether their primary features were created by erosion of land or deposition of eroded materials. Erosional coasts have developed where the shore is actively eroded by wave action or where rivers or glaciers caused erosion when the sea level was lower than its present level. Depositional coasts have developed where sediments accumulate either from a local source or after being transported to the area in rivers and glaciers or by ocean currents and waves.

Of primary interest to the ROV pilot, with regard to coastal zones, is the general classification of these zones and its affect upon general water turbidity in the operational area. Depositional coasts tend to have a higher quantity of suspended solids in the water column, thus a higher turbidity and degraded camera performance. Erosional coasts tend to possess less suspended particles, thus featuring better camera optics. Further, the depositional source will greatly affect the level of turbidity since

Table 6.5 *Ocean coverage distribution.*

	$10^6\ km^2$	*Percentage of earth's Surface area*
Continental margin	93	18.2
Deep basins	268	52.6
Total	361	70.8

mud deposited from a river estuary will have a higher turbidity, than will a rock and sand drainage area.

As stated earlier, oceans cover 70.8 percent of the earth's surface. Of that composition, the distribution between continental margins and deep-sea basins is provided in Table 6.5.

A substantial amount of scientific and oil exploration/production work is done in the continental margins with ROVs. The continental margins are, in large part, depositional features. Their characteristics are driven by run-off deposited from the adjacent continent.

Sediments are carried from the marine estuaries, then deposited onto the continental shelf. As the sea floor spreads due to tectonic forces, the sediments fall down the continental slope and come to rest on the abyssal plain. Substantial amounts of oil and gas deposits are locked in these sediments, and are the focus of exploration and production efforts. The general bottom characteristics of this shelf will be mud and sediment.

6.4 EFFECTS OF WAVE PATTERN UPON ROV OPERATION

There is a plethora of information on wave propagation and its effect upon vessel management. Please consult the bibliography for more detailed information. This section will address the effect of waves upon ROV operations.

The energy to produce a sea wave comes principally from wind, but can also be generated through some lesser factors, such as submarine earthquakes (which happen less often but are possibly devastating when they occur), volcanic eruptions and, of course, the tide. The biggest concern for wave pattern management in ROV operations is during the launch and recovery of the vehicle. During vessel operations, the hull is subject to motions broken down into six components – pitch, roll, and yaw for rotational degrees of freedom, and heave, surge, and sway for translational motion. The distance from the vessel's pivot point on any axis, combined with the translational motion from waves hitting the hull, will translate into the total swinging moment affecting the suspended weight – the ROV.

The closer the launch platform is located to the center of the vessel's pivotal point, the smaller will be the arm for upsetting the suspended weight. The longer the line is from the hard point on the launch winch, the longer the arm for upsetting the suspended weight. Add a long transit distance from the hard point to the water surface, with an over-the-side launch from a larger beam vessel, and the situation is one just waiting to get out of control (Figure 6.17). Regardless of the sea state, it is best to complete

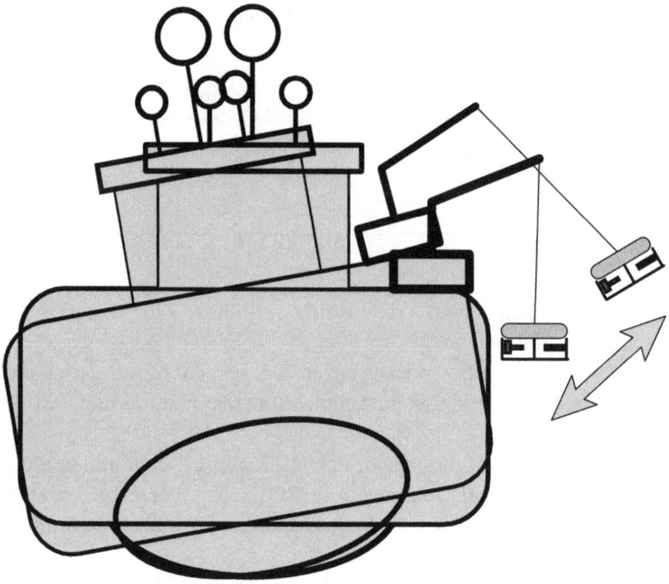

Figure 6.17 *Vehicle suspended from vessel over the side.*

this vulnerable launch operation (from the time the vehicle is lifted from the deck to the vehicle's submergence) as quickly as possible, given operational and safety constraints.

ROV operations manuals should set the maximum sea state in which to deploy the submersible. In a high sea state on a rocking vessel, the submersible becomes a wrecker ball suspended in mid-air, waiting to destroy the vessel structure as well as the vehicle, deployment system, and any bodies unfortunate enough to be in the wrecker ball's path.

If the sea state ramps up unexpectedly while the vehicle is still in the water, most operations manuals specify to leave the vehicle submerged until a sea state that allows safe recovery.

Another important wave propagation factor affecting ROV operations includes the heave component of the vessel while the vehicle is in the water. Particularly affected is the umbilical or tether length between the hard point of the launching platform and the clump weight, cage deployment platform, tether management system, or the vehicle itself. Of particular concern is the snap loading of the tether due to tether/umbilical pull with the vessel heaving. This rapid loading of the tether/umbilical can easily exceed the structural limitation of the tether and/or part conductors and communications components within the line. If this snap load parts the line, the vehicle could be lost. Larger and more sophisticated launch systems have a stress limit adjustable to the tether or umbilical loading limit. The system will slack or pull based upon the given parameters to maintain tension while avoiding overstressing the line.

Chapter 7

Environment and Navigation

In this chapter, the basics involved with operating ROV equipment are explored with an emphasis on underwater observation, navigation, and tether management.

7.1 THE 3D ENVIRONMENT

Most of us live, think, and operate vehicles in a two-dimensional world. We drive cars in 2D. We drive boats in 2D. Part of the difficulty encountered by new ROV pilots is the transition to operating, and thinking, in a three-dimensional space. The following is provided to help ease that transition.

7.1.1 Obstructions

Obstructions hazardous to ROV equipment fall into these general categories:

- *Fixed structures* – Man-made or natural structures affixed to a land mass protruding above the surface and into the water (e.g. piers, rocks, jetties, etc.)
- *Surface floating obstructions* – Vessels (anchored and otherwise), flotsam, or floating surface obstructions that protrude to some point below the water level (ships, buoys, anchor chains, drag lines, nets, etc.)
- *Objects suspended in the water column* – Objects close to neutral buoyancy (often attached to items in other classifications) subject to flow with a movement of water (e.g. fishing lines, loose netting, kelp fields, etc.)
- *Bottom obstructions* – Subsurface items anchored to the bottom of any size that can obstruct the movement of ROV equipment (e.g. rocks, bottom refuse, crab traps, subsurface structures, wrecks, etc.).

7.1.2 Hazards to Underwater Vehicles

Underwater hazards to ROV equipment generally fall into the following classifications:

- Obstructions (as classified above) to vehicle movement
- Water current flow exceeding the performance capabilities of the vehicle, causing a partial or total loss of control
- Mechanical damage to system components (such as being struck by a ship's propeller)
- Foreign object ingestion to thrusters, causing thruster performance degradation

- Chemical damage to ROV materials, causing damage or performance degradation
- Environmental, chemical or radiological damage to electronic components
- Water optical conditions causing reduced visibility
- Sharp temperature gradients causing thermal shock loading of vehicle materials and components (an example is deploying a submersible from −40°F arctic air into 32°F water while the vehicle is still cold-soaked, thus shattering the domes and plastic components).

Consideration of these environmental factors during pre-deployment operations will mitigate equipment damage and help assure mission success.

7.1.3 Underwater Optics and Visibility

The physics of underwater lighting/optics will affect the image properties produced by ROV equipment used in any underwater mission. Most people's experience (and frame of reference) with optics centers on light in air; Therefore, this section will address the changes that occur when light enters water.

Light refraction and dispersion occurs when a light beam is passed through water (Figure 7.1).

When viewing an object in water through a plane window, unless the lens is corrected, refraction causes focus error, field of view error, and distortion as follows.

7.1.3.1 *Focus error*

Rays diverging from a point at a distance from the window appear, after they have passed through the window, to come from a shorter distance.

7.1.3.2 *Field of view errors*

A ray coming to the window at any angle of incidence other than zero will leave it at a wider angle. Thus, an optical system such as a camera, which has a given angular field

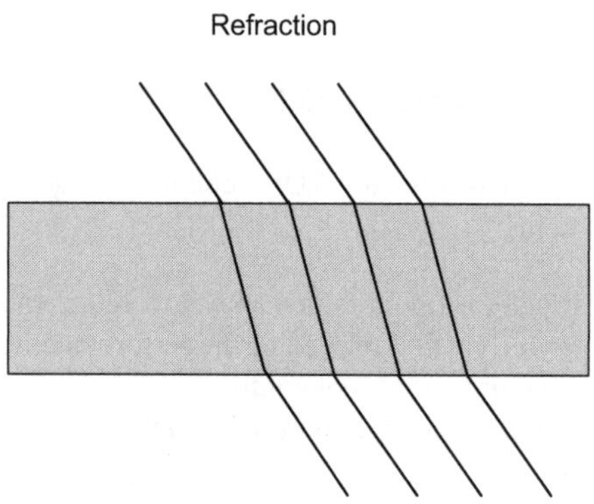

Figure 7.1 *Light refraction in water.*

of view in air, will have a small angular field in water. In addition, an object, which would produce an image of a certain size if it were in air, produces a larger image if it is in water. The lens appears to have a longer focal length, approximately 4 : 3 that in air. This is why subjects appear magnified when using a diving mask underwater.

7.1.3.3 *Distortion*

Rays from points forming a rectangular grid in water will not seem to come from a rectangular grid after they have passed into air. There will be distortion in a lens that is not corrected for underwater use (Figure 7.2).

About 4 percent of the natural light striking the surface at a normal angle of incidence is reflected away. The rest is quickly attenuated by a combination of scattering and absorption, which are discussed below.

7.1.3.4 *Absorption*

Visible light occupies the electromagnetic spectrum between approximately 400 nanometers (violet) and 700 nanometers (red). The absorption rate of water varies depending upon the wavelength (Figure 7.3). The ends of the spectrum (the red and the violet ends) are absorbed first, with the maximum penetration/lowest absorption rate in the blue/green spectrum.

Maximum penetration is gained when no particulate matter is suspended in the water (which would cause scattering), such as in the warm tropical waters of the

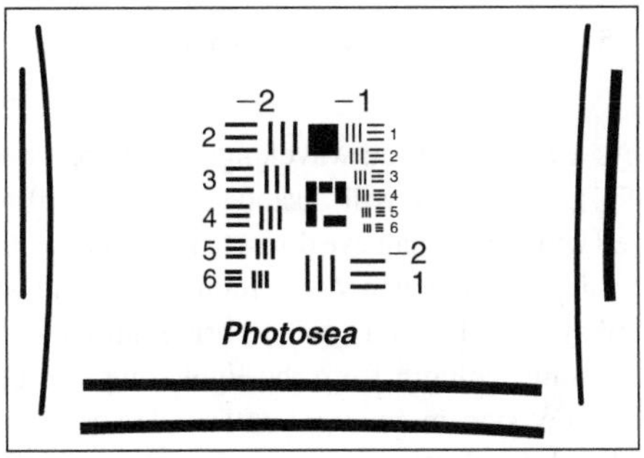

Figure 7.2 *Note the 'pincushion' distortion in this underwater photo taken with an uncorrected lens (from Wernli, 1998).*

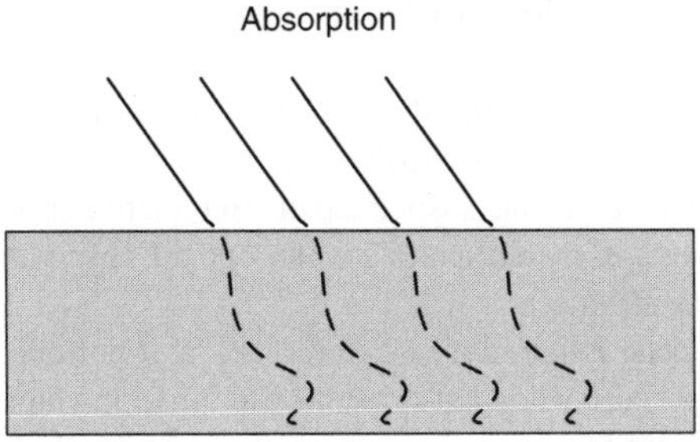

Figure 7.3 *Water absorption of visible light.*

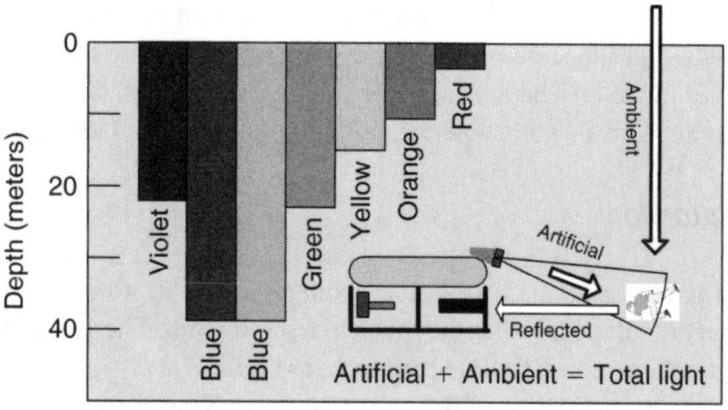

Figure 7.4 *Light penetration and total illumination of target.*

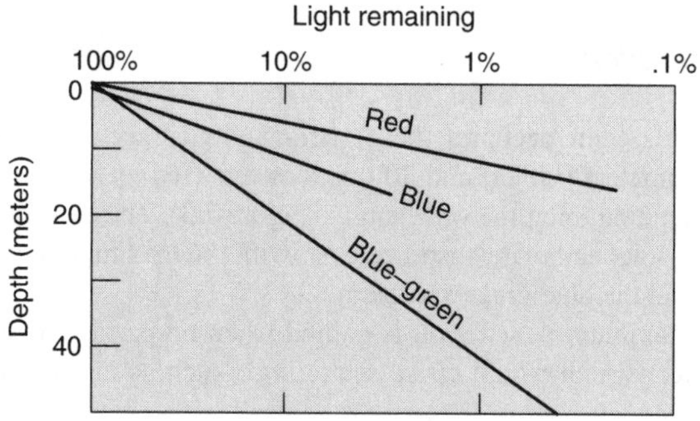

Figure 7.5 *Light absorption by wavelength.*

Pacific. As a result, the red wavelengths are absorbed within the uppermost 60 feet (Figure 7.4), the yellow will disappear within 330 feet, but green light can still be recognized with the human eye down to more than 800 feet below the surface. So, even at shallow depths, objects become monochromatic when viewed through the camera of an ROV system. The only way to bring out the color of the object of interest is to reduce the water column from the light source to the object. If that light source is aboard the ROV system, the light will need to be near the object in order to illuminate it with full color reflection.

An example of the absorption spectrum for pure water is shown in Figure 7.5.

7.1.3.5 *ROV visual lighting and scattering*

Observers agree that the absorption and scattering in clear ocean water are essentially the same as in clear distilled water, that some dissolved matter increases the absorption, and that suspended matter increases scattering. Both absorption and scattering present difficulties when optical observations are made over appreciable distances in water.

Scattering is the more troublesome, as it not only removes useful light from the beam, but also adds background illumination (Figure 7.6). Compensation for the loss of light by absorption can be made by the use of stronger lights, but in some

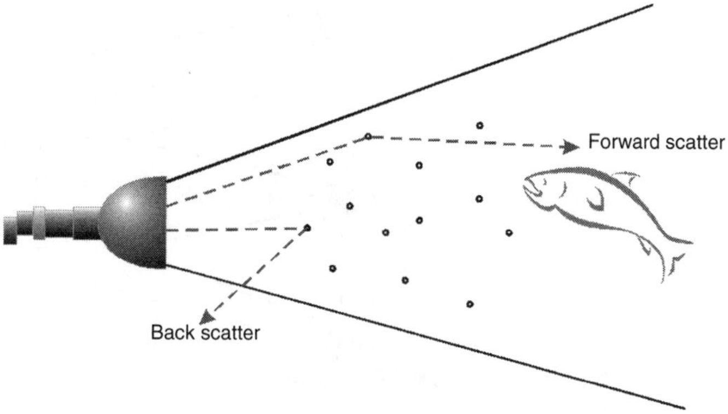

Figure 7.6 *Illustration of the scatter phenomenon (graphic by Deepsea Power and Light).*

circumstances additional lights can be degrading to a system because of the increase in backscatter. These circumstances are analogous to driving in fog; The use of 'high beam' headlights, in most cases, causes worse viewing conditions than 'low beam' headlights.

Just as driving an automobile in a fog causes reduced visibility, the lighting aboard an ROV system blinds the camera through backscattering of light hitting the particulate matter suspended in the water column. In highly turbid water conditions (such as in most harbors around the world), reduction of the lighting intensity may be necessary in order to gain any level of visibility. Another consideration to aid the viewing of items underwater is the separation of the light source from the camera so that the water column before the camera is not illuminated, thus eliminating the source of backscattered lighting (Figure 7.7).

As an anecdote to lighting, Hollywood director James Cameron accomplished optimal elimination of the water column's effect during such projects as the filming of the shipwrecks *Titanic* and *Bismarck*. This was accomplished by placing the light source aboard a completely separate manned submersible (the Russian *Mir-1* submersible) from the camera platform (the Russian *Mir-2* submersible). With small ROV platforms, the separation of lighting source from the camera point may not be possible. This may require reduced on-board lighting while taking advantage of the ambient light as back-illumination of your object of interest.

The ability to have a second vehicle like the *Mir* at hand is unlikely. Therefore, in very turbid water, using two or three lower-powered lights positioned efficiently on the ROV, instead of one higher-powered light, may help the situation.

7.2 THE NECESSITY OF ACHIEVING OBJECTIVES THROUGH NAVIGATION

A structured approach to achieving mission objectives is paramount. It is generally considered a waste of valuable field time and resources to 'Chance' upon some item of interest through the use of unstructured search methods.

The operation of the vehicle should be from known reference point to known reference point, maintaining positive navigation throughout the maneuver. Therefore,

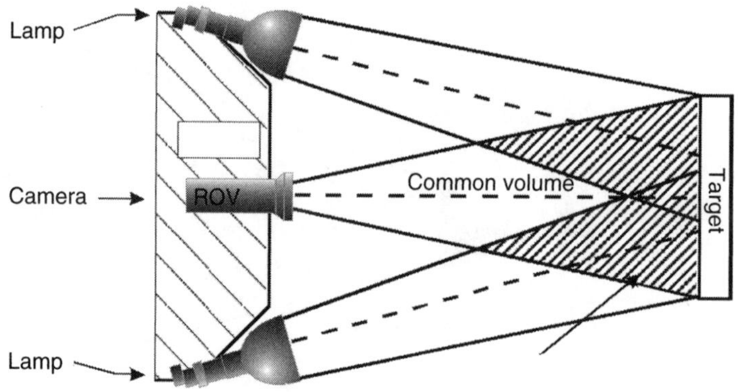

Figure 7.7 *Separation of the light source from the water column before the camera mitigates backscatter (graphic by Deepsea Power and Light).*

searches for the purposes of this manual will be conducted via the following primary navigation means:

- Visual known reference point to known reference point navigation
- Sonar navigation
- Grid search pattern navigation.

These techniques are discussed in the following sections.

7.2.1 Visual Navigation

Visual search/inspection will encompass a majority of search and inspection navigation tasks. The objective of this method is to maintain visual reference at all times with some known reference system.

7.2.1.1 *Surface navigation*

While operating on the surface in transit to the work site, it is necessary to maintain visual reference with the submersible while navigating it to a point where the submersible's camera can take over (Figure 7.8).

Another effective method of surface navigation, when it is not possible to maintain visual sighting with the submersible, is to tilt the camera upward through the water column to gain visual sighting of the structure of interest. In that way, it is still possible to maintain visual contact/orientation with your structure until underwater contact with the structure is made.

7.2.1.2 *Underwater visual search*

Once visual underwater contact is made with a known structure, a thorough search of the area may be conducted. Methods of structure inspection vary from operator to operator, but the following remain effective:

- Gain a macro view of the structure by scanning the surface, then perform a detailed inspection of the edges of the structure

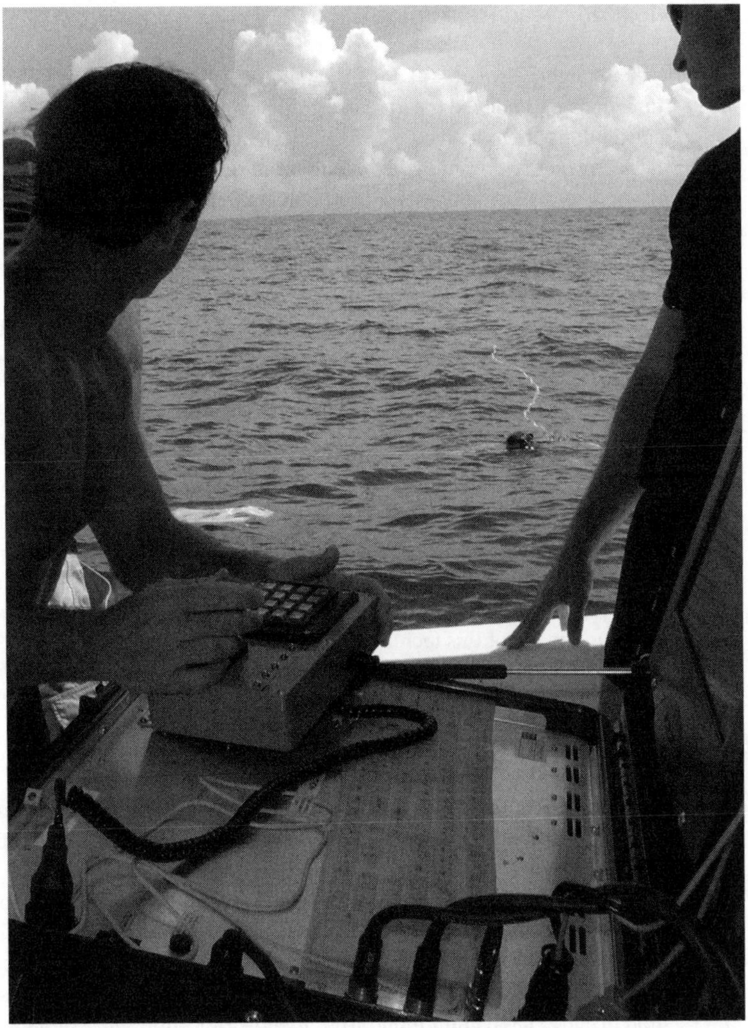

Figure 7.8 *Visually surface swimming the submersible to the hull.*

- Make liberal use of reference lines on your structure, such as longitudinal or lateral seams (e.g. ships' hulls), structural bracing (e.g. bracing between piling members for positive transition to next pile), lines of attachment points (follow rivet lines on walls for reference), etc.
- Obtain as much prior information (e.g. construction diagrams, updated engineering drawings, recent surveys, etc.) to assist in gaining assurance that the area was inspected, including all relevant items.

Extreme care must be taken during operations to maintain adequate control over tether lay. During the planning for any structure investigation, a prudent plan of action will allow for a more efficient inspection by avoiding time-consuming and annoying tether snags.

7.2.2 Navigation (Dead Reckoning)

When targets are of a known-distance and bearing from reference objects, a time/distance deduced (dead reckoning) navigation can be made. Basic time/distance calculations are quite handy in ROV navigation.

Borrowing a common technique from visual aircraft navigation, an effective method of performing straight-line navigation in near-bottom conditions is to pick out a target at the end of the visibility limit then fly to it. As that item is approached, find another item in the distance past that item in a straight line and continue toward it. Repeat this methodology as needed to navigate visually in a straight (or near straight) line in reduced visibility conditions. Stay within sight of the bottom while keeping clear of obstacles so the vehicle does not ram into obstructions unexpectedly.

7.2.3 Navigation (INS)

A solution to the mental time/distance calculation is to perform it electronically. The Inertial Navigation System (INS) senses acceleration in all planes to gain a vector from a known datum to arrive at a resolved position based upon vector movement from the known datum. INS is widely used in aircraft and submarine applications, but is just becoming of use in observation-class ROV applications. Entire books have been written about this technology. Look for more movement in this field as the cost and size factors approach the observation-class ROV threshold.

7.2.4 Navigation (DVL)

As discussed more fully in Chapter 5, a resolved velocity over bottom can be gained acoustically through use of the doppler velocity log (DVL). Once the over-ground vector is determined, an accurate time/distance calculation can be gained to geo-reference the vehicle's position. In some newer applications, use of a DVL along with Geographic Information Systems mapping applications allows for 3D estimated tracking of a submersible with periodic 'snap to grid' updated position through some other technology (e.g. GPS or acoustic positioning). As the vehicle is maneuvered in time/distance navigation mode from a known location, the estimation error per unit distance or time is increased proportional to the distance traveled. For instance, if there is an assumed 10 percent error with the DVL or the INS over distance (or time for that matter) traveled, the error will be 10 percent for that unit. If another unit of travel distance/time is processed, the error will be that 10 percent plus another 10 percent as the circle of possible/probable positions increases. The farther the vehicle progresses on dead reckoned navigation without an updated accurate positional fix, the larger grows the circle of equivalent position probability (also known as circular error probable or 'CEP').

Bowditch (2002) describes CEP as: (1) In a circular normal distribution (the magnitudes of the two one-dimensional input errors are equal and the angle of cut is 90°), the radius of the circle containing 50 percent of the individual measurements being made or the radius of the circle inside of which there is a 50 percent probability of being located. (2) The radius of a circle inside of which there is a 50 percent probability of being located even though the actual error figure is an ellipse. That is, the radius

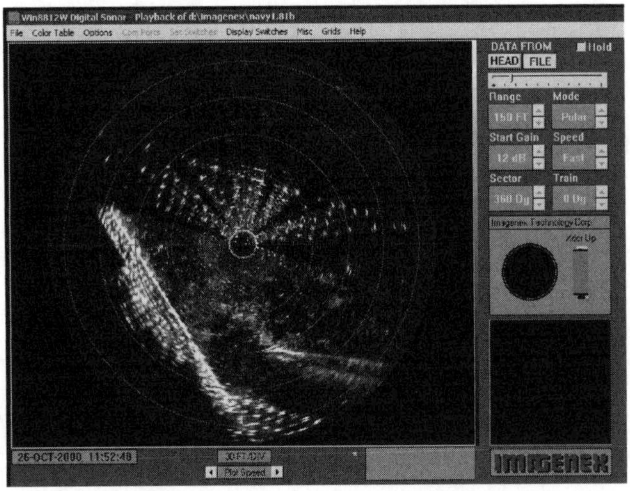

Figure 7.9 *Sonar bottom search of a slip at the Washington Navy Yard showing all bottom features within a 150-feet radius.*

of a circle of equivalent probability when the probability is specified as 50 percent. The CEP measurement is useful in navigation, weapons delivery computations, and search grid designations.

7.2.5 Sonar Navigation

For search zones with large featureless areas, sonar will prove effective in insonifying obvious items of interest, thereby quickly eliminating the areas of least concern. An example of a sonar operation that fleshes out items of interest is an 'under vessel' bottom search. In this type of search, a scan of the bottom beneath a moored vessel can be completed in less time than performing a grid search (Figure 7.9).

On a structure survey (such as a pipeline), the item can be initially located and navigated to with sonar, then transitioned to optical survey upon visual acquisition of the item.

An effective technique for scanning an area with an ROV-mounted sonar system is to fly the submersible to a fixed structure (for better sonar platform stability) and allow the sonar to generate a picture of the surrounding area.

7.2.6 Grid Pattern Navigation

When sonar is either not available or not effective (such as for an uneven bottom, which blocks sonar echo), a grid pattern must be used in order to clear an area. These are difficult in low-visibility water conditions. Some methods that may assist in achieving the grid are:

- Use of a visual grid reference mark, such as a bright-colored weighted rope, that can lay flat upon the bottom for easy, positive navigation

- Use of acoustic navigation to lay down and track transects
- Use of natural visual clues, such as position of sun or vessel shadow.

7.3 CURRENTS AND TETHER MANAGEMENT

Once the vehicle is in the water, tether management becomes critical to the success of the mission. One problem here and the mission could be a failure and/or the ROV could be lost. The critical considerations of tether management follow.

7.3.1 Currents

As stated previously, the objective of operating an ROV is to deliver a camera, or other instrumentation package, into a place where it can be of use. Under most situations, the item of interest will be submerged in a fixed location. Tidal flows can cause difficulty in flying the submersible to the work site.

The resistance to delivery of the submersible to the work site is directly related to the total drag upon the submersible and tether. As stated mathematically in Chapter 2, hydrodynamic drag upon the wetted surfaces of the submersible system (vehicle plus tether) is affected by the following:

- Density of the sea water
- Characteristic area for both the submersible and the tether
- Velocity of the submersible system through the water
- Non-dimensional drag coefficient of the system (essentially the hydrodynamic shape of the system through the water).

Increase any of these items and the submersible will have more difficulty in fighting the drag.

Drag caused by current can be mitigated by the following:

- Maintain as little tether in the water as possible in order to accomplish the task.
- When pulling tether to the work site, attempt to pull the tether into the current to present the least cross-section drag as possible.
- If possible, lay the tether across some obstruction (while assuring an easy egress) between the work site and the pilot's operating station, so that the submersible is not required to continually fight the drag while station-keeping.
- Approach a structure from the lee side to allow operation in the turbulent area downstream of the structure of interest.
- Operate at slack tide, or consider delaying the operation until slack tide, if currents prove to be above the submersible's power capability. Or invest in a more powerful ROV system.
- Make liberal use of clump weights while working on or near the bottom so that the weight takes the cross-section drag on the tether instead of the submersible's thrusters.
- When operating on the bottom, make use of the boundary layer near the bottom to stay in the calmer waters below the boundary layer.

7.3.2 Teamwork and Proper Tether Management

Proper coordination between the tether handler and the submersible pilot will do more to prevent tether management problems, as well as entanglements, than any technological solution.

An ROV pilot's success as an operator of robotic equipment will directly depend upon his ability to figuratively place his head inside the submersible. As part of that 'feel', the ROV pilot is in a position to gain a 'feel' for the amount of tether needed as well as the degree of pull being experienced by the tether at any moment. As experience is gathered, a situational 'lay of the land' orientation should be gained, allowing a mental picture of the location/lay of the tether. From this feel, the ROV pilot can direct the tether handler for best tether management.

In general, the submersible should pull the tether to the work site (as opposed to the tether pushing the vehicle). The tether handler should allow just enough slack in the water so that the submersible is not wasting thruster power pulling tether into the water. Any excess tether in the water, other than absolutely necessary to accomplish the job, is inviting tether entanglements and wasting thruster power from the drag of the excess tether.

Direct ingress/egress routes to the work site are generally preferable to multiple turn navigation due to tether friction and the higher likelihood of snags. The intended task should be planned in advance so that the best route can be chosen. The extra preparation time to choose the best route could save time later unfouling the tether.

In every ROV pilot/operator's tool kit should be additional floats and weights to compensate for the varying density of water. If an ingress route to the work site takes the submersible over a wall then under an overhang, a weight at the proper location, followed by a float near the top, will lay the tether properly and avoid unnecessary tether hangs.

Neutrally buoyant tether is best for practically all ROV operating situations – but neutrally buoyant in what water type? Usually, ROV manufacturers specify neutral buoyancy of the tether based upon fresh water at average temperature (i.e. zero salinity and 60° F water). As the temperature of the water moves down and the salinity of the water moves up, the tether becomes increasingly buoyant. This factor must be taken into consideration and compensation made in the field.

While operating ROV equipment, a negatively buoyant tether may be useful in situations such as an under-hull ship hull inspection (to keep the tether clear of hull obstructions). But negatively buoyant tether operated near the bottom will drag across any item protruding from the bottom much like a dragline.

Clear communications need to be established between the tether handler and the ROV pilot to avoid misunderstanding under critical situations. Standard terminology includes:

- 'Launch vehicle'
- [for pre-dive inspection] (Pilot) 'Lights' (Tether Handler) 'Check', etc.
- 'Pull tether *x* feet' for pulling in tether
- 'Slack tether *x* feet' for slacking tether
- 'Pull tether *x* pounds'
- 'All Stop'
- 'Hold'

- 'Pull until tension'
- 'Pull then slack in [frequency] succession'
- 'Recover vehicle'.

7.3.3 Tether Snags

Tether snags range from friction drag across an obstacle to total entanglement. The steps to clear tether entanglements are fairly universal on all ROV platforms and include the following:

1. When the submersible stops moving, the tether is snagged somewhere behind the vehicle.
2. Thrust forward slowly to bring the tether to 'taut' condition to isolate the tether between the location of the snag and the vehicle.
3. Reverse slightly to bring slack in the tether line behind the vehicle.
4. Make a 180° turn in place to locate the tether between the vehicle and the snag.
5. Follow the tether to the location of the snag.
6. Work the tether snag out visually in coordination with the tether handler.

Once the tether snag is located, the general procedure for clearing the snag will present itself. There are four general ways of clearing a snag:

1. The preferable choice is to move the submersible to the snag and clear the foul using positive control.
2. If the above is not possible, move the operations platform (i.e. move the tether from the operator's side).
3. It may be possible to move the snag or the item upon which the tether is snagged.
4. The last choice is to cut the tether either physically or via a connection point.

If it becomes necessary to leave the operations area before a snagged tether or stuck vehicle can be freed, the vehicle can be powered down, left in the water, and retrieved at a later time. Steps to accomplish this task are:

1. Power down the vehicle.
2. Pay out all of the tether.
3. Unplug the top-end or intermediate tether connection point and wrap in plastic or some other water-resistant wrapping.
4. Note the location with GPS.
5. Attach a buoy to the tether for easy retrieval upon return.
6. Advise operations that there is a tether in the water, which may pose a hazard to navigation, so that advisories can be made.

7.3.4 Tether Guides and ROV Traps

There are objects that, due to their shape and size, are more likely to snag a loose tether. Some can be used as tether guides, but some are considered 'tether traps'.

The typical tether trap has an edge with any converging angle less than 90°. This type of arrangement will allow the tether to slide naturally into the groove made by the

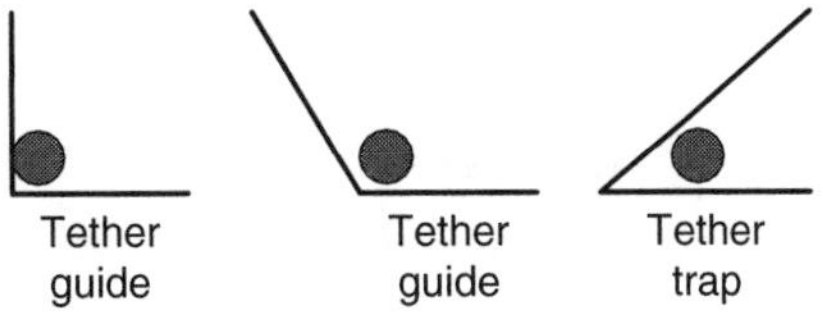

Figure 7.10 *Tether guides and traps.*

angle and, once the tether is yanked, form a friction lock on the tether (Figure 7.10). At that point, it is likely the submersible is lost.

Edges with angles of 90° or more can form a 'Tether Guide'. These types of structures are handy during operations since they allow the operator to place the tether into the groove for known placement and allow for low-friction sliding of the cable along the groove.

Any mission that involves operations in or around structures will require a choice of tether lay and guiding.

7.3.5 Clump Weights and Usage

Hold a piece of yarn in front of your mouth and blow. The yarn is blown by the air from your mouth due to the cross-section drag of the string. Take that same piece of yarn and tie a weight to the end and repeat the exercise. The yarn does not blow in the wind. The cross-section drag is now absorbed by the weight instead of the yarn itself. This is a close analogy to the need for a clump weight while operating an ROV.

The clump weight serves several purposes in ROV operations:

- It allows for orderly lay of the tether from the water insertion point to the work site on the bottom.
- It delivers the tether to a known location that can be measured above the level of the bottom so that only the tether from the clump weight to the submersible requires management.
- The submersible is only required to drag the amount of tether from the clump weight to the submersible, thereby freeing the submersible for direct work tasks (Figure 7.11).

When conducting a long bottom search or inspection task, it may be more feasible to use a clump weight to bring the tether to the bottom, then move the work platform, rather than anchor and try to swim the submersible along a transect. Consider the appropriate use of a clump weight in order to more easily complete the mission.

7.3.6 Rules for Deployment/Tether Management

A few rules for tether management follow:

- Never pull on the tether to clear a snag.
- Always follow your tether back out from the work site with the submersible to the insertion point while pulling the tether slowly back with as little free tether behind the vehicle as possible.

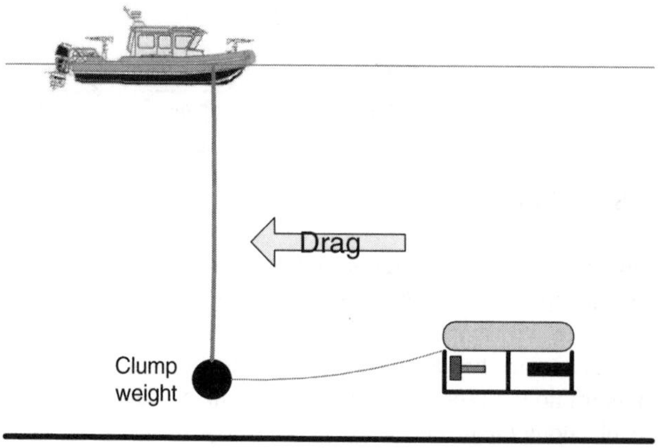

Figure 7.11 *Use of a clump weight reduces tether drag to only the length between the clump weight and submersible.*

- A clean and organized workplace is a safe and productive workplace.
- Tether turns are introduced when the submersible is turned from its original heading. Minimizing the number of tether turns while operating will enhance maneuvering. Remove all turns before submersible recovery.
- Be observant of obstacles located near the submersible that have the potential to snag the vehicle or the tether.
- If the vehicle is at the end of its tether, there may not be enough slack to allow an easy turn-around to follow the tether out. In this case, reverse direction to generate slack, then turn the vehicle around to manage tether.
- Avoid weaving in and around fixed objects, such as baffles, pipes, pilings, rocks, and anchors. When operating in possibly fouled areas, it is advisable to remain on the surface until the vehicle is approximately above the work site, then dive.
- Attention must be paid to objects standing vertically or horizontally in the water column, such as anchor chains, pilings, lines, pipes, and cables. The vehicle's tether can easily become entangled. In currents, approach such items from the downstream side.
- If operating from an anchored ship, avoid working on the upstream side of the anchor.
- When operating in an area containing obstructions or obstacles that could snag or foul the tether, the operator should endeavor to remember the route taken to get to any one position. Not only will this information be helpful on the return trip, it will be valuable in the event the tether does become snagged!
- If the tether does become entangled, do not pull the tether to free. The best course of action will present itself after careful and cautious consideration of the alternatives. During a recent operation, while performing a penetration of a wrecked aircraft at 425 feet of sea water, the submersible became entangled in the wreck a total of 16 times during an 8-hour period. Tether snags happen as a normal course of operations. Plan for them and take positive precautions to manage the tether lay throughout the work site.

Chapter 8
Homeland Security

8.1 CONCEPT OF OPERATIONS

Remotely Operated Vehicles (ROVs) provide governmental and law enforcement officials with the capability to visually inspect underwater areas of interest from a remote location. ROVs are intended to be complementary to the use of public safety and military dive teams. They can be used in conjunction with divers, or as a replacement to them, when the operation's environmental conditions allow. Factors to consider in deciding which capability to employ should include, but not be limited to, the mission objective, degree of accuracy required, current threat level, on-scene conditions, and the amount of time plus resources available.

This text covers ROV systems with submersibles weighing 150–200 pounds (excluding tether). A two-person (nominal) team is recommended to operate inspection ROVs (although single-operator functions can be performed). Experience has shown that personnel from electronics or scuba diving backgrounds are excellent candidates for assignment as the primary ROV operator. Operation and maintenance of the ROV requires an average level of electronics and situational awareness skill and experience that fits well with the background of most electronics and dive team personnel. The use of volunteer or administrative personnel also helps to reduce impacts to the department's operational schedule, as many law enforcement personnel are not assigned to dedicated 24/7 security duties.

When planning ROV operations, appropriate consideration must be given to the personal protection of the ROV team. An ROV team operating in a non-secure area under heightened threat levels may require armed escorts. Because ROV operations normally require the operators to focus on the ROV, it is strongly recommended that separate personnel carry out the task of team protection/security.

8.2 TACTICS, TECHNIQUES, AND PROCEDURES (TTPs)

The capabilities and limitations of the individual ROV systems determine the deployment procedures, since the more powerful systems allow for higher distance offset through greater currents than do their less powerful brothers. For the purposes of classifying the procedures, the systems are divided into smaller categories for ease of procedure assignment (as further defined below).

The TTPs developed for this manual are intended to be manufacturer non-specific to prevent these procedures being tied to any single manufacturer of ROV equipment. Techniques were tested on a variety of ROV systems in the small 'observation class' category to validate and gain confidence as to these procedures' applicability through a range of ROV sizes and capabilities. The systems tested were placed into three general 'observation class' categories based upon their respective sizes, weight and forward thruster output available. The size assignments were:

- *Small* – Submersible weight less than 30 pounds with forward thruster output less than 10 pounds.
- *Medium* – Submersible weight between 30 and 50 pounds with forward thruster output 10–20 pounds.
- *Large* – Submersible weight above 50 pounds and/or with forward thruster output greater than 20 pounds.

8.3 OPERATING CHARACTERISTICS OF ROV SIZE CATEGORIES

The overall size of the system partly determines the payload capacity and ability to carry larger, more powerful thrusters. Thruster output determines the vehicle's capability to deliver the submersible to a place where it can produce a useful picture while fighting currents and pulling its drag-producing tether. Generally, the larger the submersible, the more powerful the thrusters. Other control, drag, and stability considerations include the hydrodynamics of the thruster placement (affecting laminar/turbulent flow through thruster housings), vehicle stability at higher speeds and, the largest of these considerations, the diameter of the tether (circular shapes having the highest drag coefficient) being pulled behind the vehicle.

To summarize, vehicle size category determines the physical capability to maneuver into a place to accomplish the task. The viewing from that point is a function of camera coverage and distance from the item being inspected. If water clarity is 5 feet, the ROV can be no more than 5 feet away from the target. The larger the standoff (i.e. water insertion point to work area), the larger the thrust necessary to pull the tether to that workplace. The stronger the current, the larger the thrust required to overcome the parasitic drag created by the vehicle and the tether.

Conversely, while a larger ROV may be able to overcome the above-described obstacles, the larger ROV is heavier, has a larger system 'footprint' and requires more electrical power (e.g. larger generator with higher fuel consumption and space requirement, louder engine with higher emissions, etc.). Although the smaller vehicles are more easily affected by the conditions described above, they offer many beneficial trade-offs (i.e. smaller footprint, lower power requirements, more mobile, capable of getting into smaller areas).

8.4 PORT SECURITY NEEDS

For the purposes of this manual, underwater port security tasks fall under two broad categories:

- *Search and identification* of underwater targets or areas of interest. Examples may be limiting an inspection to a vessel's running gear or searching a particular location based on intelligence.
- *Search and inspection* of general areas of interest for potential threats within the security area. Examples may be conducting a search/inspection of all

the pier pilings within a specified zone prior to the arrival of a high value vessel.

Search and identification of underwater targets is accomplished through four basic steps:

1. *Research* to define the area of interest.
2. *Wide area search* with instruments and sensors.
3. *Narrow area search* with slow-speed instruments.
4. *Final identification*, visual discrimination and disposition – i.e. diver or remote camera propelled to the inspection site

ROV intervention as a productive and cost-effective means of final identification, discrimination, and disposition is analyzed in this manual. Refer to Chapter 10 for a more detailed discussion on the previous four steps.

8.5 UNDERWATER ENVIRONMENT OF PORTS

The underwater environment of ports within the USA (and the remainder of the world) vary in temperature, water clarity, operating currents, and vessel traffic. The environment in Seattle, WA (cold deep water, good visibility, moderate currents, moderate vessel traffic) varies significantly from the Port of New Orleans (warm deep water on the Mississippi river, poor visibility, high currents, moderate vessel traffic), the Port of Galveston (warm shallow water, poor visibility, low currents, low vessel traffic), and New York Harbor (cold shallow water, poor visibility, high currents, high vessel traffic, medical waste, etc.).

Accordingly, the environmental factors that determine the difficulty of completing an underwater port security task with an ROV system follow:

- *Currents* – Determine the ability of the submersible to successfully swim to, and station-keep near, a fixed object while countering these currents.
- *Water depth* – Determines the offset from the deployment (water insertion) point to the bottom for bottom clearance searches as well as proximity of bottom to vessels/moorings/anchors/piers.

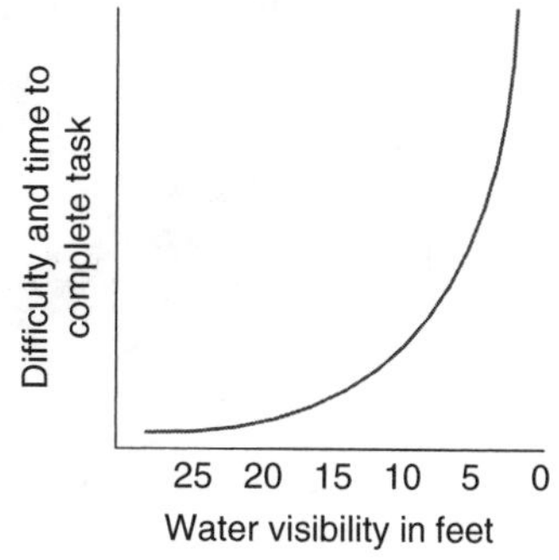

Figure 8.1 *Degree of difficulty and time required for visual inspection versus water clarity.*

- *Vessel traffic* – Determines the relative security of the inspection operation (a more secure location produces an easier inspection task).
- *Water temperature* – Determines the resources and expertise necessary for a dive team to enter that environment along with loiter time.
- *Water clarity* – Determines the degree of difficulty in completing the underwater task as well as the time to fully image the items of interest (Figure 8.1).

8.6 NAVIGATION ACCESSORIES

There are several technologies available for achieving underwater positioning and target acquisition for further investigation. Most common ROV-mounted acoustical systems involve imaging sonar and acoustic positioning.

8.6.1 Imaging Sonar

Imaging sonar is useful in underwater port security tasks in identifying items of interest by producing a sonar reflection or a blockage, which produces a sonar shadow. Imaging sonar manufacturers have been able to miniaturize the sonar unit to fit aboard practically all sizes of ROV systems. In order to get a high-resolution image of a small target at a nominal range, such as 50 feet, most manufacturers use high-frequency sonar with a fan beam in the 600–800 kHz range. This allows items protruding from the bottom to be imaged on sonar (Figure 8.2).

8.6.2 Acoustic Positioning

An acoustic positioning system calculates range from a submersible-mounted transducer to other transducers at known locations with known spacing. This permits an accurate range calculation, with adjustment for water temperature/salinity/density, by

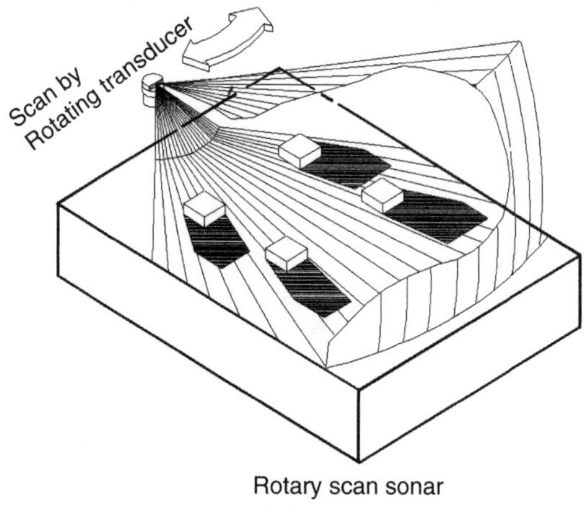

Figure 8.2 *Imaging sonar (graphic by Imagenex Technology Corp.).*

computing the one-way or round-trip timing. Bearing is resolved through triangulation of the timing differences across the transducer array (i.e. merging point of the separate lines of position). In order to resolve relative bearing to magnetic bearing, a magnetic transducer array orientation needs to be determined (easily done in software with a flux-gate compass outputting standardized data streams). In order to resolve relative location to geo-referenced location, an accurate latitude/longitude position must be determined with a GPS unit (outputting standardized data streams).

The vessel-referenced acoustic positioning system (Figure 8.4) is similar to the geo-referenced short baseline system (Figure 8.3) except that the reference is with the ship. The transducer array is placed on a measured drawing of the target vessel

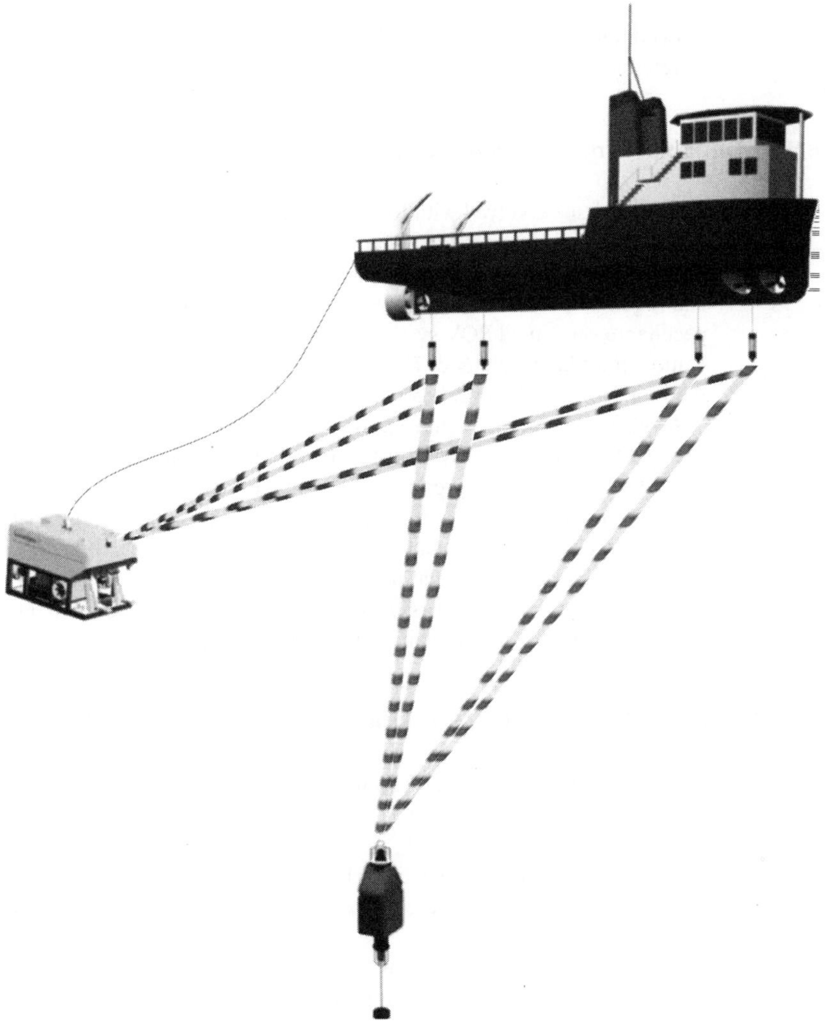

Figure 8.3 *Short baseline positioning setup (graphic by Sonardyne).*

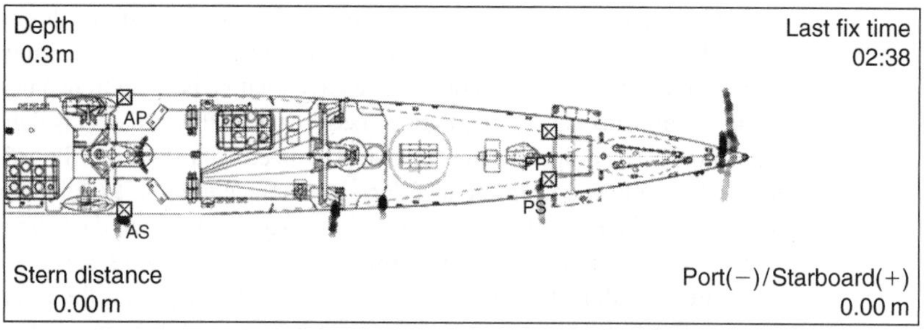

Figure 8.4 *Vessel-referenced acoustic positioning ship hull files (graphic by Desert Star LLC).*

with all transducer placements calculated based upon that scaled drawing. The frame of reference is relative to the ship.

8.6.3 Difficulties Involved with Sonar and Acoustic Positioning

The major issue involved with ROV-mounted imaging sonar is image interpretation. A basic target of interest can either be identified by acoustic reflection or by acoustic shadow. Unfortunately, image interpretation is in many cases counter-intuitive and requires special instruction in sonar theory and application. Further difficulties peculiar to very small ROV systems, due to the vehicle movement during image generation, must be understood. This so-called 'image smear' is produced from moving the vehicle/sonar platform before allowing the full image to generate.

Acoustic positioning within a port environment is problematic. A port, by its location and function, is a noisy acoustic environment. Broadband noises bounce around the water space like a ping-pong ball, causing false narrowband reception and reducing the signal-to-noise ratio of the primary positioning signal. Multi-path errors also cause difficulty (a narrowband sound bouncing between a hull and a pier wall can spoil round-trip sound calculations due to false reception).

With all of the difficulties associated with these technologies, with proper training and implementation they remain powerful tools to accomplishing port security tasks.

8.7 TECHNIQUES FOR ACCOMPLISHING PORT SECURITY TASKS

ROV-based underwater port security tasks involve two broad categories of inspections:

- Identification of targets located by other means
- Clearing an inspection area of 'suspicious items'.

It is possible to discover an item of interest while making random searches of an area. However, it has been found to be of marginal benefit to conduct unstructured searches of suspect areas. As water visibility decreases, a high degree of certainty that an area of interest has been cleared, with all suspicious items discovered, becomes

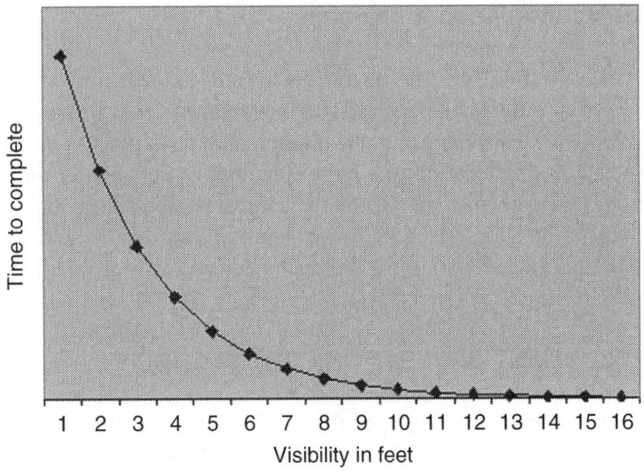

Figure 8.5 *Visibility versus time to complete inspection task.*

increasingly difficult, as does positive navigation through the search area. In order to construct a maximum risk/benefit model of search time covering the high-risk exposure, a combination of tools and techniques must be used to achieve best results. These are discussed below.

8.7.1 Hull Searches

The highest risk sections of commercial and military seagoing vessel hulls are isolated to a limited set of landmarks located on the hulls of these vessels. These targets can be located and checked in a relatively short time, thereby quickly eliminating the high-risk areas. Later, a search pattern can be implemented to image a specified percentage of the hull area.

One hundred percent visual hull coverage is exceedingly difficult to achieve due to navigational considerations as well as environmental factors, with the time–requirement curve turning nearly vertical as the water visibility nears zero (Figure 8.5). Experience has shown that the 80/20 rule applies in hull inspection missions, where 80 percent of the hull can be inspected in 20 percent of the time it would take to achieve full hull coverage. The last 20 percent of hull coverage is the most tedious and time-consuming.

8.7.2 Under-Vessel Bottom Searches

Achieving bottom clearance beneath berthing of large vessels is time-consuming without the use of imaging sonar to identify items of interest. Bottom visibility conditions become difficult due to poor lighting as well as bottom stirring from vessel traffic, ROV submersible thrusters, and silt movement from tidal flows. Results may be improved by using technology to discriminate anomalies to isolate 'items of interest' (with imaging sonar, magnetometer, and other instruments), then positively identifying these targets with an ROV-mounted camera, thus eliminating these as threats.

8.7.3 Pier/Mooring/Anchor Searches

Identification of the threat could aid in reducing the time spent searching pier/mooring/anchor areas. The further away from the high-value asset the item to be inspected resides (e.g. the placement of an explosive charge 20 pilings away from the berthing location of a large container vessel would produce minimal damage to the hull), the lower the likelihood a threat will be placed in that location. Planning the operation should take into consideration such factors and limit the time inspecting lower-risk areas or ignore them altogether.

8.8 DEVELOPMENT OF TTPs FOR PORT SECURITY

An earlier project included the development of a list of 'best practices' involving a series of underwater port security tasks, followed by testing the practices over a range of ROV equipment.

The tasks comprised the following major areas:

- Ship hull searches (pier-side and at anchor)
- Pier searches
- Bottom searches to include directly under vessels
- Day and night operations
- Inclement weather operations
- Launch/recovery operations
- Tether entanglements.

During the procedure testing, it was quickly noted that the size and power of the individual systems translated into capabilities that fell into the three general size categories (small/medium/large as stated in Section 8.2 above, see Figure 8.7).

The large ROV systems demonstrated capabilities to complete difficult tasks that required power to muscle through long offsets as well as strong wind/current combinations (Figure 8.6). The smaller systems failed under the same harsh conditions. The larger systems, however, were too bulky to be easily accommodated aboard the response boat (RB), causing difficulty in movement for the crew while underway (Figure 8.8).

The tasks tested and performed included:

- Ship hull inspections (moored/anchored)
- Pier inspections
- Simulated hazmat spills
- Simulated potable water environment.

8.9 RESULTS OF PROCEDURES TESTING BY SIZES

The power of large systems countered the lack of experience with the operators. The smaller systems took a higher degree of operator proficiency, task planning, and tether handler coordination than did the larger systems to accomplish similar tasks.

(a)

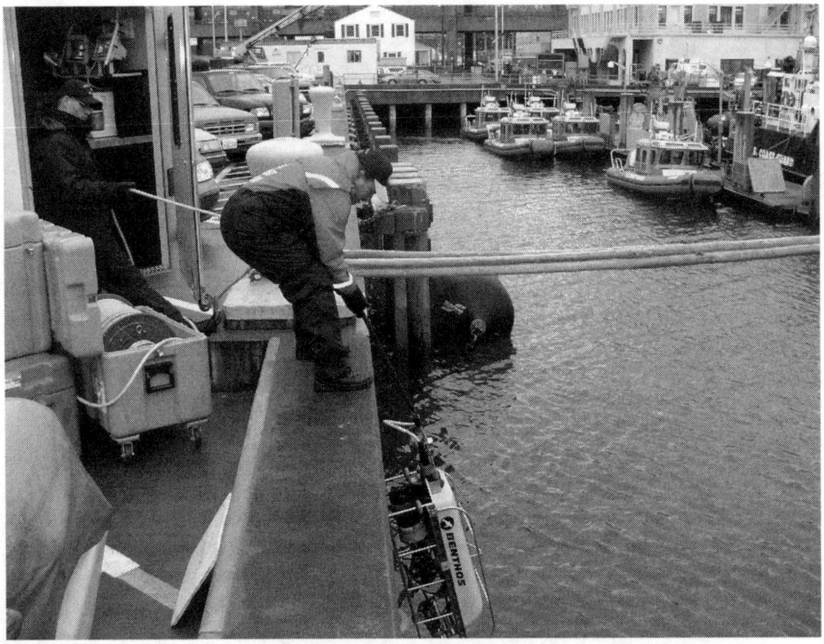

(b)

Figure 8.6 *Larger ROV systems function well with distant offset but are more difficult to handle than smaller systems.*

Figure 8.7 *Size comparison between ROV systems tested.*

(a)

Figure 8.8 *Large ROV aboard response boat showing excess green tether.*

(b)

Figure 8.8 *(Continued)*

Table 8.1 *Task completion versus current and tether offset.*

Prevailing current (kt)	*0*	*0.5*	*1.0*	*1.5*	*2.0*	*2.5*	*3+*
Small system	E	D	VD	NR	NR	NR	NR
Medium system	E	E	D	VD	VD	NR	NR
Large system	E	E	E	D	VD	VD	VD

Example task assumes 50 feet of tether offset to item of interest.

Note: The current and offset decides whether the system can accomplish the task. Visibility decides the time to accomplish this task.

The smaller systems faired better in untangling their tether and in functioning in confined spaces. The results of the tests broke down into a series of what were termed degrees of difficulty:

- E – Easy
- D – Difficult
- VD – Very difficult
- NR – Not recommended.

Table 8.1 provides a summary of the results. A detailed discussion of port security task expectations based on vehicle size is presented in Chapter 12.

As discussed in detail in Chapter 2, the ROV must produce enough thrust to overcome the drag produced by the tether and the vehicle. Although there are techniques to aid the lower-powered systems, the obvious message is that the tether drag on the vehicle is the largest factor in ROV deployment and usage. The higher the thrust-to-drag ratio and power available on the ROV system, the better the submersible will be able to pull its tether to the work site.

ROVs are being used around the world for law enforcement as well as homeland security tasks, thus protecting our ports and harbors from threats to shipping and commerce. Central to that use is the establishment of procedures and management to allow better security through cost effect and more frequent inspections.

Chapter 9

Explosive Ordnance Disposal and Mine Countermeasures

Although there will be some mines with improved capabilities, the greatest threat will be sheer numbers, rather than technological sophistication.

Lt. Gen. Rhodes and RDML Holder's doctrine paper entitled 'Concept for Future Naval Mine Countermeasures in Littoral Power Projection.'

As discussed below, the development of Mine Countermeasure vehicles has radically changed direction, driven by the threat of increased numbers vice sophistication.

9.1 BACKGROUND

A sea mine's purpose in warfare is a basic 'access denial' function for disrupting an enemy's sea navigation capability. If an amphibious landing is anticipated, a clear path to the beach requires clearing of sea mines to allow the vessel an access path. The same concept applies to shipping, since a harbor or navigation channel can be immediately shut down with any threat of sea mine presence.

Sea mining has not changed considerably since the days of the American Civil War, when Admiral Farragut declared 'Damn the torpedoes' (older term for sea mines). Sea mines are still, to this day, extremely difficult to locate and neutralize. Technology, however, is allowing for enhanced location/identification capabilities as well as safer and more environmentally friendly means of neutralizing these sea mines.

Accordingly, the development of ROVs for use in Explosive Ordnance Disposal (EOD) and Mine Countermeasures (MCM) missions is a military mission. This was both a blessing and a curse. The need for such systems was there, as was the funding, but the design specifications were such that the vehicles became large and expensive. The complexity of early MCM vehicles also lengthened their time of development. For example, the US Navy's Mine Neutralization System (MNS), AN/SLQ-48(V), took nearly two decades to move from concept to operational status (Figure 9.1). Today there are approximately 60 in the field.

The USA was not the only country building MCM vehicles. ECA of France had the workhorse of the early MCM category, the *PAP 104*, which hugged the bottom using a drag weight. Today, ECA has fielded approximately 450 PAP vehicles, 120 of which are the new PAP Mark 5 systems (Figure 9.2).

Other heavy-duty MCM vehicles include:

- Germany's Pinguin B3 (Atlas Elektronik GmbH)
- Sweden's Double Eagle line (Saab Underwater Systems)

Figure 9.1 *The US Navy's Mine Neutralization System.*

Figure 9.2 *ECA's PAP Mark 5.*

- Canada's Trailblazer line (International Submarine Engineering)
- Switzerland's Pluto line (Gaymarine srl).

All of the larger ROV MCM vehicles have the capability of delivering a small acoustically activated explosive charge to the target. The goal is for the MCM ship to locate the mine ahead of time using the ship's MCM sonar, send out the vehicle (while tracking it with the ship's sonar) to verify the target, and plant the charge. Once the charge is planted, the vehicle is retrieved and the charge (previously installed by the vehicle on the mine) is then acoustically detonated. This will, hopefully, destroy the mine by either sinking it (thus placing it outside the path of shipping), destroying its detonation capabilities (i.e. damaging the detonation mechanism or flooding the mine casing with sea water, thus rendering the explosive charge inert) or setting off the mine's warhead while maintaining a safe distance. Disarming a live sea mine is an EOD function.

An older technique of floating the moored mine has essentially disappeared. Under this method, the ROV or diver attached an explosive charge to the cable of a moored mine to sever the mooring, thus floating the mine to the surface. The mine is then destroyed with small arms fire or with the deck gun. However, the last scenario any ship captain wants is to have a mine floating somewhere on the surface. Even if the mine could be located, the mine could be lost while floating to the surface, thus exacerbating the problem. Therefore, the best mission plan is to destroy the mine in place.

To neutralize the mine in place without 'cooking off' the mine's warhead, two basic techniques are used: (1) Shaped charge/projected energy or (2) projectile. The shaped charge, while probably destroying the mine, definitely destroys the vehicle. Once the vehicle (and, hopefully, the mine) has been destroyed, the mine must be reacquired to verify that it has been neutralized (a painfully tedious process after the bottom has been stirred from the explosive charge). The projectile method allows the vehicle to be reused repeatedly, thus saving the time to reacquire the target for battle damage assessment as well as the repeated transit time for the vehicle from the launch platform to the minefield.

The long chess game of mine disposal ends when the mine has been positively characterized as inert. Under the one mine/one vehicle concept of operation, the mine must be reacquired after the mine neutralization vehicle warhead discharge to conduct a Battle Damage Assessment to verify elimination of the threat. The time and cost for the operations platform to reacquire the target is significant and may vastly outstretch the cost of the mine neutralization vehicle itself. Clearly, a reusable vehicle is preferable to an expendable vehicle from a cost and time perspective.

The change in the mine neutralization method, from explosive charge to projectile, may provide the following benefits:

- Lower environmental disruption, allowing for increased stealth
- A range of sizes available to fit the mission requirements
- Reusability of the vehicle
- Long loiter time in minefield, reducing vehicle transit time to operations area
- Immediate muzzle reload with automatic rifle mechanism
- Rapid magazine reload upon retrieval of the vehicle
- Lower risk from possible sympathetic discharge of the mine's warhead
- Environmentally friendly method of mine neutralization
- Immediate Battle Damage Assessment after delivery of the projectile to the mine
- More operational use and training due to reuse of the vehicle
- Built to standard communications protocols for rapid platform switching
- A range of weapons attachable to the vehicle, allowing scalable response
- Lower cost structure to expendable Mine Neutralization Vehicle
- Rapid prototyping with COTS components.

Further, a range of applications outside of MCM may exist, which are limited only by the warfighter's imagination:

- Torpedo tube swim-out from submarine for rapid ship's swimmer defense
- Harbor protection from swimmer attack allowing measured response with long loiter time (with both lethal and non-lethal payload)

- 'Shot across bow' ability for a range of threats encountered in law enforcement
- Usable from either submersible-powered or surface-powered vehicle
- Tele-operated or fully autonomous modes are programmable, producing a hybrid ROV/UUV.

9.2 EOD APPLICATIONS

As for EOD applications, the early ROVs didn't have the capability to counter the turbid, dynamic, near-shore environment that was the diver's domain. Little has changed in that area to date. Considerable research is being performed in this area, but if you can't see the target, you can't destroy it.

However, one area outside of the near-shore environment where EOD divers have a tough job is the inspection of ships' hulls to counter terrorist threats. Considerable research and demonstrations are being conducted to evaluate the applicability of ROVs (and AUVs) to inspect ships' hulls. By using high-resolution sonars to view the hull, advanced navigation techniques for tracking the vehicle's location while maintaining a constant distance from the hull (from several inches to a few feet), and cameras to verify the target, the burden of the EOD diver is again being reduced through robotic technology.

Other areas where the workload of the EOD diver can be considerably reduced through the use of robotic technology will be discussed in the following sections.

9.3 MCM TODAY

Advances in the miniaturization and efficiency of sensors necessary for mine countermeasures have brought the use of robotics in this mission to a new level. The MCM mission breaks down simply to 'find it' and 'kill it'. Or, in official Navy parlance, the two phases are 'Search, Classify, Map (SCM)' and 'Reacquire, Identify, Neutralize (RIN)'.

The 'reacquire' aspect of this new approach is the key to how MCM is being planned for the future. Two systems are anticipated; one to perform the SCM phase and one for the RIN phase. These two systems could actually be the same vehicle, although outfitted with a different payload.

The ability to find the mines while the operating platform is at a safe distance is critical. To solve this problem the US Navy is looking at AUVs. With small, capable sonars, excellent navigation, and the ability to store large quantities of data on the vehicle, AUVs are looking very promising to map the target area.

The US Navy used AUVs successfully during Operation Iraqi Freedom. The Naval Special Clearance Team One used the REMUS AUV (Figure 9.3) to successfully perform mine countermeasure missions (MCM). The use of the AUV allowed them to map the area, which not only reduced the time to clear the harbor, but made it a much easier job for the EOD divers.

Other companies developing AUVs for the SCM mission include Bluefin Robotics (USA), ECA SA (France), Hafmynd Ltd (Iceland), and Kongsberg (Norway).

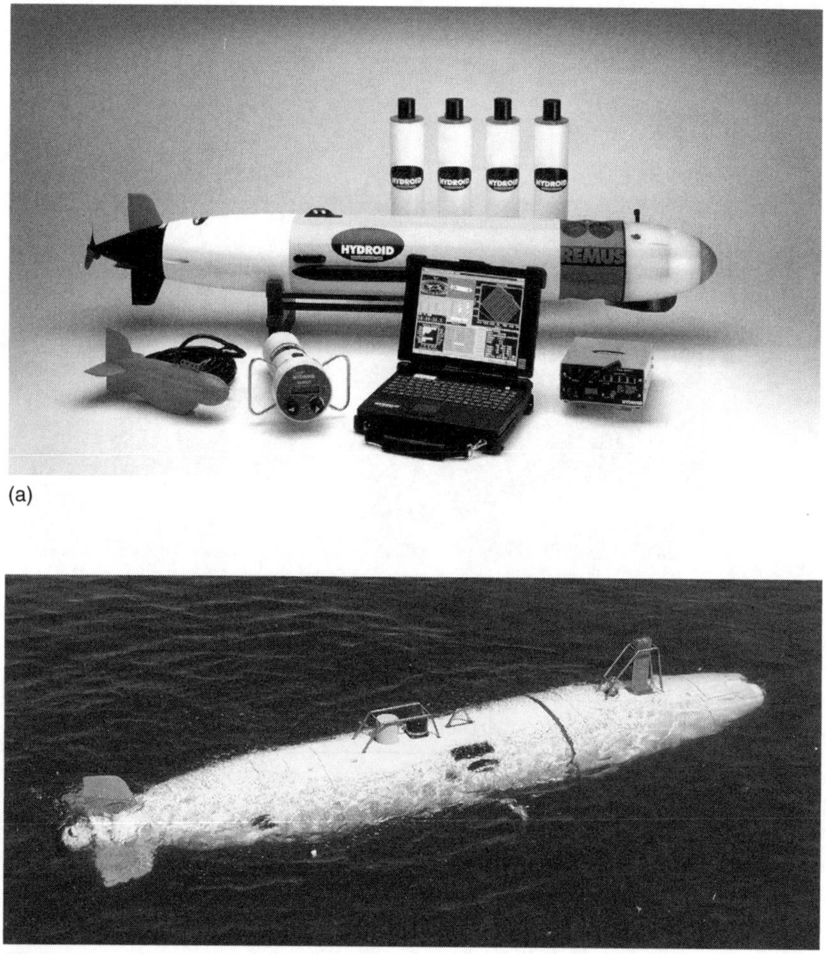

(a)

(b)

Figure 9.3 *REMUS AUV.*

9.3.1 Expendable MCM Vehicles

Of particular interest is the second portion of the mission – the RIN. Four companies that have developed a 'killer' ROV with use of projected energy follow:

- BAE System's Archerfish
- ECA SA's K-Ster
- Kongsberg's Minesniper
- Atlas Elektronik's Seafox.

An example of the method in which these systems are used is provided in Figure 9.4.

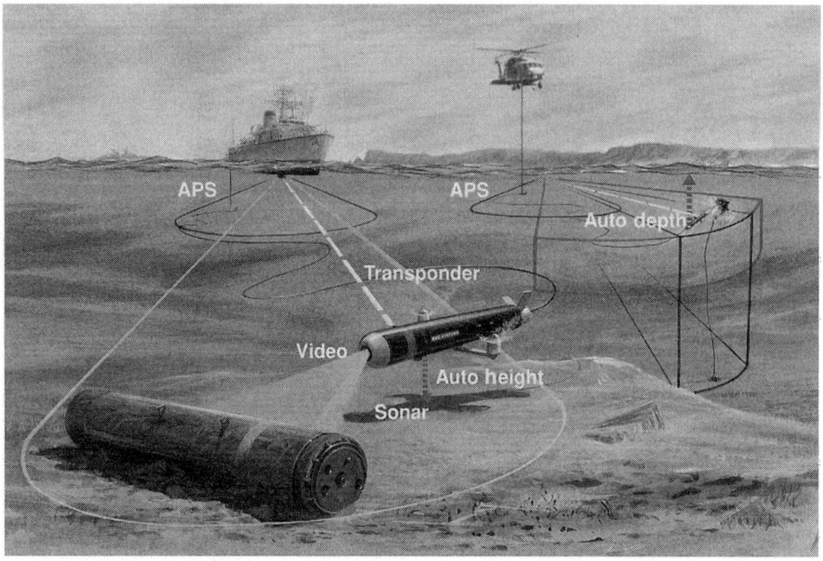

Figure 9.4 *Archerfish operational scenario (courtesy BAE Systems).*

Table 9.1 *Expendable MCM vehicle specifications.*

	Archerfish	*K-STER*	*Minesniper*	*Seafox*
Length (m)	1.05	1.45	1.8	1.3
Width (m)	0.135	0.23	0.48	0.4
Height (m)	0.135	0.23	0.17	0.4
Weight (kg)	15	50	39	40
Depth (m)	300	300	500	300
Range (m)	Unknown	Unknown	4000+	1200
Sonar	Yes	Dual freq.	Yes	Yes
TV	Yes	B&W, color	CCD	CCTV
Navigation	Trackpoint II	Yes	SBL	INS

In development within the USA is a new family of vehicles known as the 'K2' family of vehicles using projectile delivery as its primary mean of mine neutralization.

Each of these systems will be described in more detail in the following sections. However, since all are similar in concept (i.e. they use an optical communication cable, have a speed of 6–7 knots, carry cameras, sonar, navigation systems, etc.), their pertinent physical and operational specifications are provided in Table 9.1.

9.3.1.1 *Archerfish*

The Archerfish (Figure 9.5), developed by BAE systems, is a fiber-optically guided, single-shot mine disposal system (Figure 9.6) that is available in both exercise

Figure 9.5 *Archerfish (courtesy BAE Systems).*

Figure 9.6 *Archerfish prosecuting bottom mine (courtesy BAE Systems).*

and warhead variants. Twin thrusters allot it to hover and transit. Maximum speed is 7 knots.

The US Department of Defense selected a team of Raytheon and BAE Systems for the demonstration and development of the Airborne Mine Neutralization System (AMNS). The AMNS will be integrated into the US Navy's MH-60 helicopter.

9.3.1.2 *K-STER*

France's ECA SA, which has been expanding their line of undersea vehicles, has added the K-STER Mine Killer (Figure 9.7). The vehicle is an Expendable Mine Disposal

Figure 9.7 *K-STER during launch (courtesy ECA SA).*

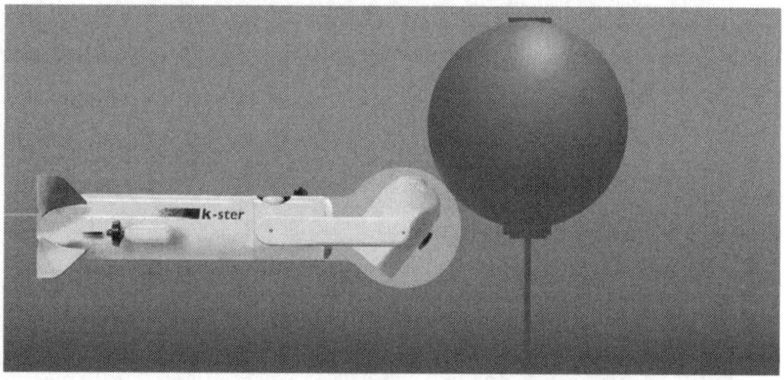

Figure 9.8 *K-STER against tethered mine (courtesy ECA SA).*

System (EMDS) that comes in both a positively buoyant, inert training variant (K-STER I) and the negatively buoyant disposal vehicle (K-STER D). The system features a unique tilting head that carries the sensors and shaped charge (Figures 9.8 and 9.9).

9.3.1.3 *Minesniper*

Kongsberg Defence and Aerospace has completed the development of the Minesniper 'one-shot' mine destructor vehicle. The Minesniper is controlled via a fiber-optic tether and can be fitted with either a shaped charge or semi-armor-piercing warhead to use against sea mines. The expendable vehicle is powered by twin rotatable thrusters that give the Minesniper the ability to rotate around its own axis. Minesnipers have been sold to the navies of Norway and Spain. It can be used from ships or helicopters.

Figure 9.9 *K-STER neutralizing tethered mine (courtesy ECA SA).*

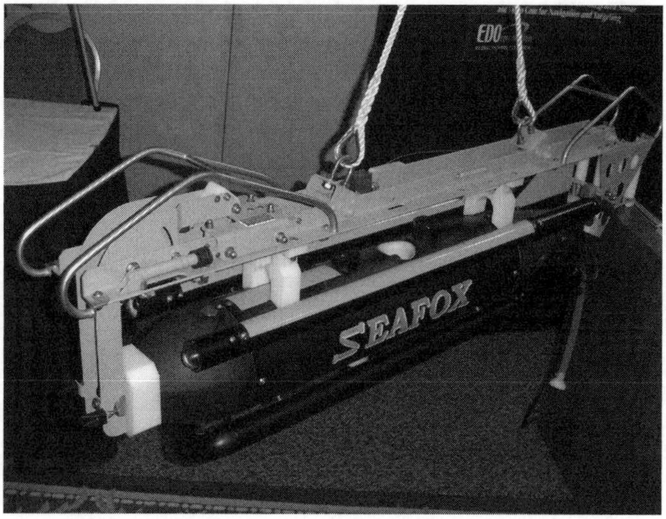

Figure 9.10 *Seafox.*

9.3.1.4 *Seafox*

The Seafox, developed by Atlas Elektronik, is a fiber-optic guided ROV that can be used for the SCM mission (Seafox I), or it can be loaded with an explosive charge and used for the RIN phase (Seafox C, for combat). The vehicle can be used against short and long tethered mines and proud bottom mines. The vehicle system is made up of a console, a launcher, and the Seafox vehicles.

The Seafox (Figure 9.10) uses its integrated sonar to reacquire the target and a CCTV camera to identify it. Once identified, the vehicle can use four independent, reversible thrusters and one vertical thruster for high maneuverability prior to firing the shaped charge. The system, which has been integrated into several navies, is capable of operating from several platforms, including dedicated MCM vessels, surface combatants, craft of opportunity, and helicopters.

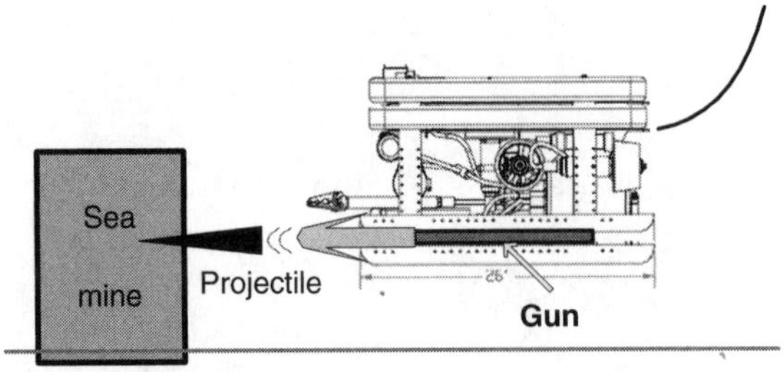

Figure 9.11 *K2 Concept drawing.*

9.3.2 Non-expendable MCM Vehicles

The K2 family of 'non-expendable' vehicles (Figure 9.11), like the expendable MCM ROVs above, is a fiber-optic or copper-guided ROV that can be used for the SCM or RIN missions. The vehicle is either surface powered or can be battery operated. The vehicle, with sensor and navigation package, is flown into the minefield, where it identifies/classifies the target then subjects the target to repeated projectiles until the target has been neutralized. Once the target has been assessed as neutralized, the vehicle proceeds to the next target.

Chapter 10

Public Safety Diving

Many of the useful applications for this technology are in the field of public safety diving. Several public safety divers (PSDs) are killed each year in the line of duty due to the hazardous environments in which they operate. Unmanned searches, as well as electronic search techniques, have the capability to significantly increase the efficiency of the search and lower the risk to field operations while enhancing the chance of success. With the right equipment, in properly trained hands, good results can be obtained with the ROV in the dive team's tool chest. This chapter will explore the basics of finding, documenting, and recovering items of interest to the PSD through the use of electronic search techniques.

10.1 PUBLIC SAFETY DIVING DEFINED

Public safety diving is normally defined as diving operations 'by or under the control of a governmental agency' for the service of the general public. This covers all aspects of search, rescue, recovery, criminal investigations, and any other functions normally covered by police departments, fire departments, and other applicable governmental agencies. Examples of public safety diving would be crime work done by the federal, state and local law enforcement agencies, search and rescue missions performed by police/fire departments, and disaster relief operations. Examples of operations excluded from this definition would be commercial inspections of public sewer outflows, inspection of bridge abutments, and other services not in the direct service of the general public.

10.2 MISSION OBJECTIVES AND FINDING ITEMS UNDERWATER WITH THE ROV

ROVs are a supplement to the PSD's tool chest. Under current technologies, ROVs are becoming much more useful and pertinent to the PSD team's operational mission, providing an enhancement to its capabilities. The function of the ROV in this long chess game of finding items of interest underwater is in the end game as a final identification tool. The ROV has limited search and some recovery capabilities. The main function of the ROV is to lower the physical risk to the PSD by putting the machine at risk in situations and environments that have been traditionally borne by the PSD.

The PSD's objectives in performing an underwater search follow:

- The isolation, securing, and defining of the search area
- The clearing of the search area of possible targets
- (Upon location of the items of interest) the gleaning of what information from the site is available, to department standards of documentation
- The disposal of the item as deemed necessary by the command structure.

The actual search technique for locating bottom targets with an ROV system is to run the search pattern across the bottom as high off the bottom as possible, based upon the water visibility, while maintaining visual contact with the bottom. The search is run either with a single camera display (which most observation-class ROV systems possess) performing a zigzag pattern across the bottom along the search line (to search on either side of the submersible) or in a straight line with multiple cameras providing wider camera coverage. It is imperative to keep visual contact with the bottom at all times while transiting the coverage area, if for no other reason than to raise the possibility of a 'lucky find'. So-called 'spaghetti searches' (unstructured look around searches within the search area) are quite popular in getting the immediate result of 'searching' and performing a task, but it has been consistently deemed that a clear, coherent, and structured clearing of the search area is the only productive means of performing an underwater search.

For an expanded explanation of public safety diving operations, refer to Teather (1994) and Linton *et al.* (1986) in the bibliography. There are also excellent training courses available to police and fire departments for training as a public safety diver in traditional scuba diving techniques.

10.3 WHEN TO USE THE DIVER/WHEN TO USE THE ROV

The ROV or the PSD should only be put into the water when a target is 'trapped' by some other means (i.e. a sonar/magnetometer target is identified or a narrow search area is identified). The general search navigation procedure should be to start at a known reference point, finish at a known reference point, and maintain positive navigation throughout the search phase so that all of the search area is positively covered with the search instrument.

For the grid search to cover a defined geographical area, search lines should be spaced to the extent of the range of the equipment (with an approximate 10 percent overlap of coverage for error margin):

- For sonar equipment – to the effective range of the system used
- For visual equipment – to the effective range of the ROV's camera or diver's sight (i.e. clarity of the water)
- For magnetometer equipment, etc.

With a trained and proficient ROV operator on the dive team's staff, all of the mundane search/identification of possible targets should be performed by the ROV up to the final identification of the item. The diver should get wet if the physical capability of the ROV system is insufficient to handle the mass/bulk of the item or if the item is of such a fragile or sensitive nature that an ROV would be inappropriate.

10.4 SEARCH THEORY AND ELECTRONIC SEARCH TECHNIQUES

An ROV can be thought of as a delivery platform for a series of sensors. Since ROVs are inherently slow-speed platforms, the full search toolbox should address the needs dictated by the operational requirements. If a very small search area is defined (e.g. a possible drowning victim was seen falling off of a dock in an exact location), an ROV with just a camera may be the only requirement if the water clarity is good. If the water clarity is bad, a scanning sonar may be needed to image the victim acoustically. If a large area is defined (e.g. a drowning victim's last seen point is unknown, thus the search boundaries cover a very wide area), other sensors decoupled from the ROV would be much more effective in covering the search area.

The basic problem with any search is that until you find the item of interest, your time is spent finding where the item 'is not' located. The objective of searching is to positively eliminate an area of interest as a target area with as high a degree of certainty as possible. The job of a public safety diver is to clear that area as quickly and efficiently as possible, given the resources available.

Finding items underwater is a chess game from opening moves to end game. The search area is defined then swept, to identify possible targets that will need to be prosecuted or classified. As introduced in Chapter 8, search and identification of underwater items is accomplished through four basic steps:

1. *Research* the operation to define the area of interest for mapping and limiting the search area. Examples of items used to outline the search area follow:
    a. Intelligence
    b. Witness interview/last seen point
    c. Survey data, prior searches, etc.
2. *Wide area search* with instruments and sensors such as:
    a. Towed or watercraft/aircraft-mounted imaging and profiling acoustics (such as side-scan sonar)
    b. Magnetic anomaly detection equipment (such as a towed magnetometer)
    c. Radiological and chemical instrumentation, etc.
3. *Narrow area search* with slow-speed instruments including:
    a. Acoustics (such as a fixed location mechanically scanning imaging sonar)
    b. Fixed or towed magnetometer
    c. Optical equipment (towed/drop/ROV-mounted camera or laser line scanner), etc.
4. *Final identification*, classification, discrimination, and disposition of the item by a human – i.e. through the diver's eyes or through the camera mounted on an ROV that is propelled to the inspection site and viewed by the operator.

The ROV's place within the PSD's tool kit is through the narrow area search and final identification phases of the search plan. ROV intervention as a productive and cost-effective means of final item identification, discrimination, and disposition is the focus of this section. The four basic steps are discussed in more detail in the remainder of this chapter.

10.4.1 Planning and Research

The planning and research segment of a search operation is by far the most important yet least practiced segment. A PSD dive team can never obtain too much information before a search. Inexpensive time spent asking questions and pouring through data could save hours of expensive, high-risk, resource-intensive time on the water.

Suppositions can be made based upon witness interviews and some basic science. Calculations of travel time and distance to impact on accidents and drowning cases help narrow the search area. For an expanded text on drowning physics and underwater crime scene investigations, Teather's work (1994) on the subject (referenced in the bibliography) is excellent.

There is a plethora of information available from many sources that all contribute to the data needed to define the search area as well as to gaining an understanding of the conditions and challenges faced while on the water. Examples of these public-domain files include:

- Geographic Information Systems (GIS) database for the search area maintained by the local township
- The US Coast Guard, State Environmental Agencies
- Local land survey companies
- The US Geological Survey
- US Board of Reclamation
- Previous searches within the (or other inter-agency) department
- Fishing maps
- Archeological data from local universities.

Data can be obtained from any number of other sources. A competent and effective PSD will spend time locating/gathering these items of information before being faced with a time-sensitive search. Examples of the types of data acquired during two prior searches, one an archeological search and the other a murder victim search, follow.

10.4.1.1 *Archeological search*

A good example of research was during a prior job cataloging the WW2 wreck sites of torpedoed tankers in the Gulf of Mexico in support of an effort with a Southern Louisiana university. Specific steps, information gathered, and issues encountered follow:

1. Petitioned and received the then-named Defense Mapping Agency (DMA) for non-submarine contacts within their database.
2. Received from the US Coast Guard's (USCG) 8th district, all submerged obstructions within their database.
3. Received from the National Oceanic Survey (NOS) their database of wrecks.

4. Received from Texas A&M University their 'Hangs' listing of all reported fishing net hangs within the geographical area in question.
5. Received from the US Minerals Management Services its archeological database of historically significant items.
6. Obtained copies of pipeline survey maps that displayed high-accuracy pipeline coordinates for positional referencing.

The data was not ready for populating the GIS system, since all of the databases were in different coordinate systems:

1. The MMS database had all information in survey block coordinates based upon North American Datum 1927 (NAD27) survey State Plane Coordinate System (since that is what the offshore oilfield leases were drawn with in the 1960s) in *x*/*y* coordinates from the reference datum point.
2. The USCG, NOS, and DMA coordinates were all in World Geodetic System 1984 (WGS84, which is the current coordinate system used by most charting systems), but the reported coordinates were sometimes as much as 70 years old, taken via dead reckoning and celestial sightings. A 'reliability quotient' had to be assigned to the positioning coordinates.
3. The Texas A&M 'hangs' listing was in Loran A 'TDs' (i.e. timing differences), which required conversion to latitude/longitude coordinates before processing.
4. The pipeline survey maps were all in NAD27. The pipeline maps had to be merged with the production platform coordinates to accurately plot the subsea pipeline track away from the production platform. The map drawings had 'shot points' where high-accuracy position readings were taken along the length of the pipeline at measured points; The remainder of the string was 'connect the dots' type of dead reckoning. Therefore, all of the coordinates had to merge to the pipeline map, so they all needed to be converted to the NAD27 coordinate system.

Once all of the data had populated the GIS map, the data points should clump onto the site of the actual wreck. As an example, the *R.W. Gallagher* was a US Merchant Marine tanker sunk in July 1942 by the German submarine U-67. The DMA, USCG, NOS, and NOAA databases had that wreck in three separate locations – dives were conducted on that 'wreck' and it was not at any of those locations. Regardless of the success of such searches, the bottom line is that without proper planning, the search team is probably destined for failure.

10.4.1.2 *Murder victim search*

A second example involved the search of a water-filled abandoned rock quarry for a murder victim who was purportedly cut in half, stuffed into two separate acid-filled sealed barrels, then rolled off a cliff into a 100-foot water depth isolated location. Although the victim was not found, the data gathered to ensure an efficient search included:

1. Satellite and aerial photos from Terraserver.
2. County survey data from county records.

3. Company records of excavation from landowner.
4. Extensive witness interviews.
5. Sheriff's investigators rolled water-filled test barrels off the cliff to test the trajectory, then marked the search area off with buoys.

10.4.1.3 *Environmental considerations*

One would prefer to swim in the warm waters of the Equatorial regions on sunny cloudless days. Unfortunately, the PSD does not have the luxury of knowing the time and choosing the place of the team's deployment. Most PSDs have within their jurisdiction aquatic areas comprised of lakes, rivers, canals, and coastal waterways where most human activities prevail. These areas are normally of low visibility and/or of a high current/tidal flow that create difficult search conditions. Visual instruments are of limited use for such environments, making the intelligent use of other sensors practically a requirement.

In theory, with low visibility conditions, items of interest should be first trapped by a non-visual sensor then held in a steady-state condition (e.g. for a sonar target, it should be held 'in sight' on the sonar screen) so the diver or ROV can positively navigate to the target. If conducting area searches, it is most convenient to use search lines of equal depth to limit the maneuvering to *x*/*y* coordinates without throwing in the additional '*z*' factor of constantly changing depth.

Smooth, flat bottom conditions are the optimum type of underwater topography for searching with sonar. Since the target can be insonified on sonar much more easily on a flat bottom, the sensors will pick up minimal 'false targets'. For rocky bottom conditions, more time must be allocated to clear the area, since the target may have fallen against or into some of the rocks. This would make identification visually or acoustically difficult.

Additional detail on environmental and bottom conditions that should be kept in mind when preparing for a search are provided in the following sections.

Types of environments

- *Rivers* – Of the search environments confronting the public safety dive team, the most difficult class is the river environment. Rivers normally drain from higher elevations and have both suspended particulate matter and high currents. Rivers also often have any number of bottom obstructions, highly concentrated surface vessel traffic, and soft, loosely consolidated or unconsolidated mud bottoms.
- *Streams* – These normally possess easy access with clear visibility and shallow depths (but possibly higher currents). Streams are normally searched visually or with shallow water, snorkeling divers.
- *Lakes* – The lake environment is probably the easiest underwater search environment due to its usual lack of currents, generally good visibility, and relative lack of concentrated surface traffic.
- *Estuaries and littoral waters* – In general, the closer the body of water is to mountainous terrain, the higher the water flow and the higher the visibility of the water (with some exceptions). In areas of generally flat terrain with field run-off, expect low visibility and difficult search conditions. The challenge of estuaries

and littoral areas is the sheer size of the search area. Covering a large search area without high-speed, high-coverage search sensors (such as a towed side scan sonar or magnetometer) will, in all likelihood, be time and/or cost prohibitive.

Bottom conditions affecting searches

- *Sea grass* – Sea grass and other bottom growth near the surface is fed by photosynthesis and suspended in the water column via air-filled pockets within the structure of the leaves. As explained in Chapter 5, air is highly reflective to sonar and will cause both a sonar echo (possibly displaying a false target) and a sonar shadow on the backside of the echo. If a significant presence of sea growth is encountered in the search area, prepare for a long and difficult search experience with plenty of false targets and fouled search equipment.
- *Rocky bottom* – This type of bottom condition generally reflects sonar signals well, causing a noisy environment for searching and numerous false targets. In order to maintain a consistent sonar display, the gain on the system must be adjusted to a lower setting. However, this sometimes masks the sonar target along with the clutter. The only true straight lines in the environment are man-made, which does help in target discrimination. Rocky bottoms are also difficult for tether management of the ROV since tethers tend to snag around rocky outcropping and get trapped in overhangs.
- *Muddy bottom* – Muddy bottoms permit easy sonar searches due to the absorption of the sonar signals, allowing a higher gain setting. Anything with a higher consistency than mud appears quickly on the sonar display. ROV systems attempting to navigate near muddy bottoms are often plagued with ineffective cameras due to low visibility. Also, when items fall into a muddy bottom condition, they are often enveloped within the mud, causing impossible visual clues and rendering them invisible to high-frequency sonar systems.
- *Sandy bottom* – Of the bottom conditions for searching, the sandy bottom is probably the easiest environment for locating and discriminating targets, both visually and with sonar. The surface tension of the sandy bottom is such that items will not normally bury in a short period of time, nor will the sonar reflectivity produce the false target data present in other less favorable conditions.

10.4.2 Wide Area Search

A wide area search is conducted with equipment and sensors mounted on or towed from a relatively high-speed platform that can cover a wide search area in a short period of time. Examples of wide area search systems include: LIDAR (light detection and ranging) or MAD (magnetic anomaly detection) equipment mounted aboard an airborne platform; Towed or boat-mounted side-scan sonar; Towed video camera and bathymetry equipment.

10.4.3 Narrow Area Search

The narrow area search is conducted with relatively immobile sensor equipment mounted aboard a fixed or slow-moving platform. Examples of narrow area search equipment include:

- Diver performing a rope-guided search
- Tripod-mounted, mechanically or electrically scanning sonar moved from location to location while covering a fixed area
- ROV performing a grid search
- Low-tech cable drags.

An earlier project involved a classic narrow area search for evidence (in conjunction with the US Bureau of Alcohol, Tobacco, and Firearms) on a suspected arson case involving a boat belonging to a high-profile politician. The boat burned within the slip of a floating dock; Therefore, investigators were certain that the clues rested on the muddy bottom, 62 feet below the surface. The visibility on the bottom of the muddy lake in the Southern USA was less than 2 feet. The search area was less then 100 feet in diameter. A tripod-mounted, mechanically scanning sonar was lowered to the bottom to image the entire search area. Two separate ROVs were used to recover all of the items of interest from the search area (over 40 items were recovered in a 6-hour time span) with the sonar used to direct the ROVs to each target.

10.4.4 Some Search Examples

Two good examples of the effect of the environment involved separate unsuccessful drowning victim searches in 2000, both of which were in rocky-bottom lake conditions. The rocky bottom caused such a large false target population that the entire search budget was expended identifying rock outcroppings. The victims were both eventually recovered, which allowed the refinement of the search techniques used.

For illustration of the search process, the efforts made on the two unsuccessful body searches are discussed below, as well as those on a successful search. These will serve as a primer to the issues involved with such an endeavor, as well as highlight the difficulties inherent in body searches.

10.4.4.1 *Drowning considerations*

Prior to conducting a mission that may involve a drowning victim, the following 'rules of thumb' should be considered:

- For drowning victims, the body has up to a 1:1 glide ratio to its landing spot on the bottom. That glide ratio must be adjusted based upon currents for an actual track over the ground. A depiction of this phenomenon is shown in Figure 10.1.
- The average adult weighs 8–12 pounds in water (for an expanded table of weights as well as an in-depth analysis of the drowning issue, see Teather's work on the subject). If the ROV is equipped with a manipulator, it should be sufficient to bring the victim to the surface if a strong hold can be maintained to the body.
- The average temperature crossover point for a body to either remain negatively buoyant or become positively buoyant is approximately 52°F (12°C). Check the water temperature to determine the likelihood that the victim is still near the last seen point.

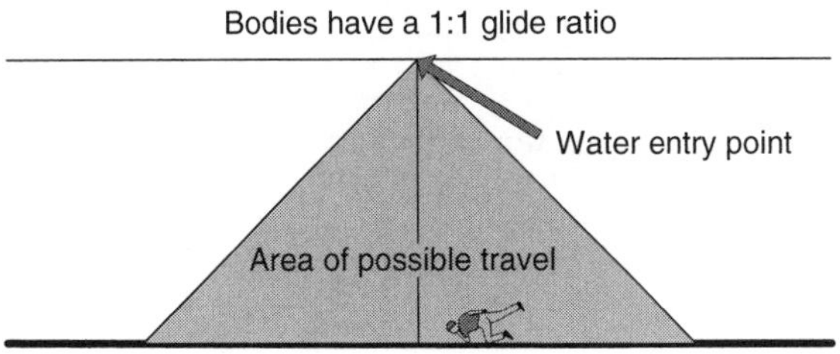

Figure 10.1 *The area of possible travel for a body from the water entry point to the bottom landing spot.*

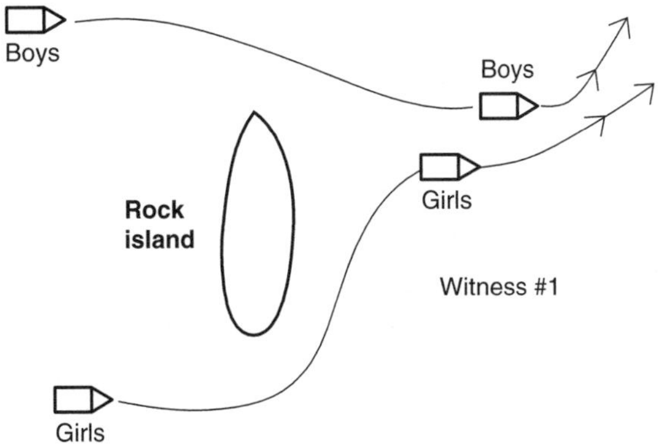

Figure 10.2 *Witness #1 version.*

10.4.4.2 *Scenario 1*

The drowning victim was a 16-year-old boy who was thrown from the back of a rapidly moving Personal Water Craft (PWC), then run over by another PWC in trail. There were two teenaged boys on the lead PWC and three teenaged girls on the trailing PWC. The conditions of the bottom of the rock-strewn reservoir in the North Central USA were very cold freshwater with an average bottom depth of 110 feet and approximately 10-foot visibility. The 'last seen' point was very confusing due to several conflicting witness testimonies. The varying depictions of the last few seconds before the victim was thrown from the PWC follow.

Witness #1 was the middle passenger on PWC #1 (girls). She stated that PWC #2 (boys) cut in front of them after they went around the south side of the rock island, then sprayed them throwing the victim into the path of their PWC (Figure 10.2). They immediately stopped and turned around to render assistance.

Witness #2 was the driver of PWC #2 (boys). He stated that they pulled alongside PWC #1 (girls) then turned left to 'spray' the girls (Figure 10.3). The victim 'fell off' into the path of PWC #1.

Witness #3 was the rear passenger of PWC #1 (girls). She stated that the boys saw them coming, then turned around to give chase (Figure 10.4). The driver of PWC #2

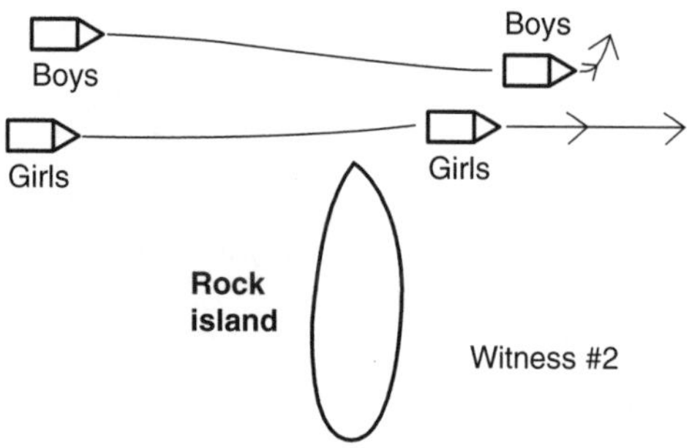

Figure 10.3 *Witness #2 version.*

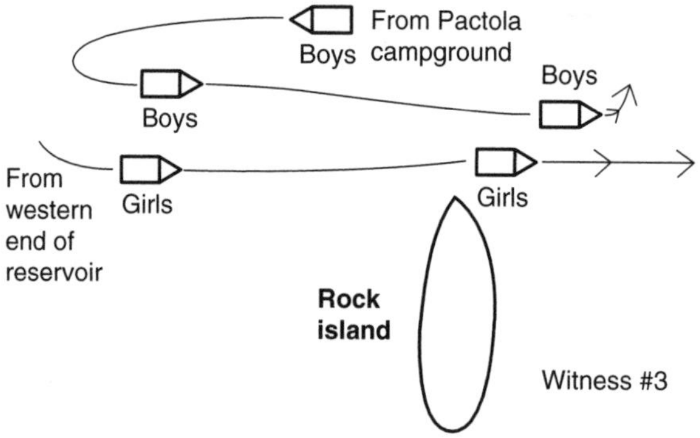

Figure 10.4 *Witness #3 version.*

(boys) did a radical left turn in front of PWC #1, throwing the victim into the path of the oncoming vehicle.

Witness #4 was the driver of PWC #1 (girls). She stated that the boys saw them coming then turned around to give chase (as with witness #3). The driver of PWC #2 (boys) did a radical and complete 360-degree turn in front of PWC #1, throwing the victim into the path of the oncoming vehicle (Figure 10.5).

There was a previous search with side scan sonar of the search area that turned up a large number of possible targets – none of which matched the shape of the victim.

The Fire Chief in charge of the operation was an experienced professional in the public safety field. The second search was much more detailed due to the tedious nature of the bottom conditions. A tripod-mounted, mechanically scanning sonar system was used, deployed from a pontoon boat with a four-point mooring. The Fire Chief had no other choice but to assume that the search area was located in the north-eastern quadrant from the rock island in the lake. He properly determined that the search lines needed to start at the island and proceed in an easterly direction until the allotted time for the search was expended. North/south search patterns were instituted proceeding in an easterly direction. The grids were spaced at 100-foot intervals. The search proceeded in a systematic fashion for 3 days, at which time the search was ended without success.

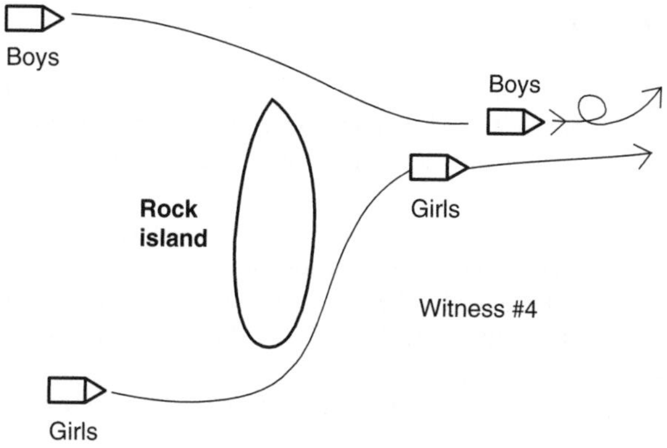

Figure 10.5 *Witness #4 version.*

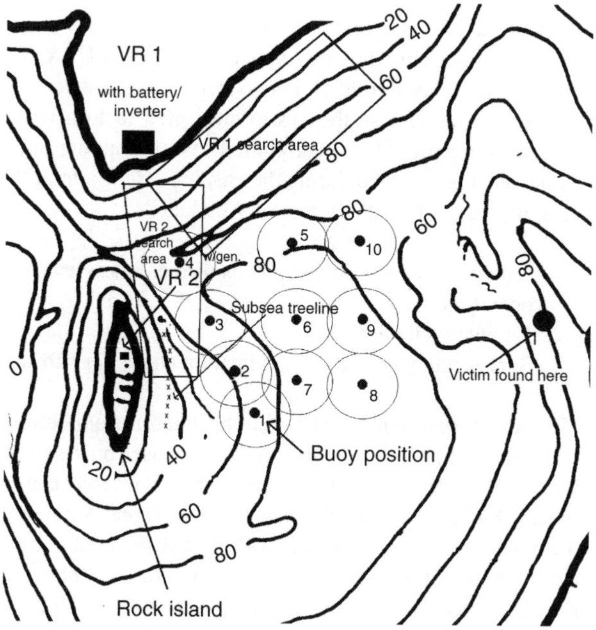

Figure 10.6 *Composite of eyewitness data.*

The victim was eventually located approximately 800 feet east of the rock island by recreational divers (Figure 10.6). The victim was found in an area approximately two search lines further from the line under evaluation when the search was called off. This is a case of an impossible PSD search due to temperature and decompression issues. The search area was difficult to define due to conflicting witness testimonies. The bottom conditions were tedious and difficult, with uneven and rock-strewn conditions causing a high number of sonar targets as well as having any number of places to trap the victim in bottom holes and crevasses.

10.4.4.3 *Scenario 2*

The victim was on a field trip in Western Canada and decided to swim across a 1-mile glacial lake. The victim drowned somewhere during the swim across. A witness last saw the victim swimming with difficulty and took a sight line on the last seen point.

Several days of searching with side scan sonar produced thousands of sonar targets. Discriminating those for evaluation and elimination took more time than the search team had budgeted for the search. The bottom conditions were difficult for divers, but acceptable for ROVs, with 25-foot visibility and an average depth of 125 feet of fresh water. The search area was simply too large due to the soft 'last seen' point. The victim was eventually recovered when the lake was drained for reservoir maintenance.

10.4.4.4 *Scenario 3*

An example of a successful drowning victim recovery is the search for the bodies of an overturned public water taxi in the Eastern USA. The accident happened on a cold, windy day during frontal system passage when the vessel capsized due to the turning moment as the vessel attempted to turn back into port. Rising wind conditions overturned the dual-hulled vessel, trapping the victims within the cabin. Six passengers drowned during the ensuing rescue attempt. Data gathered during the planning phase of the body recovery included:

1. Extensive witness interviews as well as interviews of the survivors and vessel operators.
2. Navigational charts.
3. Survey and work permit application charts for subsurface harbor construction.

The initial search focused on finding a debris field before the tedious task of body search was initiated. Initial search efforts focused on the north side (upwind side on the day of the accident) of the harbor near where the vessel capsized. After several days of searching without results, the data was reanalyzed. A trajectory of the vessel track due to wind drift after the capsizing was plotted, then the search area was redrawn based upon witness interviews of the recovery efforts. What seemed insignificant during the original search effort took on more importance – an LST ('Landing Ship Tank' or military surplus landing vessel for loading/offloading equipment in the construction industry) had been dispatched to render assistance. Once a re-estimation of the LST's position at the time of intercept of the water taxi was made, the search area was refocused on that position. Once the search was resumed in the new area, the severed canopy from the water taxi was found with side scan sonar on the first search line. The location of the debris field formed the new central datum from which the new search area could be drawn. The victims could not be (and were not) far from that point.

Within 3 days of finding the debris field, all of the remaining victims were located and recovered. They were located with side scan sonar, then reacquired with an acoustic lens sonar mounted to an ROV. An example of the detailed search capability of an acoustic lens sonar is the fact that it was able to image the stitching in the shoes of one of the victims to the shoe manufacturer's logo.

The victims were recovered with divers as the ROV held position while imaging the targets acoustically. Neither the ROV equipment nor the divers were put into the water until sonar targets were identified.

Excellent planning and research can lead to a successful search operation.

10.4.5 Final Identification

The final identification of an item of interest requires human discrimination of an image of the item. Positive classification of the target as either the object of the search, or the exclusion of the object as a possible target from the search area, achieves the objective of the task.

Once the final identification is obtained, the objective of the entire field task must be concluded. In most instances, clues as to the cause of the incident under investigation may be gleaned from the scene by taking the extra time to document the area. What other evidence can be found within the area? In what position did the victim or the evidence come to rest? How can one (or should one) recover the victim or evidence?

Extreme attention to detail in the final phase of the search and recovery operation can significantly enhance the results of the PSD operation. The quality of the findings will support court proceedings as well as fact-finding tribunals where the public safety diver could be called to testify.

Chapter 11
Commercial, Scientific, and Archeological

Since commercial, scientific, and archeological operations involve a set of 'How To' TTPs (tactics, techniques, and procedures) similar to those in Chapter 8, this chapter will address the development of a list of specific environments and operations peculiar to these applications.

As discussed in Chapter 2, vehicle geometry does not affect the motive performance of an ROV nearly as much as the dimensions of the tether. ROV systems are a trade-off of a number of factors including cost, size, deployment resources/platform, and operational requirement. The various sizes of observation-class ROV systems (as discussed in Chapter 2) have within them certain inherent performance capabilities. The larger systems usually have a higher payload and thruster capability, allowing better open-water functions. The smaller systems are much more agile in getting into tight places in and around underwater structures, making those more suitable for enclosed structure penetrations. The challenge is to find the right system for the job.

Some examples of tasks, along with the best system size selection, are provided in Table 11.1.

Operational considerations regarding various tasks are discussed throughout this book. The remainder of this chapter will deal with additional considerations more applicable to commercial, scientific, and archeological tasks.

11.1 VIDEO DOCUMENTATION

The difference between a film produced by a Hollywood production company and an amateur videographer is less about the equipment and more about the technique. What separates the amateur from the pro is the attention to detail that allows a complete portrayal of the subject matter through moving images. It is not about the image, it is about the message.

Table 11.1 *Task and ROV size matching.*

Tasking	*Best size*
External pipeline inspection	Large
External hull inspection	Medium/large
Internal wreck survey	Small
Open-water scientific transect	Medium/large
Calm-water operations	All sizes

It's been said in many different industries: 'If you didn't get it on video, it didn't happen'. The meaning of this saying indicates that video and still-camera recordings are an integral part of a professional documentation package, forming part of the deliverable at the end of a project. Part of any documentation package is a set of prolific notes documenting the entire operation from start to finish so that as few items as possible are left to guesswork or memory. Field notes, as well as audio annotation, complete the video content and report, allowing a stand-alone document that can 'tell the whole story'.

As an anecdote to professional technique, consider the freefall photographer. At a parachuting school in the Southern Central USA, a service offered to customers included freefall video and stills during each tandem skydive. With each new videographer, it took time to learn the new skills of video documentation over and over again. It was a case of having to teach skilled skydivers how to be competent cinematographers – the first few camera runs had the cameraperson with the head-mounted video camera jerking his/her head in all directions with occasional filming of the customer. Not only was the video unacceptable to the customer, in most instances it made the customer airsick just viewing the film.

The same problem applies to underwater photography. Video documentation with an ROV system is a case of an ROV technician being required to learn the skills of a cinematographer. It is important to understand the status of the underwater location under investigation, thus it's critical that an image documentation of the area/item be made that properly orients the viewer while maintaining his or her interest. All of the principles of land cinematography apply to filming underwater.

Experience gained during prior projects with some of the legends in the underwater photography business is invaluable. These legends included still photographer David Doubilet, the incredible cave cinematographer Wes Skiles, and the man whose accomplishments form the textbook of modern underwater photography, Emory Kristof, who became famous working for *National Geographic*. A few of their rules of thumb are listed below:

- Get as much footage as possible – it can always be edited.
- Keep the camera/submersible stable and still. When taking a mosaic of a subject area, stay on the starting shot for a few seconds, then slowly pan from left to right (the direction in which a book is read) and come to rest on the ending shot. Hold the ending shot still for a few more seconds then move on.
- Go from macro (to get a situational reference) to micro. If going into a structure, get as full a view as possible then go to specific items.
- Some observation-class ROVs have a camera zoom function. Try not to use the zoom function too frequently while filming, since it makes for poor subject content.
- Vary (and get as many of) the camera angles of the items being filmed as possible. It gives the capability to paste together a full video mosaic of the subject.
- Video is about gathering information in the form of moving images. Get as many close-ups as possible of the item being filmed, along with varying distance/frame content, to enhance the information.
- The job of the cinematographer is to gather image quality and content. It may be better for the operator to do the navigating and allow an observer/supervisor to direct the operation to assure full content, much like the director on a movie

set directs the camera operator. In practically all instances, it is best to have a separate notetaker to assure proper documentation of the project. Attention to detail is essential.

- Upon encountering an item of interest, leave the item in frame then count to 5 before slowly panning to the next item.
- Make all movements slow, controlled, and deliberate, understanding that the submersible is both an eye and a camera platform.
- Attempt to be consistent with the filming style. When panning, try to always pan from left to right (or right to left). When approaching an item for inspection, attempt to look all around the item for status and structure before going in for close-ups.
- Since the specialty of an ROV system is its on-station loiter capabilities, take the time to fully document the subject before moving on to another location. Do as many 'takes' as necessary to get the required shot.
- Go into a subject area with an eye for the final edit and get the footage needed for that edit in mind.
- An audio overlay will assure the video is annotated and allow for the tape to be a stand-alone document. Make sure to have an on/off switch, because an open mike can make for embarrassing playback.

11.2 HIGH CURRENT OPERATIONS

The optimum environment for operating ROV equipment is clear water, calm seas, and no current. Unfortunately, the real world intervenes with this perfect world, since one cannot wait for the weather to change in order to get the job done. It is possible to mitigate the effect of currents by doing drift work within the water column, but most commercial inspection jobs require viewing items that are anchored to the bottom or the shore.

According to Bowditch's *American Practical Navigator* (2002), horizontal movement of water is called current. It may be either 'tidal' or 'non-tidal'. Tidal current is the periodic horizontal flow of water accompanying the rise and fall of the tide. Non-tidal currents include all other currents not due to tidal movement. Non-tidal currents include the permanent currents in the general circulatory system of the oceans as well as temporary currents arising from meteorological conditions. Currents experienced during ROV operations will normally be a combination of these two types of currents.

In order to complete an underwater task successfully, work with nature to assure that the factors affecting the outcome of the operation are timed in such a way as to mitigate their effect. Planning the dive for commencement during times of slack tide (time during the reversal of tidal flows where current is at a minimum) can lessen current effects upon the operation. Also, an understanding of the dynamics of water flow over rivers, lakes, and structures will assist in taking advantage of these factors (refer to Chapter 10 for additional information on the environment).

The hydrodynamics of water flowing over an underwater structure (Figure 11.1) can have dramatic consequences on the success of the operation.

If the situation allows, approach a submerged structure from the downstream side proceeding against the current. Place the weight or managing platform (i.e. the cage

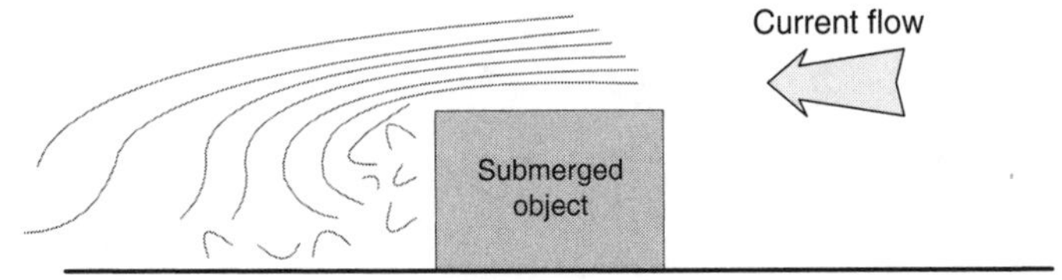

Figure 11.1 *Currents eddy on the downstream side of a submerged structure.*

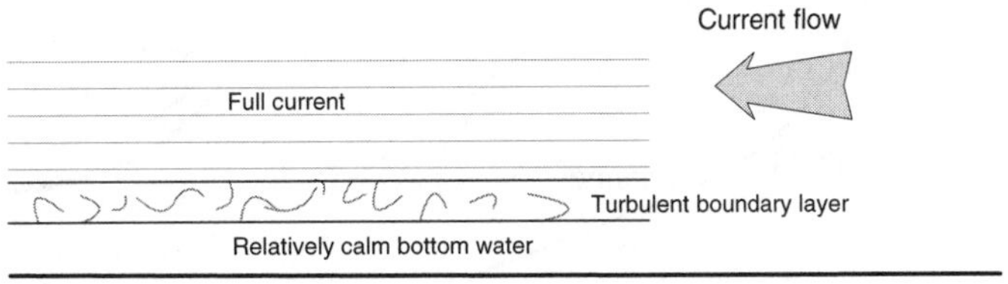

Figure 11.2 *Bottom conditions relating to current flow.*

or TMS) above the level of the structure (to lower the risk of fouling) and as close to the object/work site as possible.

In general, the deeper the water in the ocean environment, the lower the current. At the bottom is an area of water that is trapped by surface tension, holding the water immediately near the bottom at zero current. The boundary between the zero current water and the full current stream is an area of high gradient (and turbulent) velocity change called the 'boundary layer' (Figure 11.2). If the vehicle is placed below that layer, a relatively current-free operation can be accomplished. The boundary layer moves closer to the bottom as the current increases. In very high currents, where the layer is closer to the bottom than the height of the submersible, this positive effect may be negated.

Once the hydrodynamic conditions are understood, some other possible solutions to operational problems become apparent. For instance, rivers channel on the outside of a bend and bar on the inside of the turn. If the approach to the object is from the lower current side on the bar, where the vehicle can sneak beneath the boundary layer, the capability is increased to successfully complete an inspection task that otherwise would be unobtainable with a more direct approach (Figure 11.3).

11.3 OPERATIONS ON OR NEAR THE BOTTOM

By far, the most important working issue in ROV operations is tether management. In operations at or near the bottom, the objective is to reduce the amount of tether the submersible is required to pull to the work site through judicious use of the deployment cage, tether management system, or clump weight. With the weight or managing platform (i.e. the cage or TMS) forming the center point of a circle of operation, the deployment platform (e.g. the boat, dynamically positioned vessel, or drilling rig) can be moved to within the operational radius of the work site, allowing the work to be easily performed with minimal excess tether (Figure 11.4).

The deployment platform should be placed directly over the work site with the managing platform in a position to have good angular access to all points on the work

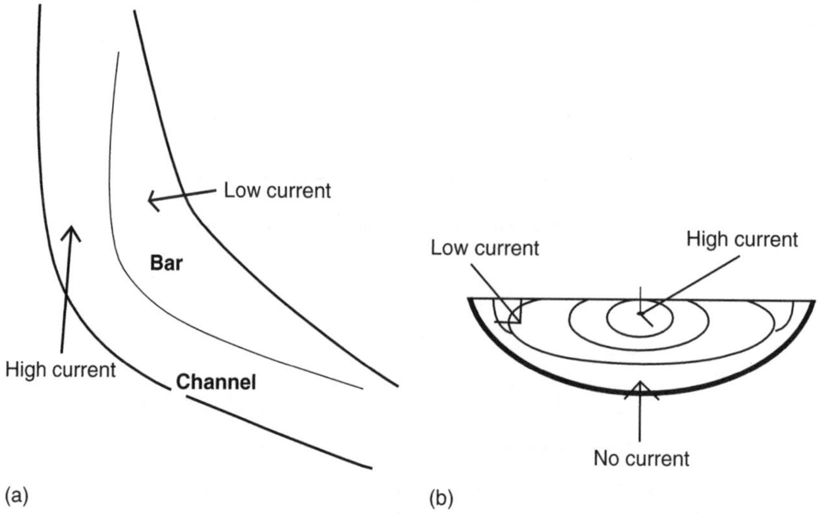

Figure 11.3 *Examples of channeling on river bends.*

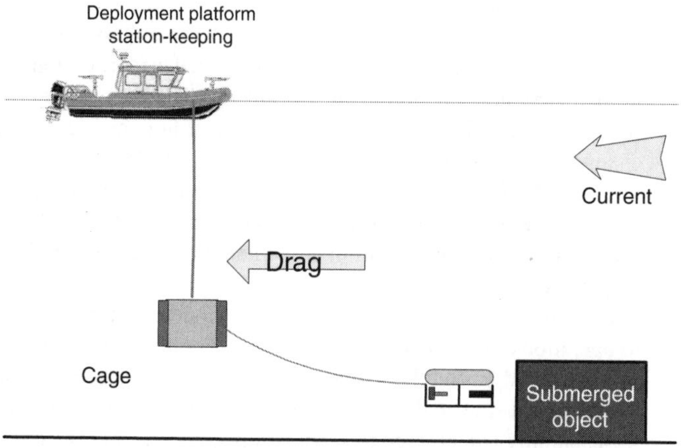

Figure 11.4 *Position of boat with respect to current and target.*

site. If direct access to a location on the job site is obstructed, the best practice, in most cases, will be to move the deployment platform, rather than risk tether entanglement upon the structure.

Perform all work operations with as little tether actually dragging the bottom as possible (to cut down on the possibility of tether hangs), then recover the submersible before moving the deployment platform or leaving the site.

11.4 ENCLOSED STRUCTURE PENETRATIONS

Probably the most exciting (and nerve racking) application of ROV technology is the enclosed structure penetration. It is within this application that the small ROV really shines (Figure 11.5). Due to the inherent dangers to divers in underwater structure

(a)

(b)

Figure 11.5 *Small ROV performing internal wreck penetration of the USS* Arizona *(photo by Brett Seymour, National Park Service).*

(c)

(d)

Figure 11.5 *(Continued)*

(a)

(b)

Figure 11.6 *Small ROV in aquaculture applications.*

penetrations (difficulty of rescue, danger of entanglement upon the structure, questionable integrity of the structure, etc.), the ROV may be the only way possible to survey the site.

Some rules of thumb gathered during many hours of enclosed structure penetration with small ROVs follow:

- Choose an entry point into the structure that has a smooth transition (to avoid tether chaffing or trapping) and allows for easy tending of the tether as close to the entry point as possible.
- Purposely choose the tether lay along the structure with a constant eye to recovery.

(c)

(d)

Figure 11.6 *(Continued)*

(e)

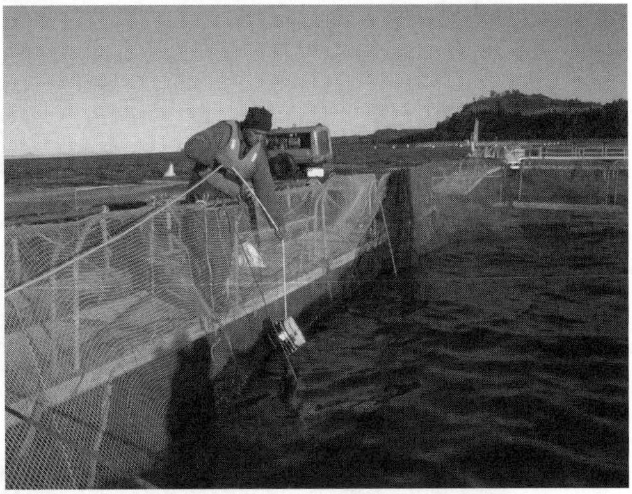

(f)

Figure 11.6 *(Continued)*

- First pull the entire length of tether into the structure, to the full extent possible in order to lay out the tether, then work back toward the entry point, keeping slack in the tether to a minimum.
- As discussed in Chapter 7, tether traps are the vehicle's biggest danger within enclosed structures. A tether naturally channels to juncture points within the structure, possibly trapping the tether in a less-than-90° wedge point. Plan the tether lay to avoid the tether being pulled into one of these traps, possibly trapping the expensive vehicle irretrievably within the structure.
- It is best to assure a neutrally buoyant tether is available for structure penetrations since tethers can be fouled on either the ceiling or the floor of the structure.

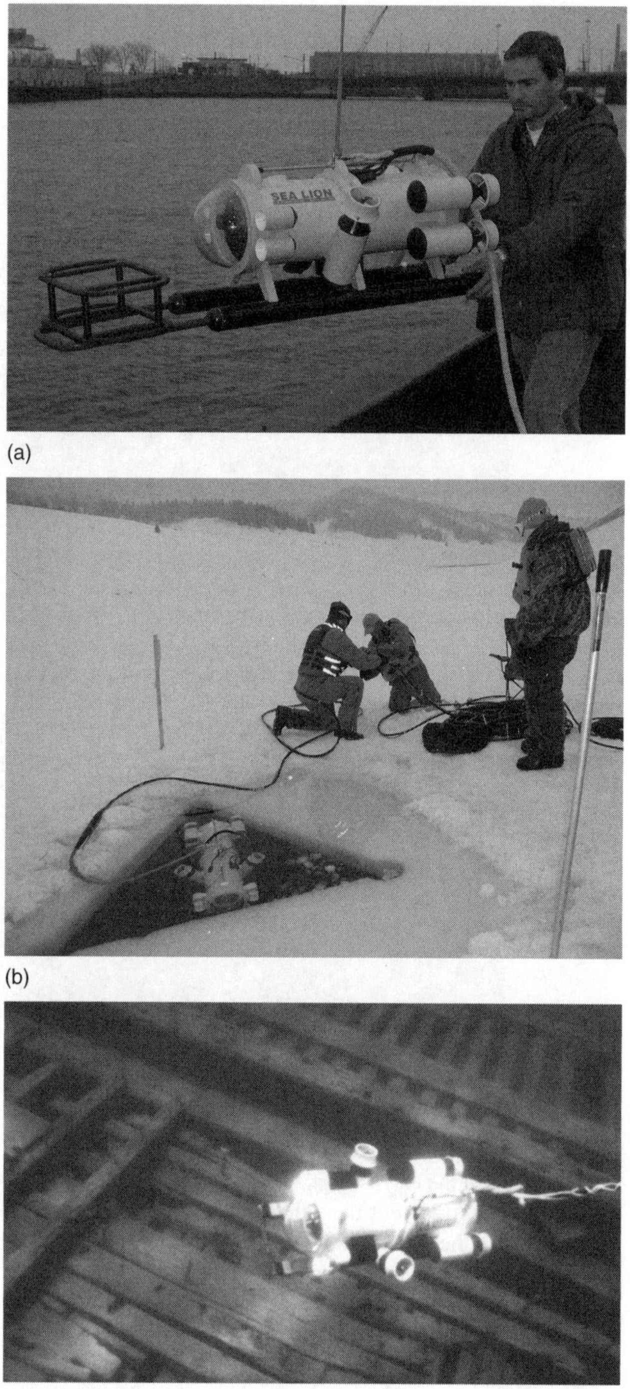

Figure 11.7 *Small ROVs are excellent for aquaculture and scientific operations.*

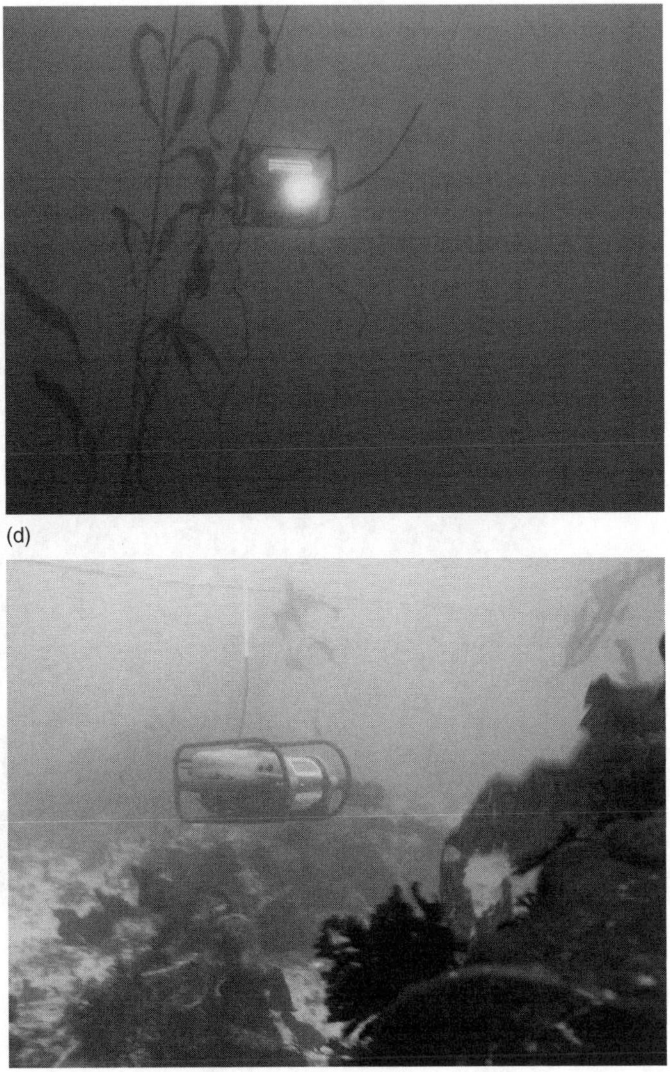

(d)

(e)

Figure 11.7 *(Continued)*

- Enclosed structures (especially old archeological sites) normally have significant silt buildup within the structure. To assure the least stirring of silt (thus obscuring visibility), maintain the vehicle close to neutral buoyancy (cheating on the positive side so that any thrusting needed is away from the bottom silt).
- When exiting the structure, fly the vehicle out adjacent to the tether while the tether handler slowly pulls the tether out of the structure.
- Never pull the tether to free a snagged vehicle. If there is any resistance to pulling the tether from the structure, slack and then reassess.

(f)

Figure 11.7 *(Continued)*

- Plan on getting into some minor snags and plan the time budget with this in mind.

When the inevitable does happen, keep calm, continue attempting to free the problem and never give up.

11.5 AQUACULTURE

See Figure 11.6. ROVs are quite useful in aquaculture applications for the following reasons:

- A small ROV does not stress the stock nearly as much as a swimmer.
- Use of an ROV allows more frequent inspections of the nets, mooring, and supports for the farm without the danger to divers.

- If configured properly, the ROV can recover 'morts' (dead fish) remotely, thus reducing disease propagation and diver bottom time.
- Reduce excess feeding (some species – like Norwegian salmon – will only feed from the water column and will not bottom feed), thus saving feed meal when the stock is finished feeding.
- Comply with regulatory reporting through video documentation of farm conditions.
- Study feeding patterns for optimization of feed delivery.
- Hydrological profile (through vehicle delivery any number of sensors to all parts of the cage) of cage for environment optimization.
- For sea cages, reduce stock loss through constant monitoring of cage integrity.

Traditionally, fish farming operations have been low-budget endeavors. With the introduction of modern management techniques, new technologies have gained wide acceptance within the industry, thus encouraging investments in robotics. ROVs are replacing the mundane tasks previously performed by divers, allowing a safer and more cost-effective aquaculture operation (Figure 11.7).

11.6 DOCUMENTATION AND DISPOSITION

The end product of any underwater operation is the documentation of the project and the final disposition of the object of interest. Some ROV systems allow digital capture of images directly onto magnetic media; Others simply allow screen capture of a frame of video to use in formal documentation.

Any final report to the customer should include a condensed version of all the notes taken during the operation, in a readable format, along with an edited video presentation of the operation.

There are many excellent inexpensive digital video-editing programs available. These enable coarse editing while in the field to allow for a deliverable report immediately at the end of the job. Also available in post-processing is a much more detailed condensation of the operation. In order to organize a formal, professional presentation to the customer, design a template for the final report that is easily adaptable to different operations and open customer base.

The format of 'Who, What, When, Where, Why, and How' allows the work to be in a journalistic format, with complete reporting of the event at an adequate level of documentation. The final report is all about transmitting the information from the underwater project in a clear, concise, accurate, and readable format. The final report becomes a calling card, showcasing the professional capabilities of the team and possibly a ticket to the next interesting assignment.

Chapter 12
Standard Operating Procedures

This section contains detailed examples of standard operating procedures (SOPs) for ROV deployments used in general underwater operations with particular focus upon port and harbor security. Every operational situation is unique, with its own peculiar set of operational requirements. These SOPs have been operationally tested and proven. Use the below procedures as a starting point for deployments. The SOPs assume an ROV operator of average proficiency under normal conditions. A very proficient operator may be able to perform items located in the 'Not recommended' category, but due to equipment performance constraints, it may require an expanded timeframe.

12.1 OVERALL OPERATIONAL OBJECTIVES

The professional ROV crew will approach any field assignment in a systematic fashion with constant focus on the achievement of the operation's ultimate goal. The overall objectives of the assignment will be accomplished through the following general tasks:

- An integral part of any ROV support operation includes a full understanding of the overall operation to be conducted as well as the ROV's role within that mission plan. As part of the planning for the field assignment, a full review of any customer-produced documentation is required in order to maintain full operational efficiency.
- Obtain a scope of work from the customer so that the appropriate equipment may be scheduled for the operation.
- Mobilize all equipment allocated and scheduled for the operation.
- Transport all equipment to the work site.
- Prepare all equipment for the individual tasks in proper order of use.
- Thoroughly brief the crew on the planned operation as well as the vessel crew and/or any necessary third party (Port Captain, Facility Owner/Operator, nearby vessels, etc.).
- Complete all pre-dive functionality tests as well as checklists for deployment.
- Deploy the equipment, locate the work site, and begin the work task.
- At the completion of the work task, recover the equipment (per detailed procedures).
- Perform post-dive functionality checks as well visual condition inspection.
- Upon completion of the assignment, demobilize the equipment and transport it back to base.

12.2 EQUIPMENT MOBILIZATION

Equipment for deployment should be readied for travel and use as follows:

- Verify all preventative maintenance and servicing has been performed to standards established by the manufacturer.
- Verify all consumables (i.e. lights, seals, o-rings) are serviceable and in good condition.
- Ensure an adequate spares package is available to meet the priority level of the mission.
- Verify all functions of the equipment before packaging it for shipment.
- Ensure the equipment is packaged adequately for travel to the work location.

The Operator is responsible for the operation of his ROV system. All tests and mobilization steps should be performed before leaving base.

Standard ROV equipment needs (pre-deployment checklist) for field deployment fall under these broad categories:

- ROV manufacturer-provided equipment
- ROV manufacturer-provided spares
- Other than ROV manufacturer-provided equipment deemed essential to the ROV system (e.g. viewing monitor, video cables, transport cases, power cables, etc.)
- Spares, adapters, and electrical wiring
- Test equipment for troubleshooting component parts.

For the first three items above, refer to the manufacturer's operations and maintenance manual for a complete listing of items needed for deployment. For the last two items above, refer to Chapter 13 of this manual for a general listing of items needed for troubleshooting, backup, and testing of ROV equipment.

12.3 OPERATIONAL CONSIDERATIONS

12.3.1 Operations On and Around Vessels

The following should be considered when conducting any ship hull inspection:

- The bridge of any vessel to be inspected must be informed prior to commencing the dive to determine if any conditions are present that are hazardous to ROV or vessel operations.
- When launching and recovering the vehicle, attention must be paid to any submerged obstructions in and around the deployment platform or vessel to be inspected.
- Where ROV operation is required within and around the operating structure of the inspection vessel, care must be taken to ensure that the tether or umbilical is not snagged or damaged.

12.3.2 Operations from the Vessel of Opportunity while Station-Keeping

The following principal points shall be noted in all ROV operations from vessels utilizing station-keeping:

- Vessel shall not be moved without the prior knowledge of the ROV operator
- The ROV shall be launched as far away from thrusters as practical
- The tether shall be manned at all times
- Vehicle buoyancy shall be slightly positive
- Good communications must be maintained between the ROV operator and the helm to ensure that the ROV operator is aware of all actions regarding the relative positions of the ROV and vessel.

12.3.3 ROV Operations with Divers in the Water

Concurrent ROV/diver operations may potentially carry an increased risk to the diver for various reasons, including possible entanglement, mechanical contact between the diver and the vehicle, and various other areas of interference. Diver safety is paramount. Observe the following guidelines during all in-water concurrent operations:

- As an integral part of any industrial operation, a pre-job briefing session should be conducted with all concerned operators. Once the customer representative, vessel or facilities operator, and vendor personnel agree upon the proposed operation, the ROV supervisor shall brief his/her crew (see Section 12.4).
- During the operational crew briefing, the diving supervisor should establish a perimeter around the diver that the ROV should not penetrate without express permission and positive approval from the diving supervisor. The ground fault interrupt system on the ROV system should be fully operational and tested.
- Consideration should be given to thruster injestation of items belonging to the diver. The possible foreign object damage to the thrusters and diver injury could be avoided through thruster guards or through properly stowing diver tooling.
- The sequence for diver/ROV deployment will directly affect the possibility of a diver/ROV umbilical entanglement. The ROV should be deployed to the dive site with the tether fully lain to the work site before the diver is cleared to the area. The diver should work then clear the area *before* the ROV is recovered. No vehicle recovery should be started unless prior approval is gained from the diving supervisor.
- Clear, open, and direct lines of communication should be maintained between the ROV operator and the diving supervisor. Emergency procedures should be fully briefed and understood before the beginning of the concurrent operation.

12.3.4 Precautions and Limitations

The following precautions should be exercised to properly protect the ROV equipment while operating within its design limitations (Wernli, 1998):

- The ROV should only be operated by approved and appropriately trained company operators.

- Piloting ROVs in excess of 8 hours within any 12-hour period subjects the operator to excessive fatigue, thus possibly compromising personnel and equipment safety. If extended periods in-water are planned, a relief pilot should be scheduled or shifts should be staggered.
- The launch and recovery phase of the dive cycle is probably the most dangerous portion of the dive. The vehicle and launch platform are subjected to an inordinate amount of risk to damage while the vehicle is suspended in air (lift from the deck until contact with the water). Weather conditions at the time of launch should not exceed company-specified limitations. This would increase the risk of damage to the equipment to an unacceptable level.
- The ROV, as well as system components, should not be operated outside of the manufacturer-specified operating parameters.
- When operating from dynamically positioned vessels, or vessels under station-keeping (as specified above), ensure that the vehicle is kept clear of thrusters and surface obstructions.
- ROV operations should not be conducted in low water visibility (less than 4 feet/1.2 m) or in uncharted and/or unstructured areas unless the vehicle is equipped with submersible-mounted imaging sonar.
- Approval from command and possibly the insurance company may be required when operating the ROV in hazardous circumstances. Examples of such circumstances include:
    1. HAZMAT spills
    2. Use of explosives
    3. Working within underwater structures
    4. Extreme dynamic environments caused by waves and/or surge.
- Ensure demobilization procedures are properly followed to include post-operation maintenance and storage procedures.

12.3.5 Night Operations and Extreme Operational Environments

12.3.5.1 *Night operations*

As discussed in previous chapters, light is subject to scattering and absorption in water. The largest negative factor in working at night is the enhanced effect of backscattering due to the lack of ambient lighting (causing the auto iris on most CCD cameras to open further, amplifying the backscatter effect). The only significant source of lighting at night to illuminate the item of interest is the submersible's lights; Therefore, the submersible may have a lower effective range of vision during night operations. The submersible should be moved closer to the item of interest while the lights are set on a slightly lower setting to counter the enhanced backscattering.

There are, however, several advantages to deploying ROV systems at night for underwater port security, including:

- Lower harbor traffic that could inhibit operations
- Easier sighting of the submersible's lights below the water, assisting in visually locating the vehicle
- Better discrimination of items by the camera due to higher lighting contrast.

12.3.5.2 *Extreme temperature operations*

For operations in Arctic and Desert conditions, special considerations are needed to protect the equipment from malfunctioning due to temperature damage.

High temperatures are especially hard on electronic components due to the printed circuit board's and conductor's normal heating, plus the local high temperature, exceeding the maximum temperature limits of the materials.

Low temperatures are especially hard on seals, o-rings, plastics, and other malleable components (most of which are consumables).

Shock temperature loading of components happens when the components are left in air and allowed to settle to the local extreme ambient temperature. Once the components are temperature-soaked to the local extreme ambient air temperature, the possibility exists of shock temperature damage. If the submersible is then operated in water with a large temperature difference from the local air, shock loading of the components will be experienced.

An example of this in the Arctic would be a cold-soaked submersible left out overnight in −40°F air, then deployed into +32°F water, causing a total shock temperature loading of 72°F. This mistake, in all likelihood, would shatter the domes and cause seal failure, with the resulting ingress of sea water into air-filled spaces of the submersible.

A similar instance happened recently in the Gulf of Mexico aboard a derrick barge during a hot, windless August day. Prior to an operation, the vehicle was left out on the steel deck without shade while the crew relaxed in the air-conditioned operations shack. When the time came to dive, the vehicle (now heated to approximately 150°F/65°C) was lowered into the 75°F/24°C water, unevenly contracting the faceplate on the lights, thus allowing salt water incursion into the light housing. Once the salt water penetrated the light housing, there was an immediate ground fault tripping the GFI (ground fault interrupt) circuit, shutting down the vehicle electronics. Upon vehicle recovery, the lighting circuit had been destroyed due to metal oxidation through the ground fault.

Practically all temperature-related issues are resolved on the submersible by gradually matching the temperature of the submersible with the temperature of the water before deploying the vehicle. Extreme temperature problems with the control console are more easily resolved since it will be with the operator, normally, in a temperature-controlled environment.

12.4 PRE-DIVE OPERATIONS AND CHECKS

12.4.1 Crew Briefing

Any team effort requires a full understanding of the operation to be conducted by all team members. Before the start of any ROV operation, a full crew briefing is essential so that efficient tasking is accomplished in order to enhance team synergies. A proper crew briefing should include the following essential elements:

- Mission tasking and/or threat outline, detailing the specific requirements of the task

- A thorough and complete explanation of the job components
- A scope of work and specific goals to be accomplished during the task
- Ingress/egress routes to the work site
- Crew positions during the task as well as specific responsibilities
- Tactics, techniques, and procedures for accomplishing each individual task as well as completion of the mission
- Specific information needed as well as methods of documentation
- All relevant information on the object of inspection, including (but not limited to) drawings, maintenance records, damage reports, survivor statements, and any other information that would assist in the work task
- Work site coordinates
- Topographical maps, bathymetry data, tide tables, underwater obstruction analysis, prior surveys, and any other environmental information that will assist in accessing the work site
- If there was any prior work done on the work site, it is imperative to thoroughly review this information (and, if possible, interview the previous crew) and job reports to better gain an insight into the current site condition
- Schedule for completion of the task objectives (best case/worst case/most likely case) with consideration given for field delays
- A thorough detail (and, if deemed necessary, drill) on emergency procedures.

The ROV supervisor should also take further steps to be fully briefed on any other conditions that would affect or interfere with his planned operations, including (but not limited to):

- Planned vessel movements
- Vessel/platform support operations (including support vessel logistics)
- Diving operations
- Drilling operations
- Pipeline lay operations
- Operation of surface and subsurface machinery
- Scheduled power generation equipment outages/changeovers.

12.4.2 Vehicle Preparation

The sequence of events from crew briefing to the contact of the vehicle with the water should be conducted in a logical and fluid flow. Once the briefing is completed, the pre-dive checklist is begun and continued until the vehicle is in the water. If there is a checklist item that is unable to be cleared, the supervisor's judgment will dictate if the items will stop the pre-dive sequence. If a long interval is encountered between the completion of the pre-dive checklist and the start of the actual dive, the checklist may require a restart from the top or from some intermediate step so that all checklist items are completed in a timely fashion.

Vehicle preparations are normally vehicle-specific, but should include at least the following items:

- Assure all functions of the vehicle are operational
- Any tooling or sensor packages should be operational and tested with data and telemetry flowing in a nominal fashion

- All video feeds from the appropriate camera packages are wired and operational
- Verify the generator has enough fuel to support the entire operation
- All controls are operational and functioning in the appropriate direction
- All video systems are fed through the appropriate systems with recording devices tested and staged (e.g. tapes labeled, videotapes/DVDs logged, notes columned, test footage verified) – if the dive is not recorded, it did not happen!!!
- All documentation packages are staged and ready for action (e.g. video overlay prepared with the appropriate job number/location/date/time)
- Crew members at their assigned stations
- Crew communication systems tested
- Bridge notified of hoist of vehicle from the deck
- Bridge notified of vehicle in the water.

12.4.3 Pre-Dive Checklist

The following pre-dive check should be carried out prior to every dive:

- Visually inspect the vehicle to ensure the propellers are not fouled, all components are secured, and there is no mechanical damage to the frame or other components.
- Check the tether for scrapes, nicks, or other visible damage. The vehicle should not be used if the tether jacket is broken through.
- Verify correct operation of thrusters. CAUTION: Most ROV manufacturers advise that the thrusters should be run for only a few seconds in air. Water forms a heat sink that cools the thrusters during normal in-water operations.
- Check that all fasteners are in place and secure.
- Ensure that the whip connectors at the electronics and tether termination cans are plugged in securely. Note: Wet connectors have a tendency to become dry over time. Use dielectric silicone grease to lubricate both sides for easy installation.
- Ensure all unused vehicle connectors are capped securely with dummy plugs. A forgotten or unsecured dummy plug can lead to serious electrical system damage.
- Ensure all surface cables are securely connected.
- Unexpected thruster movement can occur when powering up. To prevent this, turn any Auto Depth switch on the control panel to the 'Off' position, center the Manual Depth Control knob, and switch any Auto Heading selector to the 'Off' position before powering up the system.
- Test the lights, camera/manipulator/tooling functions, and thrusters. Check that adequate video is obtained and that any video recording equipment is working properly. In most cases, underwater lights are not designed for operation out of water for any length of time. After testing the function of the lights, quickly turn the lights off to prevent excessive heat buildup. (At a recent trade show booth in Paris, a sound like a gunshot made all nearby duck for cover – it was an incandescent underwater light at a nearby booth that overheated and exploded.)
- After placing the vehicle in the water, check the vehicle for ballast and trim.

12.5 Specific consideration for operational deployment of ROVs

12.5.1 General Considerations

As discussed in Chapter 3, there are varying levels of capability with differing sizes of observation-class ROV systems. In the following sections, a series of operational tasks will be listed. The listing will be accompanied with suggested methods of how to approach each task, based upon the operating characteristics of different size classifications of ROV systems.

The classification of each ROV system (small, medium, and large) is based upon the systems evaluated during procedures testing. Each is capable of performing general tasks (i.e. ship hull inspection, pier and mooring inspections, etc.) under ideal conditions. The more difficult the conditions become (e.g. higher currents, longer stand-off from work site, higher tether drag, etc.), the less likely the system will be able to pull its tether to the work site.

12.5.2 Performance Considerations

From Chapter 2, a series of drag curves were developed to demonstrate mathematically the crossover point where the net thrust produced by the submersible will no longer overcome the drag of the system (vehicle plus tether). This is the point where the system fails to deliver the submersible to the work site (which is an important objective). From those charts, the net thrust (total horizontal forward thrust less total drag of the vehicle plus the perpendicular drag of the tether through the water) available to take the vehicle to the work site can be determined. The approximate median 'tether length in the water' crossover point for the systems listed in each category are provided in Table 12.1.

Again, these figures assume the tether to be perfectly perpendicular to the water flow (Figure 12.1).

If a 50 percent reduction in drag due to the tether being less than perpendicular is assumed, Table 12.2 can be derived.

If a 75 percent tether drag reduction is assumed, Table 12.3 is derived.

With the above assumptions, the approach to an ROV underwater port security task will vary depending upon the conditions to be encountered during that task.

Table 12.1 *Crossover points for net thrust by ROV size category (per Figures 2.22–2.25).*

Velocity (knots)	*Small (ft)*	*Medium (ft)*	*Large (ft)*
0.5	250	>250	>250
1.0	75	100	150
1.5	10	30	60
2.0	0	0	20

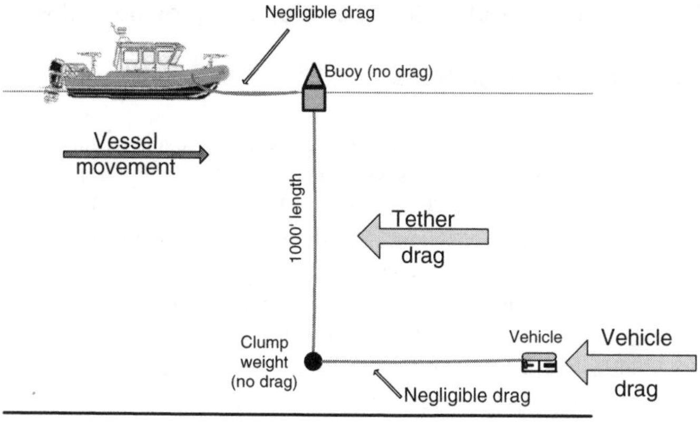

Figure 12.1 *Vehicle and tether drag at 100 percent tether drag due to perpendicular water presentation.*

Table 12.2 *Net thrust crossover points by ROV size category with 50 percent reduction of tether drag due to less-than-perpendicular tether presentation to the water flow.*

Velocity (knots)	*Small (ft)*	*Medium (ft)*	*Large (ft)*
0.5	>250	>250	>250
1.0	125	200	250
1.5	40	75	110
2.0	5	10	50

Table 12.3 *Net thrust crossover points by ROV size category with 75 percent reduction of tether drag due to less-than-perpendicular tether presentation to the water flow.*

Velocity (knots)	*Small (ft)*	*Medium (ft)*	*Large (ft)*
0.5	>250	>250	>250
1.0	200	>250	>250
1.5	75	100	175
2.0	10	15	75

12.5.3 Operational Examples

Certain maneuvers will be possible with larger, higher-performance systems that can overcome tether drag. Examples of a container ship keel run using a large and small ROV follow.

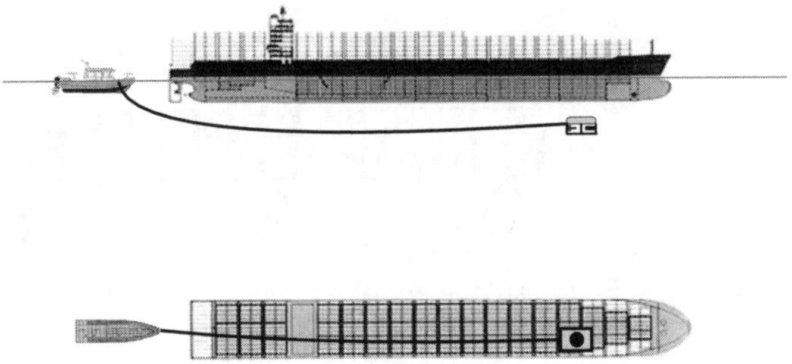

Figure 12.2 *Keel run of large cargo vessel from stern.*

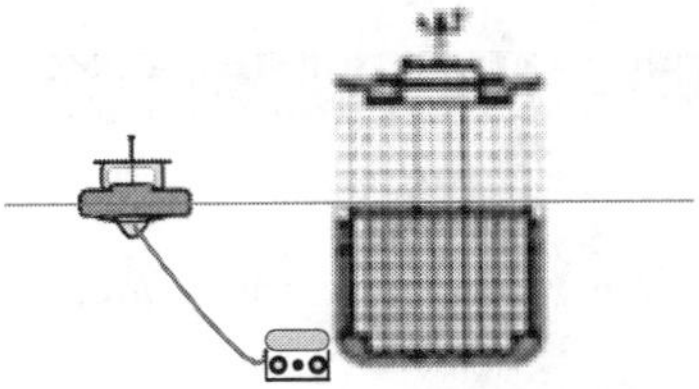

Figure 12.3 *Side scan of hull (stern view looking forward) with submersible – RB following vehicle.*

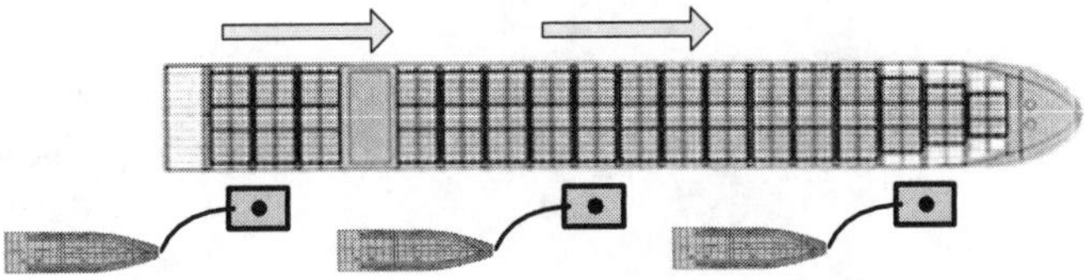

Figure 12.4 *Side scan of hull (top view) with submersible – RB following vehicle.*

12.5.3.1 *Large ROV system*

Keel run of a large container vessel at anchor while station-keeping with a response boat (RB) is shown in Figure 12.2.

Motor-by, side hull scan of large container vessel with RB following is illustrated in Figures 12.3 and 12.4.

12.5.3.2 *Small vehicle procedural modifications*

Procedural modifications to account for a smaller system include the following:

- Consider breaking the vessel inspection into smaller segments (perhaps halve, quarter, or eighth sections), so that shorter lengths of tether are in the water

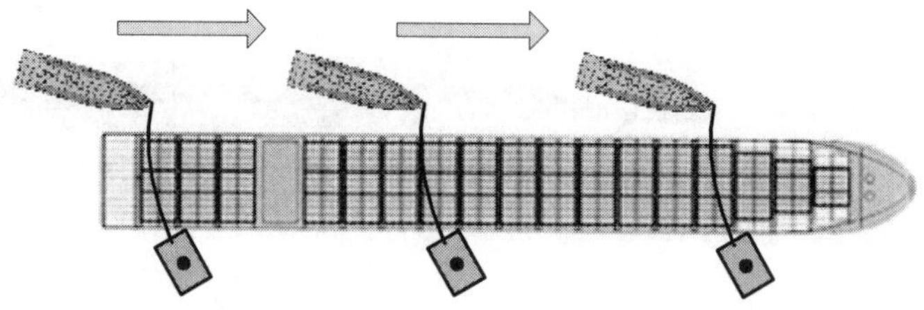

Figure 12.5 *Example of procedural change to lateral search in order to lower tether length in water (top view).*

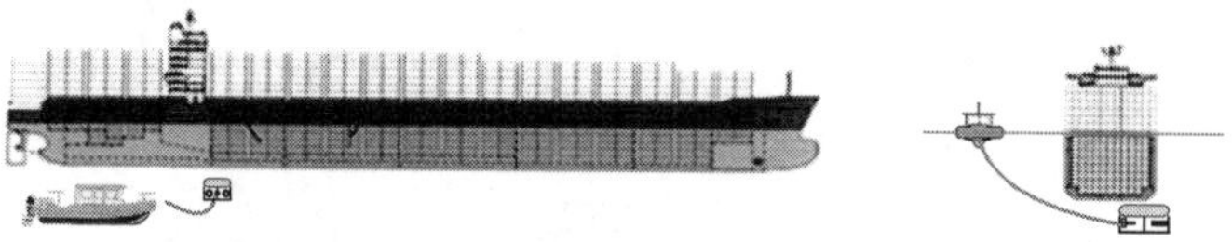

Figure 12.6 *Lateral search pattern to lower tether length in water (side and stern views).*

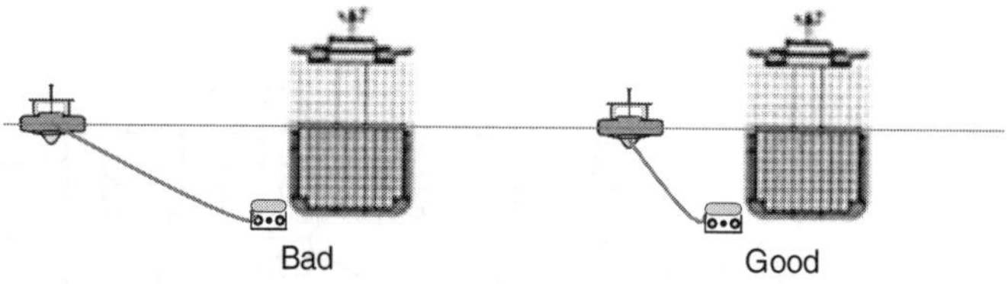

Figure 12.7 *Try to keep the vehicle as close as possible to the deployment platform.*

- Consider moving the tether handling point away from the operations platform to assist in tether mechanics (in the above example, the operator would be able to control the system from the RB, while someone aboard the inspection vessel handled the tether)
- Consider changing from a longitudinal search pattern to a lateral search pattern (as depicted in Figures 12.5 and 12.6)
- Consider the use of clump weights to cut down on the cross-section drag presented to the oncoming water flow
- Consider moving the deployment platform closer to the work site (Figure 12.7)
- Consider changing the direction of pull to the work site to a more direct pull, increasing the mechanical advantage of the vehicle's pull on the tether
- Consider delaying the operation until conditions are more favorable to the mission with the system
- Consider sourcing a higher-performance system.

12.6 TASK LIST AND GUIDELINES

The trial and error method of learning new tasks, while effective, is costly and time-consuming. The learning curve for a task is greatly accelerated if one has the benefit of others' experiences with trial and error regarding the selected task. The task listing below was developed from a menu of port and harbor security requirements. The procedures were tested to determine the best method of task approach to cut down on inefficiencies during mission planning and conduct.

The task list details a method for each general task classification. The general guidelines provide for operations in the 'best case scenario' or under perfect conditions. Along with the general guidelines is a diagram and task description for accomplishing specific tasks as the conditions become less than ideal.

Once the task has been mastered by the ROV operator, modifications to the procedure certainly should be attempted (and, if successful, fully documented) in order to continuously update and improve these TTPs. Use the matrix below as a guideline for starting the task and modify as the situation dictates.

12.6.1 Table of Task Expectations

The following categories are provided to help determine the equipment capability and level of proficiency of the operator based on the planned task.

- P – Possible with operator of average proficiency
- D – Difficult with operator of average proficiency (possible but time-consuming)
- X – Difficult or not possible with operator of average proficiency (positive outcome questionable or doubtful)
- NA – Non-applicable.

The following sizes, as described in Chapter 2, were considered in the task matrix:

- S – Small ROV
- M – Medium ROV
- L – Large ROV.

When performing the tasks described in the matrix, all items in the 'P' category are suggested methods of approaching the task with the size of equipment indicated (S, M, L). At the beginning of each task is a descriptive note suggesting the overall approach to the task. Then select the menu of procedures based upon the 'P' items within the size category.

In practically all instances, the RB is the best platform of opportunity due to the proximity of its operating platform to the water's surface, as well as its mobility and agility.

The detailed list of tasks examined in the matrix and sections to follow include:

- Ship hull inspections
- Pier inspections
- Anchor inspections

- Inspecting underwater obstructions
- HAZMAT spills
- Oil spills
- Potable water tank inspections.

All ship hull inspection tasks assume the inspection target is a 600-foot (200-meter) ocean-going cargo vessel. Sea conditions assume no current. As the current increases from zero, all ROV systems will trend from their nominal state towards the 'X' state. Also, as the length of the vessel to be inspected decreases, all nominal states tend toward the 'P' category.

12.6.2 Ship Hull, Pier, Mooring, and Anchor Inspections

In an ideal environment (perfect water clarity, no currents, 360° viewing around the submersible, large powerful ROV system, and vessel at rest), a 100 percent hull or object coverage could be completed in a relatively short time. Under ideal conditions, the following approach to each inspection task should be considered.

12.6.2.1 *Ship hull inspections*

Approach the 100 percent coverage ship hull inspection by running a series of inspection lines horizontally/longitudinally along the hull. For side hull inspections, start at the surface then descend to the water level where the submersible can still view the surface. Place the submersible in Auto Depth mode to remain at that level, then run the line from stem to stern (or reverse). Reverse direction, then descend the vehicle to a level where the last line is in sight. Run that line. Continue this method until the hull has been covered to the intersection of the bilge keel (if present). For under-hull inspections, run longitudinal lines, weighting the tether to avoid entanglement of the tether with obstructions on the hull. Continue running longitudinal lines until the hull is covered.

12.6.2.2 *Pier inspections*

Approach the 100 percent coverage pier and structure inspection on a 'voyage in/inspect out' basis. Inspect pilings on a 'bay-by-bay' basis. Swim the submersible into the bay then inspect the pilings (progressing outward) as the submersible is coming back out. Start at the surface for the first piling then run the piling toward the mud-line. Proceed on the bottom to the next piling and run it to the surface. Run on the surface to the subsequent piling and repeat.

12.6.2.3 *Anchor inspections*

Approach an anchor inspection from the surface toward the bottom, exercising caution at the lay point as discussed in previous sections of this manual.

Guidelines as conditions deteriorate – for the 'less than ideal' conditions, the following matrix of alternate approaches in varying conditions and situations is provided.

Task	Vehicle Size S	M	L
Pier/Mooring/Anchor/Hull Search from RB			
Tied off to structure			
Note: When inspecting a moored vessel (time and logistics permitting), it is best to tie off to the structure to which the vessel is moored so that the RB is stable for deployment and out of the path of other port traffic. It is best to approach a vessel from the bow for the main hull inspection then reposition to the stern for running gear inspection. This will avoid the potential tether entanglement in the running gear as the tether drags past while inspecting forward sections of the hull.			
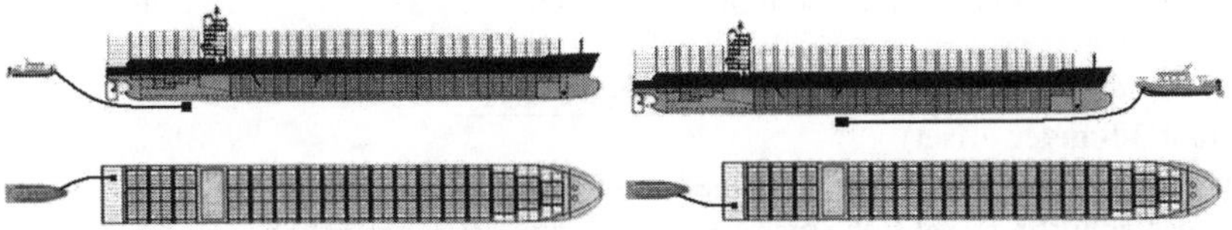			
(a) Run bilge keel or keel	X	D	P
(b) Inspect running gear to stuffing block	P	P	P
(c) Inspect sea chest(s)	P	P	P
(d) Inspect secondary thrusters (bow and laterals)	P	P	P
(e) Inspect through-hull fittings	P	P	P
(f) Inspect bulkhead/pilings	P	P	P
(g) Run anchor chain	D	P	P
(h) Acoustically/visually search bottom under vessel	P	P	P
Station-keeping			
Note: It is possible to inspect a vessel at anchor with the ROV system operated from the RB while station-keeping. As discussed above, the larger systems are capable of some time-saving maneuvers running the submersible while following with the RB. These are very efficient maneuvers, but should be conducted with caution. The approach for smaller systems is to break the vessel inspection into segments, then run inspection lines with the minimum wetted tether following the guidelines established in the reference drag tables above. When inspecting anchors, begin on the surface and inspect down the chain toward the anchor.			

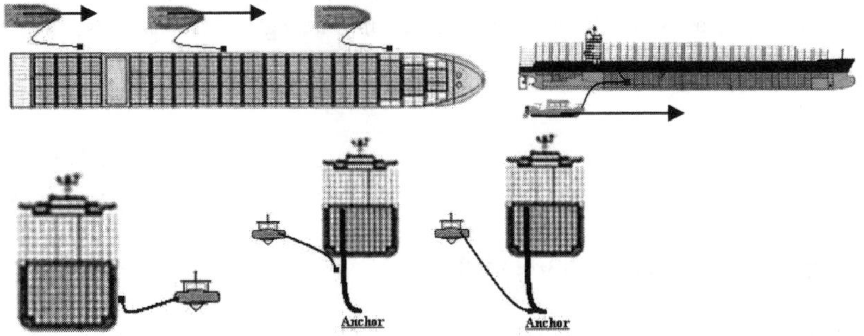

(a) Run bilge keel or keel	X	D	P
(b) Inspect running gear to stuffing block	P	P	P
(c) Inspect sea chest(s)	P	P	P
(d) Inspect thrusters	P	P	P
(e) Inspect through-hull fittings	P	P	P
(f) Inspect bulkhead/pilings	D	P	P
(g) Run anchor chain	D	P	P
(h) Acoustically/visually search bottom under vessel	P	P	P

Tied off to different structure and swimming to object (100-foot/30-meter offset)

Note: In some instances it may be necessary to tie-off to a different structure from the vessel to be inspected with some distance offset. This method of operation complicates the inspection task and may be impossible with some smaller systems. Consider moving the deployment platform closer, handling the tether from the vessel or delaying the task until logistical conditions improve.

(a) Run bilge keel or keel	X	D	P
(b) Inspect running gear to stuffing block	X	D	P
(c) Inspect sea chest(s)	X	D	P
(d) Inspect thrusters	X	D	P
(e) Inspect through-hull fittings	X	D	P
(f) Inspect bulkhead/pilings	X	D	P
(g) Run anchor chain	X	X	D
(h) Acoustically/visually search bottom under vessel	X	D	P

Visual and Acoustic Hull Inspections from Vessel

Deployed from vessel with RB handling tether

Note: In some instances (such as with an overhanging bow or stern structure), mechanical advantage may be gained while operating the ROV system from the vessel to be inspected by having the tether tended from the RB.

(a) Run bilge keel or keel	X	D	P
(b) Inspect running gear to stuffing block	P	P	P

(c) Inspect sea chest(s)	P	P	P
(d) Inspect thrusters	P	P	P
(e) Inspect through-hull fittings	P	P	P
Deployed from RB with tether handled from vessel			
Note: As stated above, in some cases it may be beneficial to separate the deployment platform from the location of the tether handler. If this is deemed advisable, either swim the vehicle to the inspection platform for retrieval of the tether or physically walk the vehicle to the vessel to be inspected. Then deploy the submersible over the side and commence the inspection.			
(a) Run bilge keel or keel	X	D	P
(b) Inspect running gear to stuffing block	P	P	P
(c) Inspect sea chest(s)	P	P	P
(d) Inspect thrusters	P	P	P
(e) Inspect through-hull fittings	P	P	P
Visual and Acoustic Hull Inspections from Dock			
Deployed from dock with RB handling tether			
Note: In some instances (such as the inspection of the seaward side of a moored vessel), some mechanical advantage may be gained by tending the tether from the RB.			
(a) Run bilge keel or keel	X	D	P
(b) Inspect running gear to stuffing block	P	P	P
(c) Inspect sea chest(s)	P	P	P
(d) Inspect thrusters	P	P	P
(e) Inspect through-hull fittings	P	P	P
Deployed from RB with tether handled from dock			
Note: In some instances (such as the inspection of the shore side of a moored vessel), some mechanical advantage may be gained by tending the tether from the dock.			
(a) Run bilge keel or keel	X	D	P
(b) Inspect running gear to stuffing block	P	P	P
(c) Inspect sea chest(s)	P	P	P
(d) Inspect thrusters	P	P	P
(e) Inspect through-hull fittings	P	P	P
Stationary deployment from RB			
Note: For some smaller moored vessels, some advantage may be gained by attempting to swim the entire hull from a single point. However, this method is, in most cases, inefficient.			
(a) Sub swimming entire ship from single spot – bow	X	D	P
(b) Sub swimming entire ship from single spot – amidships	D	P	P
(c) Sub swimming entire ship from single spot – stern	X	D	P

Minimal tether in water from RB			
Note: This method requires coordination and practice, but in most instances is the most efficient.			
(a) Station-keeping with RB moving along with sub	X	D	P
(b) Move RB from stationary point to point then deploy	D	P	P
Longitudinal/lateral searches from RB			
Note: Longitudinal inspections are more time-efficient than are lateral searches since it requires much less effort to reposition for each search line. However, lateral searches allow for better control of the tether due to less wetted surface on each search line. These two considerations must be balanced based upon the conditions prevalent at the time of the tasking.			
(a) Longitudinal search from bow	X	D	D
(b) Longitudinal search from stern	X	D	D
(c) Lateral search	P	P	P
Section searches (quarter/halve vessel) versus landmark/ high-exposure searches			
Note: All hull inspections are greatly enhanced with possession of a diagram of the hull to be inspected displaying prominent features on the hull. With a full diagram of the hull, feature-based visual searches will ensure coverage of that section of the hull inspection.			
(a) Obtain 'as built' drawings and perform landmark/high-exposure searches	P	P	P
(b) Full coverage search	NA	NA	NA
(c) Convenience of breaking vessel down into sections versus full ship	NA	NA	NA
Adrift and tied to vessel hull search			
Note: During tidal flows, it may be beneficial to place the submersible in the water then scan as the submersible drifts by. This allows the submersible to drift (and look) downstream facing down-current or attempt up-current vehicle swim while tied off to the stern of the vessel to be inspected. Caution should be exercised when doing drift-by scans due to the possibility of the submersible snagging on items in the water column or attached to the vessel to be inspected.			
(a) Drift in current from bow to stern while conducting hull scan	P	P	P
(b) Tie off to bow then do drift scan with sub facing down-current	D	D	D
(c) Tie off to stern in current and do up-current search	X	D	P

Pier/Mooring/Anchor/Hull Search from Both Shore and Vessel			
<u>Shore to structure/vessel</u>			
Note: The least resource-intensive hull search scenario is with the ROV deployed from shore while conducting a hull or structure inspection. This will form a high incidence of hull inspection scenarios. In areas of low tidal flow, this is a simple and efficient method of hull and structure inspection.			
(a) Run bilge keel or keel	X	D	P
(b) Inspect running gear to stuffing block	P	P	P
(c) Inspect sea chest(s)	P	P	P
(d) Inspect thrusters	P	P	P
(e) Inspect through-hull fittings	P	P	P
(f) Inspect bulkhead/pilings	P	P	P
(g) Run anchor chain	NA	NA	NA
(h) Acoustically/visually search bottom under vessel	P	P	P
<u>Vessel to same vessel</u>			
Note: Inspections of hulls while positioned aboard the same vessel is convenient. It may, however, be difficult to inspect the underside of the vessel.			
(a) Run bilge keel or keel	X	D	P
(b) Inspect running gear to stuffing block	P	P	P
(c) Inspect sea chest(s)	P	P	P
(d) Inspect thrusters	P	P	P
(e) Inspect through-hull fittings	P	P	P
(f) Inspect bulkhead/pilings	P	P	P
(g) Run anchor chain	NA	NA	NA
(h) Acoustically/visually search bottom under vessel	P	P	P
<u>From vessel swimming to other vessel (100-foot offset)</u>			
Note: Positioning the operations platform from one vessel while inspecting another is possible. The larger the offset, the more limited is the submersible's ability to perform its inspection task. If this scenario is presented, consider moving the tether handling function to either an RB or to the vessel to be inspected.			
(a) Run bilge keel or keel	X	D	P
(b) Inspect running gear to stuffing block	X	D	P
(c) Inspect sea chest(s)	X	D	P
(d) Inspect thrusters	X	D	P
(e) Inspect through-hull fittings	X	D	P
(f) Inspect bulkhead/pilings	X	D	P
(g) Run anchor chain	X	D	D
(h) Acoustically/visually search bottom under vessel	X	D	D
Visual and Acoustic Hull Inspections from Vessel and Shore			
<u>Deployed from vessel with shore-based handling of tether</u>			
Note: This is a convenient and effective method of performing an inspection upon the shore side of a moored vessel.			
(a) Run bilge keel or keel	X	D	P
(b) Inspect running gear to stuffing block	P	P	P

(c) Inspect sea chest(s)	P	P	P
(d) Inspect thrusters	P	P	P
(e) Inspect through-hull fittings	P	P	P
(f) Inspect bulkhead/pilings	P	P	P
(g) Run anchor chain	NA	NA	NA
(h) Acoustically/visually search bottom under vessel	P	P	P
Deployed from shore with tether handled from vessel			
Note: This method is effective for inspecting the seaward side of a moored vessel while operating from shore. Consider this method if access to the vessel to be inspected can be gained.			
(a) Run bilge keel or keel	X	D	P
(b) Inspect running gear to stuffing block	P	P	P
(c) Inspect sea chest(s)	P	P	P
(d) Inspect thrusters	P	P	P
(e) Inspect through-hull fittings	P	P	P
(f) Inspect bulkhead/pilings	P	P	P
(g) Run anchor chain	NA	NA	NA
(h) Acoustically/visually search bottom under vessel	P	P	P
Stationary deployment on shore and from vessel			
Note: This method is not the preferred method of performing a ship hull inspection since there is no mechanical advantage gained. This may be necessary due to limited access to deployment points.			
(a) Sub swimming entire ship from single spot – bow	X	D	P
(b) Sub swimming entire ship from single spot – amidships	D	P	P
(c) Sub swimming entire ship from single spot – stern	X	D	P
Minimal tether in water			
Note: In all instances of ROV operations, the best vehicle movement will be gained with minimal tether in the water, thus minimizing the tether drag to which the vehicle is subjected. Consider this as a primary method of ROV vehicle deployment.			
(a) Tether handler moving along with sub	P	P	P
(b) Tether handler moving from stationary point to point then deploy	P	P	P
Longitudinal/lateral searches from vessel and from shore			
Note: Longitudinal pulls along the hull of any vessel involve long tether pulls with considerable tether drag for the length of the hull. Lateral searches, however, provide less drag across the hull's surface through shorter transects of the hull surface.			
(a) Longitudinal search from bow	X	D	P
(b) Longitudinal search from stern	X	D	P
(c) Lateral search	P	P	P

12.6.3 Pier Inspections

Pier inspections are performed by all sizes of ROV systems. The most efficient method demonstrated for performing pier inspections is with straight pulls into the work site. One bay of pilings should be inspected, then the submersible should be recovered (or swam out of the piling structure), before performing the next bay. Best results have been achieved by the 'look-down then swim-down' followed by the 'look-up then swim-up' piling scan method. Inspect down on one piling to the bottom, swim to the next, then scan upward toward the surface (Figure 12.8).

12.6.4 Inspecting Underwater Obstructions

Navigation to, then inspecting of, underwater obstructions involves a peculiar set of operational problems not associated with other underwater port security tasks:

- The object may not have a positive link to the surface (such as a wreck, object dropped overboard, or other underwater obstruction)
- The identification dimensions of this obstruction may not be known
- Navigation to this obstruction may be difficult without the use of acoustics
- Significant unseen entanglement hazards could (and in all likelihood will) be present
- Station-keeping above the obstruction may be difficult without dynamic positioning equipment
- Keeping the submersible in visual sight of the object could be difficult or impossible with any measure of tidal flow.

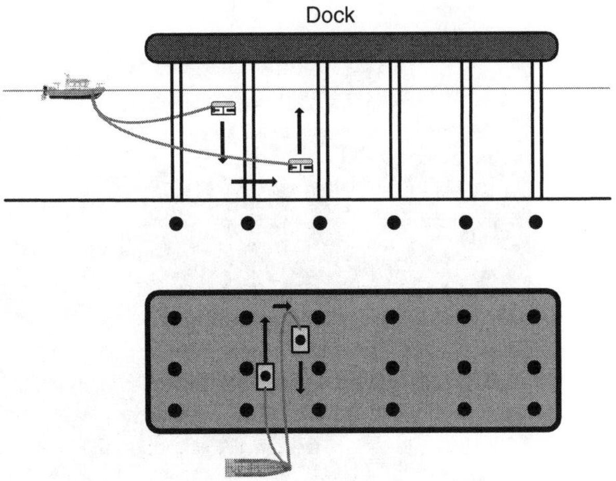

Figure 12.8 *Pier inspections from both a top view and side view.*

Some procedural modifications to compensate for the above factors while operating on and around underwater obstructions follow:

- Make liberal use of clump weights to keep the managed tether length to a minimum, while keeping the surface tether away from vessel thrusters. Assure that the clump weight stays above the level of the obstruction to reduce the risk of clump weight entanglement with the obstruction.
- Approach the obstruction with the vehicle from the leeward (or down-current) side, moving up-current.
- Consider use of a down-line, anchored to or near the wreck, to maintain positive navigation to the bottom.
- Avoid crossing over the obstruction to minimize risk of tether entanglement.
- The location and survey of an underwater obstruction without the use of scanning sonar may be exceedingly difficult in low-visibility conditions.

12.6.5 Other Non-Standard Operations

12.6.5.1 *HAZMAT spills*

ROV systems are effective in performing inspections of hazardous substance spills. The time and resources required to place a diver into these substances are considerable. In order to protect the vulnerable components of the submersible, a general idea of the substance into which the vehicle will dive is needed. Upon recovery of the vehicle, adequate decontamination is required (Figure 12.9).

Figure 12.9 *ROV contaminated during an oil spill survey.*

The suggested method of penetration into a HAZMAT environment is to determine the length of tether needed for the operation, then quarantine all parts of the submersible/tether combination from the insertion point to the submersible. A suggested decontamination method follows:

- Prepare the submersible for deployment into the HAZMAT substance
- Mark the tether at the point of entry (or point of quarantine) to the submersible
- Obtain (or create) an open container of decontaminate large enough to handle both the submersible and tether upon recovery of the vehicle
- Perform work then recover vehicle
- Thoroughly immerse the vehicle and tether in decontaminate until cleansed to EPA (or controlling authority) guidelines
- Rinse all parts of decontaminate and perform scheduled maintenance.

Protecting the environment at the site is critical. It is paramount for both personal safety, as well as environmental protection, to assure that all safety cleanup protocols are followed. This will allow timely and assured site recovery, preserve community health, and prevent any liability for the team.

12.6.5.2 *Oil spills*

ROV systems are effective as a quick responder to oil spills. The technique for inspecting the oil spill location is to get below the oil sheen without obstructing the domes, seals, and lights. For oil spills, the following are suggested guidelines:

- Find a location upstream of the oil slick to deploy the vehicle. If there is no current flow, place a deployment vessel (tubular object such as a garbage can or pipe), evacuate all oil from the tube, and deploy the vehicle through the clear area.
- Locate the source of the spill.
- Upon completion of the mission, recover the vehicle, clean with solvent, and perform scheduled maintenance.

12.6.5.3 *Potable water tank inspections*

Small ROV systems do quite well in the low flow, quiet environment of a potable water tank. The following steps are suggested when performing a potable water tank inspection:

- Determine the length of tether needed to perform the inspection of the potable water tank.
- Thoroughly clean all components of the vehicle with decontaminate in accordance with the American Waterworks Association (or other controlling authority) from the quarantine point on the tether to the vehicle.
- Obtain some type of container (a clean cardboard box, clean barrel, or other large container) large enough to hold the tether and the vehicle after cleaning. This will form the quarantine area for all insertion/extraction operations.

- Upon completion of the mission, perform scheduled maintenance with consideration that the potable water tank could contain chlorine (or other chemicals) affecting the seals and o-rings of the vehicle. Signs of chemical damage to o-rings and seals may include whitening, flaking, and drying of wetted areas.

12.7 POST-DIVE PROCEDURES

12.7.1 Post-Dive Checklist

A post-dive inspection should be carried out after every dive:

- Visually inspect the vehicle following each dive to ensure no mechanical damage has occurred.
- Check the propellers for any fouling.
- Visually check through the ports to ensure that no water has entered the camera, thruster, or electronics housings.
- Inspect the tether for cuts, nicks, or kinks in the outer shell.
- Rinse the vehicle and tether in fresh water if it has been operated in salt water.
- Check all vehicle functions again before power-down.
- Store the tether and vehicle properly for the next use. Refer to specific manufacturer's instructions for system storing procedures.

12.7.2 Demobilization of Equipment

Once the mission is completed, ensure the equipment is packaged adequately for travel from the work location. Upon return to the base, perform the following steps to demobilize the equipment for storage:

- Ensure all systems functions are performing and operational.
- Verify that the system is completely free of salt water. If there is any presence of salt-water residue, rinse completely with fresh water and dry before storing.
- Perform preventative maintenance in accordance with manufacturer-specified guidelines.
- Ensure all o-rings, seals, joint, and turn points are greased and packaged for storage.
- Store in accordance with manufacturer's recommendations.

Chapter 13

Servicing and Troubleshooting

All ROV systems share the same basic operating characteristics and maintenance needs. This chapter contains an outline of troubleshooting and preventative maintenance that should be performed on a regular basis. These procedures are not a substitution for manufacturers' suggested operating and maintenance procedures. These are guidelines to supplement manufacturer-specific instructions. A major source for this section is Wernli (1998).

13.1 MAINTENANCE

Equipment maintenance forms a vital part of a safe and efficient ROV operation. Properly maintained systems can achieve substantially reduced downtime. System schedules ranging from simple pre-dive checklists through detailed planned maintenance procedures must therefore be used to attain and maintain the highest possible standard of operating efficiency.

All work should be undertaken in compliance with supplier's/manufacturer's recommendations. Each ROV system type is provided with a full set of vehicle manuals and vendor subsystem technical information to enable efficient maintenance and re-ordering of system spare parts.

ROV system maintenance is divided into the following main areas of documentation:

- Operations and maintenance manuals and drawings
- Suppliers manuals and drawings
- Catalogs of equipment.

The above includes the following key subsections:

- Vehicle maintenance procedures
- Subsystem maintenance procedures, i.e. video cameras, sonar system, tools/motors
- Detailed repair and maintenance procedures (found in the specific ROV Operations Manual).

All equipment shall be suitably labeled to indicate its operational status, on arrival at the deployment base, in accordance with company procedures.

Every manufacturer of ROV equipment has a set of maintenance standards peculiar to their respective equipment. It is the responsibility of the ROV maintenance supervisor to assure that the individual manufacturer's maintenance schedule is meticulously followed. A sample maintenance schedule is provided below in the 'Operational forms' (section 13.5).

13.2 BASICS OF ROV TROUBLESHOOTING

13.2.1 Basics of an ROV System

There is not much 'high technology' in designing, manufacturing, and producing an ROV system. Every year, MATE (Marine Advanced Technology Education Center) hosts an ROV competition for high school and college students who put together their own operational ROV system. Educators in British Columbia have put forth a book on how to build an ROV in a garage out of hardware store parts (Bohm and Jensen, 1997). The difficulty is in producing a commercially viable and reliable system that can take the abuse of fieldwork and produce results.

The basic parts of a free-swimming, observation-class ROV system follow:

- Submersible
- Tools and sensors
- Tether
- Power supply
- Controller
- Viewing device.

Essentially, an ROV is a camera generating a video signal mounted in a waterproof housing with electric motors attached to a cable. Practically all of the vehicles use common consumer industry standard commercial off-the-shelf (COTS) components.

There are a few items on the system that require computer processing power, including the sonar, the acoustic positioning system, some instrument packages, controls to run the motor driver boards or manipulator, and telemetry from the submersible to the surface. Practically all of these are located on 'easily changeable' printed circuit boards. The only real non-electronic challenge for an ROV technician is working with the o-rings, seals, and tight machining tolerances needed to complete a waterproof seal on the submersible.

A simplified schematic of an ROV submersible control system is shown in Figure 13.1.

From these basics (with the use of manufacturer supplied schematics and drawings), one should be capable of performing basic troubleshooting in the field to complete the mission requirements.

13.2.2 The Troubleshooting Process

Effective and efficient troubleshooting requires gathering clues and applying deductive reasoning to isolate the problem. Once the problem is isolated, one can analyze, test, and substitute good components for suspected bad ones to find the particular part that has failed.

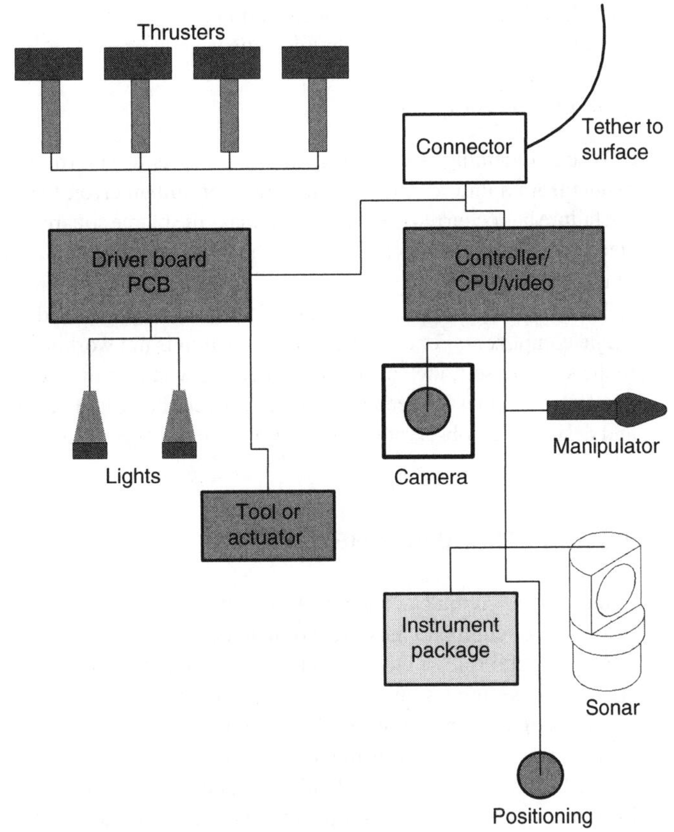

Figure 13.1 *Basic submersible schematic.*

The use of general test equipment (such as a digital multi-meter) or special test equipment (such as an NTSC pattern generator and an oscilloscope) can speed the analysis, but for many failures, deductive reasoning can suffice. Once it is determined whether the problem is electronic or mechanical, deductive analysis changes to intelligent trial and error replacement. Reducing the number of suspected components to just a few, then using intelligent substitution, is the fastest way to identify the faulty device.

In general, follow these steps when the ROV system fails:

1. Obtain the service manual for the ROV system in question.
2. Observe the conditions in which the symptoms appear.
3. Make note of as many of the symptoms as possible.
4. Use senses to locate the source of the problem.
5. Retry.
6. Document the finding and test results.
7. Assume one problem (when multiple symptoms are present, troubleshoot the easiest one first).
8. Diagnose to a section (fault identification).
9. Consult the troubleshooting chart of the manufacturer's maintenance manual.

10. Localize to a stage (fault localization).
11. Isolate to a failed part (fault isolation).
12. Repair.
13. Test and verify.

When something goes wrong, the first step is to determine whether the trouble results from a failure, a loose connection, or human error. Once it is confirmed that the failure has occurred (i.e. the operator lays the blame squarely on the ROV), the next step is to determine which portion of the ROV system is not operating – mechanical or electrical.

Then, step by step, partition each section into stages and track the trouble to a single component. For example, if one function is not working, the problem could be in the switch itself, the connector/connecting wires, or the electronic circuitry of the controller. For these procedures to be effective, a basic understanding of the operation and design of an observation-class ROV system is necessary.

13.3 TOOLS AND SPARES FOR FIELD WORK

In the field, anything can happen. A good work routine with some simple spares will help solve or eliminate most field problems.

As a general rule, a clean workplace is a safe and productive workplace. If there are items unsecured, someone will naturally trip on them. If an electrical connection is frayed, it will short out and bring down the entire system (requiring extensive troubleshooting and/or trips to the hospital).

The following is a general field tools and spares listing, which could assist any ROV pilot/technician in outfitting a field pack for servicing ROV equipment:

- A good Swiss Army type knife or one of the newer 'multi-tool'.
- A good folding pocket/diving knife (although many industrial settings have gone to a 'no knife' policy, for obvious reasons, in favor of high-quality scissors).
- A roll of electrical tape (black vinyl type) as well as a roll of high-quality duct tape.
- Plenty and varying lengths of cable ties.
- Video cable adapters:
  - BNC to RCA
  - RCA to BNC
  - BNC Tee
  - Spare adapters for as many types of connectors as anticipated
  - Spare video cables.
- A pocket magnetic compass (for calibrating vehicle compass).
- A pocket magnifying glass (for the gray-haired technicians losing their eyesight).

The following electronic accessories are also recommended for the field toolkit:

- A 50-ft, 12/3 extension cord.
- A minimum of one high-quality power bar (with surge protection).
- At least one A/V three-way amplifier (common Radio Shack part #15-1103).

- A four-way A/V selector (common Radio Shack #15-8250).
- An amplified audio/video selector (common Radio Shack #15-1951).
- A video RF modulator (common Radio Shack #15-1283A – for those times when the TV monitor doesn't have RCA-type video inputs).
- A video isolation transformer, an example of which is North Hills, FSCM 98821, Wideband Transformer, Video Isolation, Model 1117C, 10 Hz–5 MHz.
- A small container with at least one of every cable adapter Radio Shack carries. Suggestion: If Radio Shack or some other electronics store has an adapter that isn't in the toolkit, buy it. Stocking them in the kit is priceless when needed.
- A small variety of video cables, both RCA and BNC types. It is important to have a variety of the standard lengths available at any electronics retail outlet.
- Make up a 100-ft video cable extension cord. It may be handy to have one that has a male RCA plug on one end and a female RCA plug on the other, then use the adapters in the kit for any changes in format.
- A small digital multi-meter is essential. The biggest problem with the multi-meter is power. Always keep spare batteries.

In addition to the above, carry a small toolbox with a variety of small hand-tools, pliers, wire cutters, 6-inch (or metric equivalent) adjustable 'crescent' wrench, multi-bit screwdriver, soldering iron, solder, a small roll of wire, Allen wrenches (both standard and metric), and virtually anything else that will fit in the box.

13.4 STANDARD PREVENTATIVE MAINTENANCE CHECKLIST

The items in Table 13.1 are for general guidance and should not supplant the specific directives of any manufacturer-specific maintenance schedule. These maintenance items are typical of a proper manufacturer's preventative maintenance checklist that should be provided with the ROV system upon delivery from the manufacturer.

13.5 OPERATIONAL FORMS

Often after the field operation is completed, field operations personnel are required to substantiate findings made while on duty. For operational and legal consideration, documentation of operations should be meticulously kept. The following pages provide sample forms that should assist in documentation of ROV operations to include:

- ROV Pre-Dive Operations Check
- ROV Post-Dive Operations Check
- ROV Dive Log (two pages).

Table 13.1 *Maintenance schedule.*

Maintenance action	*Check every*			
	Pre-dive	*Post*	*50 h*	*500 h*
Check wires, cables, and hoses for wear and damage.	X	X		
Check for loose or missing hardware. Repair or replace as necessary.	X	X		
Check sacrificial anodes for deterioration.	X	X		
Visually inspect the vehicle to ensure that the propellers are not fouled, that all components are secured, and there is no mechanical damage to the frame or other components.	X	X		
Check the tether for scrapes, nicks, kinks, or other visible damage. The vehicle should not be used if the tether jacket is broken through.	X	X	X	X
Check that all fasteners are in place and secure.	X	X		
Check ballast weights.	X	X		
Ensure the tether connection at the vehicle is plugged in all the way, and that the tether termination can is well secured by the flotation.	X	X		
Power-up test: Test all system functions.	X	X	X	X
Check for water intrusion through the ports.	X	X	X	X
Drain and check the oil in the thruster cone end.			X	X
Replace thruster seals and inspect propeller shaft. Replace when grooved.			X	X
Store properly.	X	X	X	X

ROV Pre-Dive Operations Check

Dive No:	Date (Y/M/D):
Dive Location:	Start Time:
Operator:	Stop Time:

Check Conducted	**Initials**	**Comments**
Visually inspect the vehicle to ensure that the propellers are not fouled, that all components are secured, and there is no mechanical damage to the frame or other components.		
Check the tether for scrapes, nicks or other visible damage. The vehicle should not be used if the tether jacket is broken through.		
Check that all fasteners are in place and secure.		
Check ballast and trim adjustments.		
Ensure that both whip connectors at the electronics can and tether termination can are plugged in all the way. Also check that the tether termination can is clamped securely between the lower bracket and flotation.		
Ensure all unused vehicle connectors are capped securely with dummy plugs. A forgotten or unsecured dummy plug can lead to serious electrical system damage.		
Ensure that all surface cables are properly connected.		
To prevent unexpected thruster operation, switch the Auto Depth switch to 'OFF', center the Manual Depth Control knob, and switch Auto Heading 'OFF' **before powering up the system.**		
Power up the system. Test the lights, camera functions & thrusters. Check that good video is obtained and that any video recording equipment is working properly. **Caution: refrain from running the thruster and/or lights any longer than a few seconds out of the water.**		

ROV Post-Dive Operations Check

Dive No:	Date (Y/M/D):
Dive Location:	Start Time:
Operator:	Stop Time:

Check Conducted	Initials	Comments
The vehicle should be visually inspected following each dive to ensure that no mechanical damage has occurred.		
Check the propellers for any fouling.		
Visually check through the ports to ensure that no water has entered the camera housings.		
Inspect the tether for cuts and / or nicks or kinks in the outer shell.		
A fresh water rinse is required if the vehicle has been operated in salt water.		
Re-check all vehicle functions. To prevent unexpected thruster operation, switch the Auto Depth switch to 'OFF', center the Manual Depth Control knob, and switch Auto Heading 'OFF' **before storing the system.**		
Store the tether and vehicle properly for the next use. Refer to specific manufacturer's recommendations for tether storing procedures		

ROV Dive Log

Vessel: ________________ No. of Sheets: ________________

Location: ________________

Date: ________________

Dive No: ________________

Operations Crew: ____

____

____

Conditions:

____

____

____

Purpose of Dive:

____

____

____

____

Dive Job Summary:

____

____

____

____

____

____

____

Total Wet Time: ____________________________________

Dive Log Complete By: ____________________ Signature: ________________

ROV Dive Log

Dive No: ____________________ Sheet: _______ of: ________

TIME	OPERATION

Chapter 14
Putting It All Together

Throughout this manual, the technology applicable to observation-class ROVs was discussed, along with it application in various missions. The crowning achievement is when these newly acquired skills come together to successfully find an underwater target. In this chapter, the steps to accomplish this final task will be examined in detail along with the most common tools in the underwater technician's tool chest – the side-scan sonar and the ROV deployed aboard a small boat.

14.1 ATTENTION TO DETAIL

In the underwater business, there is no so-called 'silver bullet' (i.e. there is no single piece of equipment that can be used to solve all operational situations). Finding things underwater is more a function of gathering input from many different sources, putting them together to form the most likely conclusion, then repeatedly testing that conclusion until it is positively proven or disproved.

An oversight can cause a wasted mission through simple inattention to detail. A conversion of feet to meters can throw the projected location of the search area away from the known target. A mistake while choosing the map projection can waste an entire day of searching. Choosing the wrong operating voltage can destroy equipment. Careless wiring and arrangement of equipment can cause serious injury and even death while in the field.

Attention to detail is paramount while performing any field task.

14.2 TRAINING AND PERSONNEL QUALIFICATIONS

The knowledge requirements for operators, repair technicians, and tasking personnel are:

- Operators are required to know the theory and application of ROV technology.
- Repair technicians are required to understand the operations and maintenance of the components for the ROV system.
- Tasking personnel need only concern themselves with the capabilities and limitations of an ROV inspection system in use for various missions.

This manual has provided the basics in these areas. However, a training program should be established and a record kept in the personnel file of those involved to ensure that each individual, and thus the team, is ready for the task at hand.

14.3 EQUIPMENT SETUP CONSIDERATIONS

Some considerations while setting up equipment aboard a small boat are needed to properly and efficiently operate the equipment in a safe and productive manner. Time spent initially during the setup phase will pay dividends repeatedly while underway through a tidy, clean, and efficient working environment.

Equipment setup considerations include:

- Both side-scan sonar and ROV equipment use considerable lengths of cable. Attempt to use only one piece of equipment at a time while completely stowing the second piece of equipment until needed.
- Run and stow all cables and wires away from travel spaces. A tripping hazard will repeatedly catch personnel off guard, possibly causing serious injury or death.
- The mission equipment should never interfere with the boat's operational equipment. Blocking access to engine compartments, anchor lockers, dock lines, and any boat safety equipment is a danger to both the vessel and the mission.
- Set up all equipment with consideration to the order in which it will be needed. Completely stow equipment when not in use.
- The figure '8' flaking of cables has shown repeatedly to be a very efficient method of mechanical tether management and is usually preferred to the use of a tether reel. A tidy workspace is a safe and efficient workspace.
- To satisfactorily view computer screens and video monitors, a location away from sunlight is needed, such as an enclosed cabin or a tarp over the monitors. Also, a comfortable temperature-controlled area will assist in eliminating the 'physical need' to end the operation due to discomfort.
- A power source separate from the vessel's generator is preferred. Assure that the exhaust from the generator and/or the vessel's exhaust does not vent near the enclosed work area.

14.4 DIVISION OF RESPONSIBILITY

The captain of the boat is responsible for the vessel's safe and efficient operation. The equipment operator is normally responsible for the mission. The captain of the boat is the final authority regarding all operations aboard the vessel. It is proper protocol to gain permission before deploying any equipment that will affect the operation of the vessel. It is also proper protocol to keep the vessel operator completely informed of all of the team's intentions and planned tasks.

The mission specialist and ROV team are guests aboard a vessel of opportunity. Work along with the crew within the vessel management structure in order to get the maximum out of the equipment and to achieve the mission objectives.

Before accepting a demanding assignment aboard a vessel of opportunity, qualify the captain and crew for the mission at hand. Such tasks as station-keeping in a difficult sea state or maintaining a survey line during a side-scan search are paramount to completing the mission. Many of the operational problems can be solved before leaving the dock by screening the boat crew to ensure they are properly qualified. A fishing boat crew may be the best crew for hauling in a large catch, but maintaining a tight-tolerance survey demands another skill set.

14.5 BOAT HANDLING

Many failed operations can be traced directly to a simple matter of boat handling. Unless the equipment can be deployed consistently onto the location of the target, the mission, in all likelihood, will fail. Also, without the deployment platform maintained in a steady and stable state over the top of the work site, the entire operation may become a complete waste of time and resources.

14.5.1 Side-Scan Sonar Operations

To perform a proper side-scan sonar survey of an area, survey lines must be followed and tow fish altitudes must be maintained within fairly tight tolerance to achieve area coverage with a high degree of certainty. A high-quality GPS receiver with the capability for survey line input, as well as course deviation indication, is very helpful. Complete the survey of the entire area in the survey phase before attempting to switch to the identification phase of the mission (with diver or ROV). Many operations have been inefficiently run by stopping the survey to look at each suspected target (only to discover it was not the proper item). The switchover time between equipment can be considerable. Consult the equipment manufacturer's performance specifications to obtain the proper tow speed and altitude requirements. And do not be afraid to request corrective action of the boat operator if those parameters do not meet the mission requirements.

14.5.2 ROV Operations

It is paramount to have the ROV deployed over a steady work location. If the boat drifts off location, the ROV operator will be required to repeatedly reacquire the target (once the submersible is dragged off the target), frustrating efforts to complete the mission. If an ROV operation over a target site is to be performed properly, either multi-point anchoring over the site or dynamic positioning will be necessary. In some situations, a skilled boat operator can keep the vessel steady enough to get the identification done. But as the wind and sea state worsens, the ability to keep the vessel on-station becomes increasingly difficult. For shallow-water operations, a three- or four-point anchoring system directly over the work site (or a jack-up barge) is recommended to complete the work task.

14.6 MARKING THE TARGET(S)

To mark a target for positive identification, the following steps should be performed:

- Complete the side-scan sonar survey of the entire area, then retire to a location where a complete analysis of the data can be made. The targets can then be identified, classified, and prioritized. Once the targets are identified, further investigation can be performed.
- Fabricate a sonar reflector to place next to the target of interest. Sonar targets are highly (sonar) reflective weighted anchors connected to the surface via a line and

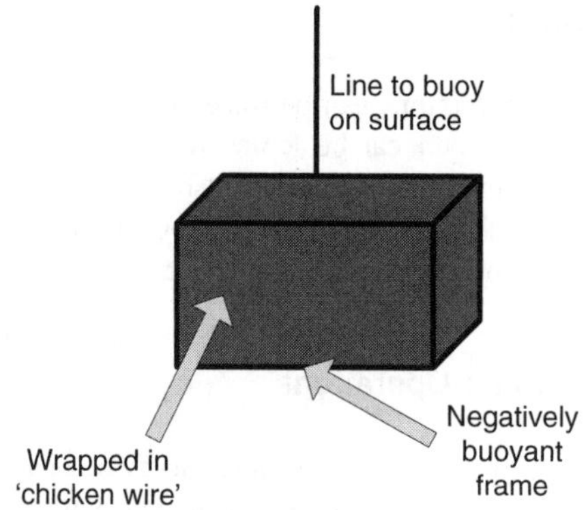

Figure 14.1 *Sonar reflecting target.*

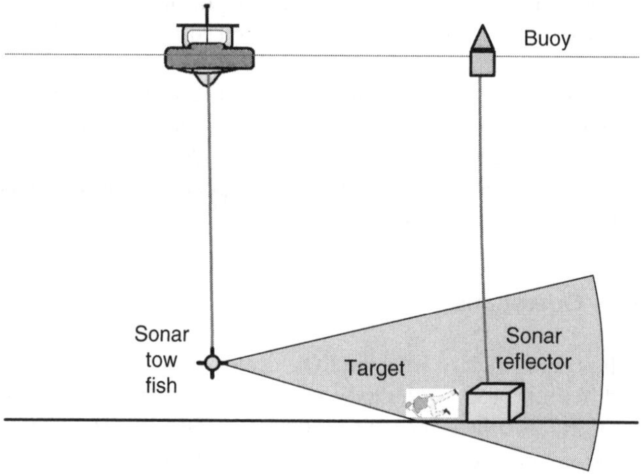

Figure 14.2 *Positioning of the sonar reflector.*

buoy (Figure 14.1). The objective of the sonar target is to give the ROV or diver a direct visual path to the target on the bottom via the buoy line. A suggestion for a low-cost sonar target is: Fabricate an approximately 3-foot (1 meter) cubed metal box structure wrapped in chicken wire, which reflects high-frequency sonar waves at all angles of incidence.

- Once the decision is made to positively identify a sonar target, proceed to the coordinates of the sonar target, drop the sonar reflecting target (make sure the buoy line is well secured with minimal slack to avoid accidental propeller entanglement), then re-scan the area to determine the range and bearing of the sonar target to the reflector. Continue to move the sonar reflector until the reflector is as close as possible to the target (Figure 14.2).
- Anchor the boat (Figure 14.3), then swim the submersible on the surface to the buoy line and follow the line down to the sonar reflector. From there it should just be a range bearing steer to the sonar target.

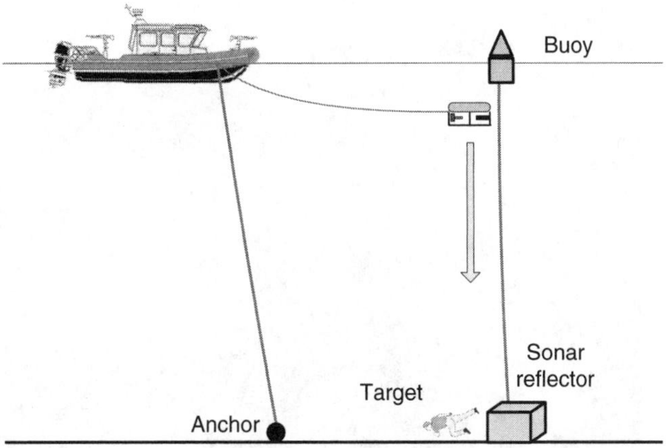

Figure 14.3 *ROV follows the buoy line to the target.*

14.7 METHODS FOR NAVIGATING TO THE TARGET

The method described in Section 14.6 is the simplest method of ROV/sonar navigation to targets on the bottom. Although tedious and time-consuming, it does maintain the operational objective of 'sustained and controlled environment' in that it allows for relatively immovable visual reference points throughout the process. This method has been proven to be effective.

Other methods of navigating to a known target on the bottom are with the use of a mechanically scanning tripod-mounted sonar, in addition to the ROV mounted sonar.

14.7.1 Tripod-Mounted Sonar/ROV Interaction

Just as an aircraft can be navigated to a landing area with ground-based radar, an ROV can be navigated to a target while tracking the target and submersible from a fixed location.

After the tripod sonar has been lowered onto the bottom and has acquired the target, the submersible is launched, then it follows the sonar line down to the bottom. Once on the bottom, the submersible determines the orientation of the sonar head, then flies in the direction of the target. The tripod sonar operator, after locating the submersible on the sonar, can guide the submersible operator to the target (Figure 14.4).

14.7.2 Sonar Mounted Aboard the ROV

The target can also be located via the on-board scanning sonar system (Figure 14.5). But be warned – the perspective of the target from the side-scan sonar will be drastically different than that viewed from the submersible. From the side-scan sonar perspective (i.e. from above), a drowning victim will give the shape and appearance of a human form. From the submersible's perspective, the body may look like a log, a pipe, a rock, or any of a number of items depending upon the viewing angle.

(a)

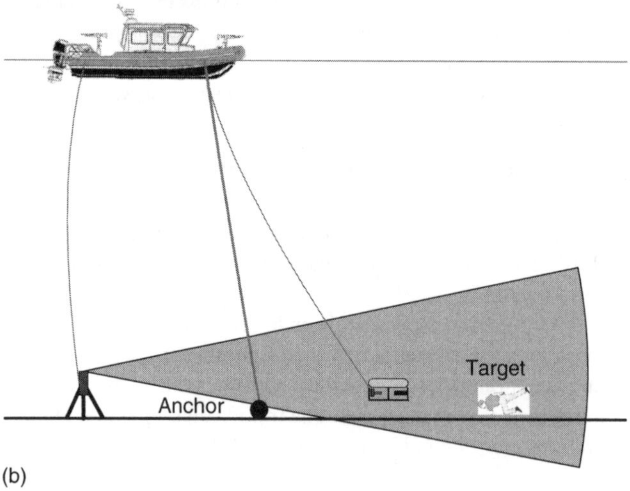

(b)

Figure 14.4 *Fixed location tracking of ROV and target.*

14.8 SONAR/ROV INTERACTION

A technique for gaining an orientation on the bottom is to find either a flat rock or a solid location on the bottom to plant the submersible to gain a good vantage point to

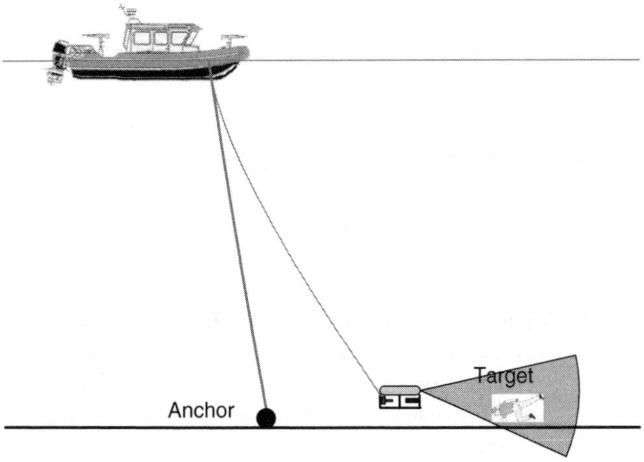

Figure 14.5 *ROV mounted sonar.*

image all around. A mud bottom makes this 'planting' a bit more permanent, since it may require an excess amount of thrust to 'unstick' the submersible.

Once a location is found, put the submersible on the bottom so that it becomes a stable platform from which to build an image of the surrounding area. The image is generated as the sonar head scans around the vehicle, isolating the target from the surrounding terrain. To better identify the target, it may be required to move the vehicle to another location to rescan the area.

The target is then brought to zero bearing (i.e. directly ahead of the vehicle), and is maintained in view while the vehicle is navigated to the target. Once the target is located and positively identified, the vehicle can remain on-location for as long as necessary. If the vehicle is equipped with a manipulator or some form of physical attachment device, it is best to hook on to the target to maintain visual contact until the next step in the process is initiated.

Appendix A

TEST QUESTIONS AND ANSWERS

The following test questions are true/false and multiple choice. The answers are provided on separate pages that follow.

1. Which of following is *not* characteristic of an ROV?
   (a) The submersible's camera allows for transmission of real-time image generation viewable on the surface or operating platform.
   (b) ROVs can perform either a human-navigated or a pre-programmed search grid.
   (c) ROVs can be operated free of a hard-wire link to the surface or operating platform.
   (d) ROVs can receive their power from either the surface or from on-board battery packs.

2. Which of the following is *true* with reference to ROV handling and stability?
   (a) ROVs normally have full freedom of underwater flight, including pitch, roll, yaw, and heave.
   (b) Best stability of the ROV platform is achieved with the center of buoyancy and the center of gravity spaced as far apart as possible, considering the size of the submersible.
   (c) Observation-class ROV systems are best known for their large size and high-powered hydraulically actuated thrusters.
   (d) The company normally operates work-class ROV systems in their port and harbor security missions.

3. The ROV submersible used by the company in its underwater port security mission should be ballasted in the following manner:
   (a) Slightly positively buoyant so that the vertical thruster does not fight the submersible's inherent buoyancy.
   (b) Very positively buoyant so that the submersible will float to the surface in case of power failure.
   (c) Slightly negatively buoyant so that the vertical thruster does not stir the bottom silt and reduce visibility.
   (d) Very negatively buoyant so that the vertical thruster functions like a helicopter, keeping the submersible from hitting the underside of the hull.

4. Assuming constant temperature, when going from a salt-water environment to a freshwater environment with a fixed ballast ROV, ballast weight will need to be added to maintain the same level of buoyancy.
   (a) True.
   (b) False.

5. Which of the following is *true* with reference to light absorption in water?
   (a) The highest absorption rate of light in water happens in the blue–green color spectrum.
   (b) The maximum water penetration distance is achieved in the red color spectrum.
   (c) Radio waves have no absorption in water and can penetrate to full ocean depth.
   (d) The only way to bring out color, even within a few feet of the surface, is with artificial lighting from the submersible.

6. The best means of mitigating the negative effects of backscattering of light from the ROV submersible's lighting system to the item of interest is to move the light source as far away from the camera as possible.
   (a) True.
   (b) False.

7. A tether management system (TMS) is used to protect the submersible from damage upon deployment and to absorb the cross-section drag on the umbilical/tether from currents. What item is used to achieve the cross-current drag reduction for an ROV system without a TMS?
   (a) Clump weight.
   (b) Control station.
   (c) Power source.
   (d) Display station.

8. Normal sea water is made up of how many parts per thousand of dissolved salts?
   (a) 10 PPT.
   (b) 23 PPT.
   (c) 34 PPT.
   (d) 45 PPT.

9. According to the TTP Manual, there is no small ROV that can be considered effective in currents over:
   (a) 2 knots.
   (b) 3 knots.
   (c) 5 knots.
   (d) 10 knots.

10. Acoustic positioning works on which of the following basic principles to resolve range and bearing?
    (a) Frame of reference exclusion.
    (b) Water density change rate.
    (c) Material reflectivity.
    (d) Sound propagation and triangulation.

11. The use of only two beacons produces a correct position resolution to two distinct locations. This is known as:
    (a) Baseline ambiguity.
    (b) Sound absorption.

(c) Sound refraction.
(d) Multi-path.

12. A line of position is?
    (a) A point or location.
    (b) A line of equal distance from a reference point.
    (c) A rope used for station-keeping.
    (d) A distance measuring error.

13. In order to determine the acceptance of a sound signal for use in acoustic positioning, which of the following signal discrimination techniques are used?
    (a) Thresholding and filtering.
    (b) Absorption and refraction.
    (c) Attenuation and reflection.
    (d) Beacon and transponder.

14. Which of the following is *not* a type of acoustic positioning technology?
    (a) USBL.
    (b) ABL.
    (c) SBL.
    (d) LBL.

15. What is multi-path?
    (a) Alternate ship channel routes.
    (b) A transducer malfunction.
    (c) Light reflection at the transducer.
    (d) Different receptions of the same sound signal from different sound paths.

16. The vessel-referenced ship hull inspection acoustic positioning system is what type of positioning technique?
    (a) LBL.
    (b) SBL.
    (c) USBL.
    (d) LUSBL.

17. Which of these signals is the only true distance measurement signal in acoustic positioning?
    (a) The second multi-path signal.
    (b) The baseline ambiguity signal.
    (c) The refracted signal.
    (d) The line-of-sight signal.

18. Which of the following is applicable with reference to acoustic positioning environment of ports and harbors?
    (a) Low noise/high reverberation.
    (b) High noise/low reverberation.
    (c) High noise/high reverberation.
    (d) Low noise/low reverberation.

19. To adjust the acoustic receiver for a harbor environment, which of the following settings would be appropriate?
    (a) Low receiver sensitivity/long time delay for position sampling.
    (b) High receiver sensitivity/short time delay for position sampling.
    (c) Low receiver sensitivity/short time delay for position sampling.
    (d) High receiver sensitivity/long time delay for position sampling.

20. Operating acoustic positioning equipment in close proximity to underwater structures significantly increases the instances of multi-path phenomenon.
    (a) True.
    (b) False.

21. The ship hull inspection acoustic positioning system can be used on ships, dock, dams, and other relative-referenced structures?
    (a) True.
    (b) False.

22. Which of the following has the highest sonar reflectivity index?
    (a) Mud.
    (b) Wood.
    (c) Sand.
    (d) Rock.

23. With regard to sonar propagation, which of the following statements is *true*?
    (a) The higher the sound frequency, the farther the sound propagation distance.
    (b) Sonar is not affected by the temperature of the water.
    (c) Shadowgraph effect on sonar displays is the area of highest reflectivity.
    (d) The higher the sound frequency, the higher the attenuation, thus a lower effective range.

24. Which of the following would require the highest gain setting on the sonar display in order to view detail of the surface?
    (a) Wood.
    (b) Sand.
    (c) Mud.
    (d) Steel.

25. The preferred method of clearing an ROV tether snag is?
    (a) Pull slowly and firmly in order to clear the snag.
    (b) Swim the vehicle along the length of the tether to determine the source of the snag then unfoul with the submersible.
    (c) Launch dive team to recover the vehicle.
    (d) Cut the tether.

26. Tether turns are insignificant to the operation of a port security ROV system.
    (a) True.
    (b) False.

27. Depending upon the threat level, efficiency gains by searching only high-risk areas of ships can be achieved.
    (a) True.
    (b) False.

28. Which of the following locations on a ship hull below the water line would best hide a parasitic attachment from water drag?
    (a) Inside the sea chest.
    (b) Attached to the keel.
    (c) Attached to the bilge keel.
    (d) Attached to the rudder.

29. Operation of the thrusters of a shaft-drive ROV thruster system after propeller fouling will?
    (a) Burn the motor or driver board.
    (b) Shut down the camera system.
    (c) Disable the rear camera.
    (d) Not affect the operation of the system.

30. Isolation and diagnosing of ROV system faults are in which two broad categories?
    (a) Acoustical and optical.
    (b) Mechanical and electrical.
    (c) Visual and radiological.
    (d) Usual and unusual.

ANSWERS

1. Which of following is a *not* characteristic of an ROV?
    c) ROVs can be operated free of a hard-wire link to the surface or operating platform.

2. Which of the following is *true* with reference to ROV handling and stability?
    b) Best stability of the ROV platform is achieved with the center of buoyancy and the center of gravity spaced as far apart as possible, considering the size of the submersible.

3. The ROV submersible used by the company in its underwater port security mission should be ballasted in the following manner?
    a) Slightly positively buoyant so that the vertical thruster does not fight the submersible's inherent buoyancy.

4. Assuming constant temperature, when going from a salt-water environment to a freshwater environment with a fixed ballast ROV, ballast weight will need to be added to maintain the same level of buoyancy.
    b) False.

5. Which of the following is *true* with reference to light absorption in water?
    d) The only way to bring out color, even within a few feet of the surface, is with artificial lighting from the submersible.

6. The best means of mitigating the negative effects of backscattering of light from the ROV submersible's lighting system to the item of interest is to move the light source as far away from the camera as possible.
    a) True.

7. A tether management system (TMS) is used to protect the submersible from damage upon deployment and to absorb the cross-section drag on the umbilical/tether from currents. What item is used to achieve the cross-current drag reduction for an ROV system without a TMS?
    a) Clump weight.

8. Normal sea water is made up of how many parts per thousand of dissolved salts?
    c) 34 PPT.

9. According to the TTP Manual, there is no small ROV that can be considered effective in currents over:
    b) 3 knots.

10. Acoustic positioning works on which of the following basic principles to resolve range and bearing?
    d) Sound propagation and triangulation.

11. The use of only two beacons produces a correct position resolution to two distinct locations. This is known as:
    a) Baseline ambiguity.

12. A line of position is?
    b) A line of equal distance from a reference point.

13. In order to determine the acceptance of a sound signal for use in acoustic positioning, which of the following signal discrimination techniques are used?
    a) Thresholding and filtering.

14. Which of the following is *not* a type of acoustic positioning technology?
    b) ABL.

15. What is multi-path?
    d) Different receptions of the same sound signal from different sound paths.

16. The vessel-referenced ship hull inspection acoustic positioning system is what type of positioning technique?
    a) LBL.

17. Which of these signals is the only true distance measurement signal in acoustic positioning?
    d) The line-of-sight signal.

18. Which of the following is applicable with reference to acoustic positioning environment of ports and harbors?
    c) High noise/high reverberation.

19. To adjust the acoustic receiver for a harbor environment, which of the following settings would be appropriate?
    a) Low receiver sensitivity/long time delay for position sampling.

20. Operating acoustic positioning equipment in close proximity to underwater structures significantly increases the instances of multi-path phenomenon.
    a) True.

21. The ship hull inspection acoustic positioning system can be used on ships, dock, dams, and other relative-referenced structures?
    a) True.

22. Which of the following has the highest sonar reflectivity index?
    d) Rock.

23. With regard to sonar propagation, which of the following statements is *true*?
    d) The higher the sound frequency, the higher the attenuation, thus a lower effective range.

24. Which of the following would require the highest gain setting on the sonar display in order to view detail of the surface?
    c) Mud.

25. The preferred method of clearing an ROV tether snag is?
    b) Swim the vehicle along the length of the tether to determine the source of the snag then unfoul with the submersible.

26. Tether turns are insignificant to the operation of a port security ROV system.
    b) False.

27. Depending upon the threat level, efficiency gains by searching only high-risk areas of ships can be achieved.
    a) True.

28. Which of the following locations on a ship hull below the water line would best hide a parasitic attachment from water drag?
    a) Inside the sea chest.

29. Operation of the thrusters of a shaft-drive ROV thruster system after propeller fouling will?
    a) Burn the motor or driver board.

30. Isolation and diagnosing of ROV system faults are in which two broad categories?
    b) Mechanical and electrical.

Bibliography

Benthos, Inc. *StingRay Mk II Operations and Maintenance Manual*. Website for Benthos, Inc. is located at: http://www.benthos.com/.

Bohm H. and Jensen V. *Build Your Own Underwater Robot*. West Coast Words, 1997. ISBN 0-9681610-0-6.

Bowditch N. *The American Practical Navigator*. National Imagery and Mapping Agency, 2002. ISBN 1-57785-271-0.

Burcher R. and Rydill L. *Concepts in Submarine Design*. Cambridge University Press, 1994.

Busby R.F. *Manned Submersibles*. Office of the Oceanographer of the Navy, 1976.

Deep Sea Power and Light. '*Frequently Asked Questions*' website page at: http://www.deepsea.com/faq.html.

Department of the Army, Corps of Engineers. *Coastal Engineering Manual*. Publication number EM 1110-2-1100, 2002.

Desert Star Systems LLC. Various Operational Manuals for the Dive Tracker line of acoustical positioning systems. Reprinted with permission. Located on the web at: http://www.desertstar.com/.

Duxbury A.C. and Alison B. *An Introduction to the World's Oceans*, 5th edition. William C. Brown, 1997. ISBN 0-697-28273-2.

Edge M. *The Underwater Photographer*, 2nd edition. Butterworth-Heinemann, 1999. ISBN 0-240-51581-1.

Everest F.A. *The Master Handbook of Acoustics*, 3rd edition. TAB Books (a division of McGraw-Hill), 1994. ISBN 0-8306-4438-5.

Fondriest Environmental. Informational website page on environmental sensors at: http://www.fondriest.comlparameter.htm. Reprinted with permission.

Giancoli D.C. *Physics*, 3rd edition. Prentice-Hall, 1991. ISBN 0-13-672510-4.

Gray R. Light sources, lamps and luminaries. *See Technology*, 45(12): 39–43, December 2004.

Huang H.-M. *Autonomy Levels for Unmanned Systems (ALFUS) Framework*, Volume I: Terminology, Version 1.1 National Institute of Standards and Technology Special Publication 1011, September 2004.

Imagenex Technology Corporation. Sonar theory from their website located at: http://www.imagenex.comlsonar theory.pdf. Reprinted with Permission.

Inuktun Services, Ltd. *ROV Seamor Operations Manual*. Website for Inuktun Services, Ltd. is located at: http://www.seamor.com/.

Joiner J.T., editor. *NOM Diving Manual*, 4th edition. Best, 2001. ISBN 0-941332-70-5.

Jukola H. and Skogman A. *Bollard Pull*. Paper presented at the 17th International Tug and Salvage Convention, ITS 2002, 13–17 May 2002, Bilbao, Spain.

Kongsberg S.A.S. *Introduction to Underwater Acoustics*, 2002.

Linton S.J. et al. *Dive Rescue Specialist Training Manual*. Concept Systems, Inc., 1986. ISBN 0-943717-42-6.

Loeser H.T. *Sonar Engineering Handbook*. Peninsula, 1992. ISBN 0-932146-02-3.

Medwin H. and Clay C. S. *Fundamentals of Acoustical Oceanography*. Academic Press, 1998. ISBN 0-12-487570-X.

Milne P.B. *Underwater Acoustic Positioning Systems*. E. & F. N. Spon Ltd, 1983. ISBN 0-419-12100-5.

Moore J.E. and Compton-Hall R. *Submarine Warfare: Today and Tomorrow*. Adler & Adler, 1987. ISBN 091756121X.

Olsson M.S. et al. ROV lighting with metal halide. White Paper on Metal Halide Technology displayed on the Deep Sea Power and Light Web Site at: http://www.deepsea.com. Reprinted with permission. Revision 1 dated 27 January 2000.

Outland Technology, Inc. *ROV Model Outland 1000 Operations Manual*. Website for Outland Technology, Inc. is located at: http://www.outlandtech.com/.

Remotely Operated Vehicle Subcommittee of the Marine Technology Society. Educational website page at: http://www.rov.org/student/education.cfm.

Remotely Operated Vehicles of the World, 7th Edition, Clarkson Research Services Ltd., 2006/2007. ISBN 1-902157-75-3.

Seafriends.org. *Underwater Photography – Water and Light*. From their educational web page located at: http://www.seafriends.org.nz/phgraph/water.htm.

Segar D.A. *Introduction to Ocean Science*. Wadsworth, 1998. ISBN 0-314-09705-8.

Sonardyne International Ltd. *Acoustic Theory*. Website page at: http://www.sonardyne.co.uk/theory.htm.

Teather R.G. *Royal Canadian Mounted Police Encyclopedia of Underwater Investigations*. Best, 1994. ISBN 0-941332-26-8.

Thurman H.V. *Introductory Oceanography*, 7th edition. Macmillan, 1994. ISBN 0-02-420811-6.

Urick R.J. *Principles of Underwater Sound*. McGraw-Hill, 1975. ISBN 0-07-066086-7.

US Geological Survey. Educational website page on water science at: http://ga.water.usgs.gov/edu/earthhowmuch.html.

Van Dorn W.G. *Oceanography and Seamanship*, 2nd edition. Cornell Maritime Press, 1993. ISBN 0-87033-434-4.

Waite A.D. *Sonar for Practicing Engineers*, 3rd edition. John Wiley, 2002. ISBN 0-471-49750-9.

Wernli R. *Operational Effectiveness of Unmanned Underwater Systems*. Marine Technology Society, 1998. ISBN 0-933957-22-X.

Wilson W.D. Equation for the speed of sound in sea water. *Journal of the Acoustic Society of America,* 1960.

Index